History, Philosophy and Theory of the Life Sciences

Volume 30

History, Philosophy and Theory of the Life Sciences is a space for dialogue between life scientists, philosophers and historians—welcoming both essays about the principles and domains of cutting-edge research in the life sciences, novel ways of tackling philosophical issues raised by the life sciences, as well as original research about the history of methods, ideas and tools, which constitute the genealogy of our current ways of understanding living phenomena.

The series is interested in receiving book proposals that • are aimed at academic audience of graduate level and up • combine historical and/or philosophical and/or theoretical studies with work from disciplines within the life sciences broadly conceived, including (but not limited to) the following areas: • Anatomy & Physiology • Behavioral Biology • Biochemistry • Bioscience and Society • Cell Biology • Conservation Biology • Developmental Biology • Ecology • Evolution & Diversity of Life • Genetics, Genomics & Disease • Genetics & Molecular Biology • Immunology & Medicine • Microbiology • Neuroscience • Plant Science • Psychiatry & Psychology • Structural Biology • Systems Biology • Systematic Biology, Phylogeny Reconstruction & Classification • Virology The series editors aim to make a first decision within 1 month of submission. In case of a positive first decision the work will be provisionally contracted: the final decision about publication will depend upon the result of the anonymous peer review of the complete manuscript. The series editors aim to have the work peer-reviewed within 3 months after submission of the complete manuscript. The series editors discourage the submission of manuscripts that contain reprints of previously published material and of manuscripts that are below 150 printed pages (75,000 words). For inquiries and submission of proposals prospective authors can contact one of the editors: Charles T. Wolfe: ctwolfe1@gmail.com Philippe Huneman: huneman@wanadoo.fr Thomas A.C. Reydon: reydon@ww.uni-hannover.de

Pierre-Olivier Méthot
Editor

Philosophy, History and Biology: Essays in Honour of Jean Gayon

Springer

Editor
Pierre-Olivier Méthot
Faculté de Philosophie
Université Laval
Quebec City, QC, Canada

ISSN 2211-1948 ISSN 2211-1956 (electronic)
History, Philosophy and Theory of the Life Sciences
ISBN 978-3-031-28156-3 ISBN 978-3-031-28157-0 (eBook)
https://doi.org/10.1007/978-3-031-28157-0

This Springer imprint is published by the registered company Springer Nature Switzerland AG
The registered company address is: Gewerbestrasse 11, 6330 Cham, Switzerland

Jean Gayon (1949–2018)

This book is dedicated to Elisabeth, Hélène, Jean-Philippe and Violaine

Editor's Note

Pierre-Olivier Méthot[1]

This new collection of essays is a translated, revised and expanded version of *Philosophie, histoire, biologie: mélanges offerts à Jean Gayon*, first published under the editorship of Francesca Merlin and Philippe Huneman at Éditions Matériologiques in the "Sciences et philosophie" series (Merlin and Huneman 2018).[2] I am grateful to Marc Silberstein, the director of Éditions Matériologiques, for his permission to republish several essays from the original book.

The 2018 volume collected papers presented during a conference organized by Francesca Merlin and Philippe Huneman to honour Jean Gayon's career at the École normale supérieure in Paris on 13–15 March 2017. This second edition differs in two ways: first, several chapters are either new contributions or substantially revised versions written in view of a broader Anglo-Saxon readership.[3] Many of these chapters have been presented on the occasion of a roundtable organized by Philippe Huneman, Michel Veuille, Victor Petit and Elisabeth Valsecchi Gayon at the École normale supérieure in Paris on 13 December 2018 and during a thematic session on Gayon's work organized by Pierre-Olivier Méthot and Francesca Merlin at the biannual conference of the International Society for the History, Philosophy, and Social Studies of Biology (ISHPSSB) held in Oslo on 7–12 July 2019.

Second, since the publication of the original French version of the volume, a book of interviews with Jean Gayon by Victor Petit was published as *La connaissance de la vie aujourd'hui* (Gayon and Petit 2018), a nod to Georges Canguilhem's *La connaissance de la vie* (2008 [1952]). This book was translated the next year as *Knowledge of Life Today: Conversations on Biology* (Gayon and Petit 2019), but

[1] Université Laval. Email: pierre-olivier.methot@fp.ulaval.ca

[2] Jean Gayon's *Darwin et l'après-Darwin* has been republished in a revised version at Éditions Matériologiques in 2019 in the *Histoire des Sciences et des Techniques* series.

[3] In this book, 11 chapters are translations (Méthot, Brenner, Duchesneau, Pradeu, Veuille, Martínez-Contreras, Tirard, Chazal, Forest, Giroux, Huneman); 4 were originally published in English (Depew, Suárez-Díaz, Sloan, Burian); and 5 chapters are entirely new (Loison, Hodge, Wolfe, Engel, Rheinberger).

without the informative biographical chapters of the original French.[4] Both editions, however, include the complete list of publications by Gayon. Several contributors to the present book have referred to these new sources as well as to Gayon's unpublished papers and research notes collected in the "Fonds Jean Gayon" at the Institut Pasteur in Paris.[5] Thus, this volume provides a broad overview and in-depth analyses of Gayon's research as well as fresh perspectives on his path as a philosopher and as a historian of science. Since Gayon himself considered his work to fall within the remit of philosophy, biological sciences and history of science, the original title of the book was kept: *Philosophy, History and Biology*.[6]

This book not only honours Jean Gayon's legacy, but also makes some inroads into the historical development of the history and philosophy of biology, a field to which he contributed in so many ways. As a person and as a scholar, Jean Gayon inspired many of us, including the editor of this volume. I hope this book will encourage others to actively take up research on Gayon's scholarship, to situate his contributions in relation to the intellectual milieus he interacted with, and to draw on the methods and on the insights he put in front of us for future research in the history and philosophy of the life sciences.

For giving me access to her late husband's papers, for her friendship and for the wonderful conversations we had over the years about Jean's life and work, I thank Élisabeth Valsecchi Gayon. I am also grateful to Jean and Élisabeth's children: Hélène, Jean-Philippe and Violaine. This book is dedicated to them and to Élisabeth. At Springer, I want to thank Chris Wilby and Ram Prasad Chandrasekar for their assistance in preparing the manuscript, as well as Philippe Huneman, Thomas A. C. Reydon and Charles T. Wolfe for welcoming this new book in *the History, Philosophy and Theory of the Life Sciences* series. I am especially grateful to the authors for their contribution and for their patience in seeing their work published. Many thanks also go to Francesca Merlin and Philippe Huneman for trusting me with the edition of this volume; to Henry Dicks for the translation and revision of several chapters; to Alex Powell for language editing; and to Maxime F.-Giguère and Loïc Latouche-Simard for their work as research assistants at Université Laval. Finally, as always, a special thank you to Cindy Glauser for her patience, her support and her constant encouragements in pursuing this editorial project.

Funding from the Social Sciences and Humanities Research Council for Canada Research Chairs is gratefully acknowledged.

[4] The interviews for *La connaissance de la vie aujourd'hui* took place between 2010 and 2013 when Gayon was the director of the Institut d'histoire et de philosophie des sciences et des techniques.

[5] For recent works drawing on Gayon's unpublished papers, see Petit (2019) and Méthot (2022). A workshop celebrating the 150 years of Darwin's *Descent of Man* (1871), held on 5–6 July 2021 at the Institute for Advanced Studies (Paris) and organized by Michel Veuille, Philippe Huneman, Victor Petit and Élisabeth Valsecchi Gayon, provided another occasion to reflect on Jean Gayon's legacy in the history and philosophy of biology.

[6] "L'ensemble de mes travaux s'est développé au sein d'un triangle dont la philosophie, les sciences (notamment la biologie) et l'histoire des sciences constituent les sommets" (Gayon 2018, p. 1).

References

Canguilhem, G. (2008) [1952]. *Knowledge of Life* (transl. S. Geroulanos & D. Ginsburg). Fordham University Press.

Gayon, J. (2018). Avant-propos (pp. 1–5). In J. Gayon, & V. Petit (Eds.), *La connaissance de la vie aujourd'hui*. ISTE Editions, Ltd.

Gayon, J., & Petit, V. (2018). *La connaissance de la vie aujourd'hui*. ISTE Editions, Ltd.

Gayon, J. (2019) [1992]. *Darwin et l'après-Darwin: une histoire de l'hypothèse de sélection naturelle. Nouvelle édition revue et corrigée par Françoise Parot*. Éditions Matériologiques.

Gayon, J., & Petit, V. (2019). *Knowledge of life today: Conversations on biology*. ISTE Ltd. & Hoboken, John Wiley & Sons.

Petit, V. (2019). Jean Gayon (1949–2018): Itinéraire d'un darwinien. *Revue d'histoire des sciences, 72*(2), 349–380.

Merlin, F., & Huneman, P. (Eds.). (2018). *Philosophie, histoire, biologie: mélanges offerts à Jean Gayon*. Éditions Matériologiques.

Méthot, P. -O. (2022). Jean Gayon face à Georges Canguilhem: une remontée en trois temps. In C. Cherici, M. Arminjon, & P. -O. Méthot (Eds.), *Le normal et le pathologique: des catégories périmées? Textes réunis en hommage à Jean Gayon* (pp. 267–306). Éditions Matériologiques.

Contents

Chapter 1
Introduction: Jean Gayon (1949–2018), Philosopher and Historian of the Life Sciences

Pierre-Olivier Méthot and Philippe Huneman

1.1 An Overview of Jean Gayon's Life and Work[1]

Professor of History and Philosophy of Science at the University of Paris 1–Panthéon-Sorbonne since 2000, former director of the Institute of History and Philosophy of Science and Technology (IHPST) of the CNRS, Jean Gayon (1949–2018) died on April 28th 2018 following a long illness that he faced with determination and courage.

Like Gaston Bachelard and Georges Canguilhem, the founding figures of the French tradition in history and philosophy of science, Jean Gayon added to his philosophical training a scientific education to analyze his subject matter seriously: the biological sciences. As the author of around 300 publications, Jean Gayon wrote on several subjects, from Jean Piaget's epistemology to model organisms, and from the history of eugenics to medical enhancement. But most of his research was in the history and philosophy of contemporary biology, with a focus on the conceptual foundations of the theory of evolution, the history of genetics in France, and the concept of heredity.

Born in Saint Maur des Fossés (in the Paris area) on June 15th 1949, Jean Gayon obtained his high school degree ("*baccalauréat*") in 1966. Following this, he went on to study philosophy at university level at the Catholic Institute of Paris after a very brief period at the prestigious Lycée Henri-IV ("*hypokhâgne*") that left him unsatisfied. The following year at the Sorbonne, in an environment more suited to his own temperament, he received an education centered on the history of philosophy in

[1] This introduction is based on Méthot (2022) and Huneman (2018).

P.-O. Méthot (✉)
Université Laval, Quebec City, Canada
e-mail: pierre-olivier.methot@fp.ulaval.ca

P. Huneman
CNRS, IHPST, UMR 8590, Paris 1 Panthéon-Sorbonne, Paris, France

P.-O. Méthot (ed.), *Philosophy, History and Biology: Essays in Honour of Jean Gayon*, History, Philosophy and Theory of the Life Sciences 30,
https://doi.org/10.1007/978-3-031-28157-0_1

the Western philosophical tradition. In addition to playing the organ, the young Gayon learned philosophy, which he described "as an immense palace of thoughts deployed in time, where [he can] infinitely [wander]" (Gayon & Petit, 2018, p. 22). At the age of twenty, a year after the events of May 1968, he began a master's degree at the Sorbonne, attending lectures in history of science by Suzanne Bachelard and Georges Canguilhem. Though he had learned some philosophy of science, the philosophical approach he then discovered with Canguilhem and S. Bachelard differed from the education he received a few years earlier based on the history of philosophy and the study of classical authors. Thus, after the course on "medical ideology in the 19th century" with Canguilhem, he wrote, "[...] the path was marked out: this approach made sense to me, and counterbalanced the internal and arduous, but so comfortable, exegesis of the great philosophical systems" (Gayon & Petit, 2018, p. 25).

Following the defense of his master's thesis on the relation between formal and transcendental logic in Kant (1970) written under S. Bachelard, and then having obtained the "*agrégation*" in philosophy (1972), Jean Gayon taught at the Lycée Maurepas (Yvelines) and put into practice the recommendation that Canguilhem gave to his students: to learn about the results and methods of contemporary science before considering a PhD thesis in philosophy of science. Unlike his fellow students Anne Fagot-Largeault (1989) and Anne-Marie Moulin (1991), Gayon specialized in the life sciences, though he always had an interest in medicine.[2] After completing his compulsory military service compulsory military service (1973–1974), he obtained a *Diplôme d'études universitaires* in geology and biology at the University of Paris 6 (1974–1978) followed by a master's degree in plant biology (1981). These studies of the biological sciences left a deep mark on him:

> [I] cannot think about subjects involving biochemistry, genetics, evolution, without remembering that I have dissected flowers, arenicolids, oysters and mouse embryos, that I have analyzed tomato extracts, performed spectrographic analyses of these extracts, counted snail populations in the woods, identified lichens, mosses, antinafes, crossed and counted drosophila, and did endless mathematical statistics exercises. (Gayon & Petit, 2018, p. 30)

The 11 years he spent teaching philosophy, learning biology, and attending to the needs of his growing family—a situation he compared to a "marathon"—left deep traces in his body: to keep up he took some anti-depressant drugs that, in retrospect, he thought may have caused some of the cancers he developed in later life (Gayon & Petit, 2018, p. 27).

Deeply imbued with the lessons of biology, Jean Gayon admitted to having "gradually [unlearned] philosophy" (1989). But if the sciences fascinated him to the point of making him unlearn philosophy, his studies of biology did not lead to a conversion to science; eager to return to philosophy after this immersive experience, he phoned Canguilhem at the end of his master's degree, in the summer of 1981, to ask for his advice for the continuation. Canguilhem informed him that, having retired in 1971, he was no longer supervising theses and recommended that he approach Michel Foucault. As Gayon wished to work with a philosopher "more in phase with

[2] In the early 1980s, Gayon considered writing his Ph.D. thesis on the epistemological aspects of cancer disease (Gayon & Petit, 2018, p. 30).

current scientific knowledge" (Gayon & Petit, 2018, p. 31), Canguilhem directed him towards his former student, François Dagognet, then an influential professor of philosophy in Lyon and President of the jury of the "*agrégation*", who was mainly known for his writings on Louis Pasteur (1967) and on the history of natural history (1970) in the tradition of Bachelard and Canguilhem.

Back then, Gayon had not yet turned his back on a career in biology; he even added to his training a *Diplôme d'études approfondies* (DEA) in evolutionary genetics. Working under the direction of Claudine Petit—a former student of Georges Teissier, a French biologist who contributed to the evolutionary synthesis—it was during this final scientific year that he heard, for the first time, of the epistemological problems raised by the theory of evolution and that he discovered the "Modern Synthesis". Petit, in fact, had offered him to undertake a Ph.D. thesis in population genetics under her direction (Gayon & Petit, 2018, p. 30). His advanced training in evolutionary genetics, however, brought him back definitively to his philosophical roots.

As he was starting his doctoral studies with Dagognet, Jean Gayon accompanied his wife Elisabeth to the United States in 1983 on a study trip to New York, Boston, and Washington as a Fulbright student. In Boston, he met with Everett Mendelsohn, director of the Department of the History of Science at Harvard University. Upon his return, he organized a three-month stay at Harvard the next year to consult the correspondence of the biologists who were part of the evolutionary synthesis. During these few summer months (July–September 1984), he met with Ernst Mayr and visited the biologist and historian of population genetics William Provine. Thanks to Provine he made the acquaintance of philosopher Richard Burian, who became his best friend and closest colleague and introduced him to the philosophy of biology community. Upon his return to France, Gayon obtained a position of associate professor in the department of philosophy at the University of Burgundy in Dijon (1985–1997). He later became professor at the University of Paris 7 (1997–2001) and at the University of Paris 1–Panthéon-Sorbonne (2001–2018).

In 1989, under the direction of Dagognet, Gayon defended his thesis on *La théorie de la sélection: Darwin et l'après-Darwin*—a title suggested by Dagognet (Gayon & Petit, 2018, p. 35). Extremely well documented, this work explores how the "hypothesis" of natural selection put forward by Darwin was confirmed several decades later, following a complex path that went through the biometricians (Karl Pearson and Francis Galton), the early Mendelians, population genetics, and empirical studies. Gayon's thesis ends with an epistemological analysis of Motoo Kimura's "neutralist theory" and shows how it extends rather than challenges Darwinian selectionism. In 1992, Gayon published his thesis as *Darwin et l'après-Darwin: une histoire de l'hypothèse de sélection naturelle dans la théorie de l'évolution.* The book, which complemented some authoritative monographs such as Provine's *The Origins of Theoretical Population Genetics* (1971), was well received. The American philosopher of biology Marjorie Grene, whom Gayon met during his several stays in the United States, praised *Darwin et l'après-Darwin* as "a uniquely rich and significant work" and noted "it is imperative that [it] be made available in English" (Grene, 1994, pp. 161, 163). An English version was published at Cambridge University Press as *Darwinism's Struggle for Survival* in 1998 (Gayon, 1998). Read and commented by philosophers and historians of science as well as by biologists, it is a classic in the history and philosophy of biology.

As a philosopher of science, Jean Gayon's legacy on teaching and research in philosophy at university level is considerable. More than leaving his mark on the philosophy of biology in France, he genuinely founded this field by combining the French tradition of historical epistemology with an Anglo-Saxon approach, at a time when philosophy of science was mostly the philosophy of logic, physics, and mathematics. Like his fellow historian and biologist Michel Morange, Jean Gayon inspired doctoral students and early-career researchers, not by creating a school but by helping them to explore their own research questions about the life sciences or medicine.[3] Unsurprisingly, a large part of the research community in history and philosophy of biology in France (and beyond) has learned from his work. Likewise, biologists interested in the conceptual foundations of biology, its history, or in the ethical or political issues it raised often turned to Jean Gayon. Whether it is the issue of eugenics, the notion of race, the history of evolution, or the debate between nature and nurture, Gayon knew how to construct a systematic and nuanced problematic from which students could develop their own way of thinking.

Inseparably a historian and a philosopher of biology, he drew on the combined philosophical and historical heritage of Georges Canguilhem and Jacques Roger (without being their student in the traditional sense of the word) as well as on Anglo-Saxon philosophy and history of science, a field in which he immersed himself during his research stay in the United States.[4] Jean Gayon left a fruitful legacy in these two disciplines. At the most general level, he showed how some core concepts of biological sciences, such as species, heredity, natural selection, fitness, adaptation, and function are the result of complex historical layers of meanings, which are the site of multiple and often unresolved scientific and even philosophical controversies. Among many other things, Gayon has taught us that since Mendel, the notion of heredity, which was first a legal ("inheritance") and a medical term ("hereditary diseases") was brought into biological science: no longer a "force" that influences us from the distant past, heredity has become a statistical phenomenon (Gayon, 2000).

In *Darwinism's Struggle for Survival* (1998), Gayon explored the major crisis Darwin's hypothesis of natural selection underwent before it was progressively corroborated in the 1930s–1950s by the combination of population genetics, statistics, and experiments on laboratory flies. Evolution, as modeled by population genetics, Gayon argued, is based on a non-reversible temporality, distinct from that of Newtonian mechanics to which it is often compared.[5] As to the concept of "gene", torn between the disjointed pieces of DNA of molecular biologists and the abstract characterization of the early geneticists, Gayon claimed it may be at best a useful fiction (see Gayon, 2001). Another lesson Gayon taught us is that contemporary forms of eugenic practices are much more complex and diffuse than State-based

[3] On Morange, see Loison and Pradeu (2021). See also Pradeu, Huneman, and Giroux, this volume.

[4] On Gayon and Canguilhem, see Petit (2019) and Méthot (2022). See also Méthot, Brenner, Duchesneau, Loison, and Sloan, this volume.

[5] For an example, see Gayon's review of Elliott Sober (Gayon, 1989).

Nazi eugenics. Finally, and perhaps most importantly, Gayon has taught us that to understand a concept includes not only having a clear-cut definition of it, but also grasping its history and uses, reactivating the controversies that animate it, and identifying who lays claim to it and to what ends.

In the history of science itself, Gayon's research on the French school of genetics, initiated with Richard Burian and Doris Zallen, and continued with Laurent Loison, has shed new light on the success of the school of molecular biology led by François Jacob, Jacques Monod, and André Lwoff. Gayon and his colleagues have shown that instead of a French mystery, whereby the success of Jacob, Monod, and Lwoff debunked a mistaken legacy of Lamarckism, it was the intellectual and institutional legacy of Claude Bernard and Louis Pasteur's experimental physiology that enabled French researchers to draw on emerging molecular biology as a major tool, even though classical genetics had thus far remained marginal.[6] Adding to the philosophical method he embraced, Jean Gayon also explored normative questions such as those raised by human enhancement or by the concept of race, which he placed in a long historical perspective. In this task, the historian provides the philosopher with fresh material and critical distance to examine the present, though Gayon is aware that philosophy of science can do without the history or science.[7]

Following Alistair C. Crombie and Ian Hacking, Jean Gayon reflected on the notion of "style" in science and on its uses for historians of science. In retrospect, his own "style" as a historian and philosopher of science could be defined by his constant concern to build bridges between philosophy and science, between "historical epistemology" and Anglo-Saxon analytic philosophy of biology, between history of science and philosophy of science—and between people with different philosophical outlook (Petit, 2019). In fact, although Jean Gayon never resolved (in a way that would have satisfied him) his dual heritage—Gaston Bachelard and Thomas Kuhn, Georges Canguilhem and David Hull, which also extends beyond those influential figures, as this volume demonstrates, his own career as a historian and a philosopher goes a long way toward guiding us in the time to come.

Jean Gayon left us far too soon; but his contributions will long remain an essential reference point to understand contemporary development in the life sciences for philosophers, historians, and sometimes, biologists.

1.2 Overview of the Chapters

The first section of the book ("*Historical Epistemology and Philosophy of Biology*") examines the relation between historical epistemology and philosophy of biology in the works of Jean Gayon in terms of methods, approaches, and styles. Following in the wake of the "French style" in epistemology, Jean Gayon is also one of the

[6] See the chapter by Suárez-Diaz, this volume.

[7] On the importance of Lakatos in Gayon's work, see the chapter by Loison, this volume. On Gayon's critique of historical epistemology, see the chapter by Méthot, this volume.

most well-known representatives of philosophy of biology in France. In the light of this double philosophical heritage, **Chap. 2**, by Pierre-Olivier Méthot, traces Gayon's intellectual journey and explores why he naturally appears as a "Canguilhemian" outside of France—a view against which he defends himself. Attending to the connexions that Gayon established with philosophers of biology, Méthot brings to light the growing tension that inhabits Gayon, from the middle of the 1980s onwards, and which led him to identify some of the limits inherent to the matrix of historical epistemology, but without disowning his first philosophical tradition. In **Chap. 3**, Anastasios Brenner examines the reception of French philosophy of science and considers the contributions of Jean Gayon in the context of the philosophy of biology. Discussing the French tradition, Brenner notes how some of its most distinctive characteristics have emerged belatedly within the Anglo-American tradition. He concludes with a reflection on the fragile state of this dialogue which Gayon contributed to establish between French and Anglo-American philosophy of science. Then, focusing on *Darwinism's Struggle for Survival* (1998 [1992]), Gayon's most well-known scholarly contribution, François Duchesneau, in **Chap. 4**, shows how Jean Gayon can be characterized as a philosopher of biology who takes a historical approach to analysing epistemological questions. To support this claim, Duchesneau examines the methodology used by Gayon in *Darwinism's Struggle for Survival*, a book which he presented as "a scenario of rational reconstruction" (Gayon, 1998, p. 3). In this book, Duchesneau notes, Gayon took a "developmental perspective" on concepts and theories that he complemented with a detailed historical analysis of the theories that succeeded Darwin's own work.

Continuing the examination of Jean Gayon's methodological approach, Thomas Pradeu, in **Chap. 5**, suggests that Gayon's work carries out a synthesis between three fields: the history of science, the philosophy of science, and the life sciences. Pradeu argues that the philosophy of biology—a field long characterized by its emphasis on evolutionary theory—could strongly benefit from this synthetic view in order to develop a richer and broader analysis of contemporary biological sciences. Finally, in **Chap. 6**, Laurent Loison introduces a different perspective on Jean Gayon's methodological approach. Using the example of Darwin's theory of pangenesis, which Gayon rarely examined despite his interest in the history of heredity, Loison argues that one of the reasons for this omission may be that it did not fit with the "structuralist approach" Gayon expounded, an approach inspired by the examples of Imre Lakatos and Pierre Duhem and which, according to Loison, set Gayon's work apart from both philosophy of biology and French historical epistemology.

The second section of the book ("*Perspectives on Jean Gayon's Contributions to the History of Evolutionary Theory and French Genetics*") turns to Jean Gayon's contributions to the history and philosophy of evolutionary theory and to the emergence of French genetics. In **Chap. 7**, David Depew begins by considering how Gayon's encounters with American evolutionary biologists and philosophers of biology in the mid-1980s transformed his initial project of justifying the Darwinian research tradition in France. Conversely, Depew notes that Gayon taught his American colleagues how important French physiological genetics was to the Darwinism of the modern evolutionary synthesis and, with Richard Burian, insisted on the compatibility of the synthesis with postwar French molecular genetics. Depew also reflects on Gayon's institutional role in the creation of an international learned society in which historians and philosophers of biology work together. Following up on this, Edna Suárez-Diaz, in **Chap. 8**, argues that Jean Gayon and his colleagues' works on the history of molecular genetics in France set the groundwork for a critique of the "diffusionist" view of knowledge, and observes how their approach insists on place, movement, and local research traditions. According to Suárez-Diaz, Gayon and his colleagues contributed to our current narrative of how Mendelism and the molecular approach to heredity made their arrival to France, and how they were accommodated within existing disciplinary fields, traditions, and institutions, which challenged the "spreading" view of the development of science.

In **Chap. 9**, Michel Veuille highlights the importance of the studies of Jean Gayon and his colleagues in reconstructing the history of evolutionary theory in France after 1859 and of genetics after 1900. Following that, he concentrates on the work of Philippe L'Héritier and Georges Teissier, two pioneers of experimental population genetics in France. Veuille identifies the zoological roots of L'Héritier and Teissier's work and traces the intellectual legacy (and the relative isolation) of their school up to its collapse in the 1970s when a new generation of geneticists rebuilt international connections with geneticists. Moving on to another field, in **Chap. 10** Jorge Martínez-Contreras explores the critical analysis of what Jean Gayon called the "modes" of cultural evolution. These "modes", for Gayon, are research programs inspired by a shared paradigm, each of which seeks to connect biological, social, and physical sciences. Given the growing interest in cultural evolution, Martínez-Contreras argues, reflecting on Gayon's critical appraisal regarding the strengths and weaknesses of these modes is especially important and timely. Finally, going back to Gayon's analysis of Darwin and Wallace about the support for natural selection in *Darwinisn's Struggle for Survival*, Jon Hodge, in **Chap. 11**, revisits Wallace's 1858 theory of evolution and its relation to heredity. Hodge suggests that Gayon's historiography could be enhanced by giving Wallace a more positive role. In so doing, Hodge provides a different reading of Wallace and calls into question Gayon's attribution of a group-selectionist view to the biologist.

The third section ("*Case-studies on Biological Concepts and Theories*") concerns issues in biology and philosophy from the eighteenth to the nineteenth century, to which Jean Gayon has contributed in several of his publications. In **Chap. 12**, Phillip Sloan draws on Edmund Husserl's phenomenology for the solution to the "species problem" in the philosophy of biology. Situating this problem in relation to the concepts of "intentionality" and the "life-world", Sloan argues that two different intentional "thematizings" of the life world (one diachronic the other synchronic) provide a more satisfactory interpretation of the species problem than the usual "individualist" versus "class" conceptions of species. A larger goal of his chapter, however, is to highlight the importance of Gayon's work in fostering a dialogue between continental and analytic approaches to the philosophy of biology. In the following **Chap. 13**, Charles Wolfe explores the problem of the birth of biology and the nature of vitalism as a related set of issues, and asks what kind of answers the approach known as "historical epistemology of the life sciences" can tell us about these historiographical and philosophical issues. Contrasting Jean Gayon's and Georges Canguilhem's view on these matters, one being more on the side of the philosophy of biology and the other more historicist (although strongly normative at times), Wolfe reflects more broadly on how to conduct historical epistemologies of the life sciences. Stéphane Tirard, in **Chap. 14**, takes us back to Jean Gayon's analysis of "acquired character", a notion often attributed to Lamarck. Gayon's work on this topic, Tirard reminds us, was conducted during a collective project he ran along with Pietro Corsi, and Gabriel Gohau (1998–2000) in Paris, which led to the book: *Lamarck, philosophe de la nature* (Corsi et al., 2006). In his contribution, Tirard dissects the argument in Gayon's chapter on the history of the inheritance of acquired characters, and shows how Gayon clarified the oft-misunderstood role of Lamarck in the construction of this notion and how several authors contributed to it during the first half of the nineteenth century.

Continuing with nineteenth century biology, Gérard Chazal, in **Chap. 15**, addresses the issues raised in the wake of the French translation of *On the Origin of Species* (1859) by Clémence Royer in 1862. Revisiting her translation and notes, Chazal shows how Royer defended the theory of evolution against creationism, while also developing a view of organic evolution that was ideological, prompting Darwin to request another French translation of his work. Overall, Chazal argues, Royer's work belongs to the history of Darwinism's "struggle for survival" that Jean Gayon analyzed in his book (1998). This section ends with an analysis by Denis Forest of John Stuart Mill's view of natural selection and the design argument. In **Chap. 16**, Forest explores why Stuart Mill gives more credit to the design argument than to the Darwinian explanation of biological adaptations, while also recognizing how Darwin "has opened a path of inquiry full of promise, the results of which none can foresee" (Mill VII [1973], p. 499).

Collecting testimonies of colleagues, friends and students of Jean Gayon, the final section ("*Jean Gayon: Personal Memories*") has a more personal tone. Opening with Richard Burian's personal view of his work with Gayon on the history of French genetics, **Chap. 17** brings us back to the beginning of their lasting collaboration in 1984 before expanding through the next three decades. Burian illustrates key aspects of Jean's approach to understanding complex histories involving conceptual conflict and conceptual change and his efforts to explore an enormous range of topics in both breadth and depth. Unfortunately, the major book on the history of French genetics they envisaged was never completed. Next, in **Chap. 18**, Élodie Giroux provides an insider's view of Jean Gayon's philosophical style through her own experience as a doctoral student. Giroux recalls how Gayon's work represents a "unique synthesis" of analytical Anglo-American philosophy of science and French historical epistemology, stressing Gayon's belief in a "genuine dialogue" between them, a view she experienced herself during her graduate studies between Paris, Boston, and Berlin. In **Chap. 19**, Pascal Engel considers a rare text by Jean Gayon in which he takes up the topic of agriculture in Flaubert's *Bouvard et Pécuchet*. In this article, Gayon explores the knowledge of biology and agriculture in Flaubert's writings. He also suggests that Flaubert's interest in Pouchet's views on spontaneous generation is connected to romantic theories of life and matter, and to his lifelong interest in the nature of stupidity. Turning to Philippe Huneman's long relation with Jean Gayon, not as a professor or thesis supervisor but as a thesis examiner, **Chap. 20** brings out other dimensions in Gayon's work and personality, who was, for Huneman, "at once a 'major contemporary', a colleague, a friend, and an inspiration". Huneman traces his own journey as a philosopher of evolutionary biology, and describes his debt to Jean Gayon. Finally, in **Chap. 21**, Hans-Jörg Rheinberger brings this volume to a close with a personal note about his first encounters with Jean Gayon in the context of a conference of the International Society for the History, Philosophy, and Social Studies of Biology, an International Colloquium in honor of Marjorie Grene, and the "Cultural History of Heredity" project at the Max Planck Institute for the History of Science in Berlin. In this chapter, Rheinberger depicts Jean Gayon as a generous scholar, "always willing to share insights, even if unpublished".

References

Corsi, P., Gayon, J., Gohau, G., & Tirard, S. (2006). *Lamarck, philosophe de la nature.* Presses universitaires de France.

Dagognet, F. (1967). *Méthode et doctrines dans l'œuvre de Pasteur.* Vrin.

Dagognet, F. (1970). *Le catalogue de la vie.* Vrin.

Fagot-Largeault, A. (1989). *Les causes de la mort: histoire naturelle et facteurs de risque.* Préface de Georges Canguilhem. Vrin.

Gayon, J. (1989), Épistémologie du concept de sélection (analyse critique de E. Sober, *The nature of selection*). In P. Jacob (Ed.), *L'Âge de la science*, vol. 2, *Épistémologie* (pp. 201–227). Odile Jacob.

Gayon, J. (1998). *Darwinism's struggle for survival: heredity and the hypothesis of natural selection* (Cobb, M., Trans.). Cambridge University Press.

Gayon, J. (2000). From measurement to organization: A philosophical scheme for the history of the concept of heredity. In P. Beurton, R. Falk, & H.-J. Rheinberger (Eds.), *The concept of the gene in development and evolution: Historical and epistemological perspectives* (pp. 69–90). Cambridge University Press.

Gayon, J. (2001). La mort du gène. In L. Sfez (Ed.), *L'utopie de la santé parfaite* (pp. 379–394). Presses universitaires de France.

Gayon, J., & Petit, V. (2018). *La connaissance de la vie aujourd'hui.* ISTE Editions.

Grene, M. G. (1994). Review: Darwin et l'après-Darwin by Jean Gayon. *Journal of the History of Biology, 27*(1), 161–163.

Huneman, P. (2018). Jean Gayon (1949–2018). *Revue d'histoire des sciences, 71*(2), 311–317.

Loison, L., & Pradeu, T. (Eds.) (2021). *La biologie au défi de l'histoire. Mélanges offerts à Michel Morange.* Éditions Matériologiques.

Méthot, P.-O. (2022). Jean Gayon face à Georges Canguilhem: Une remontée en trois temps. In C. Cherici, M. Arminjon, & P.-O. Méthot (Eds.), *Le normal et le pathologique: Des catégories périmées? Textes réunis en hommage à Jean Gayon* (pp. 267–306). Éditions Matériologiques.

Mill, J. S. (1973) [1843]. A system of logic. In *Collected works of John Stuart Mill* (Vols. VII and VIII). Routledge.

Moulin, A.-M. (1991). *Le dernier langage de la médecine: Histoire de l'immunologie, de Pasteur au Sida.* Presses universitaires de France.

Petit, V. (2019). Jean Gayon (1949–2018): Itinéraire d'un darwinien. *Revue d'histoire des sciences, 72*(2), 349–380.

Provine, W. B. (1971). *The origins of theoretical population genetics.* University of Chicago Press.

Part I
Historical Epistemology and Philosophy of Biology

Chapter 2
From Historical Epistemology to the Philosophy of Biology: A Look at Jean Gayon's Intellectual Journey

Pierre-Olivier Méthot

Abstract The academic path of Jean Gayon (1949–2018) follows in the wake of the "French style" in epistemology, but he is also one of the first representatives of philosophy of biology in France. In the light of this double philosophical heritage, this chapter re-examines the relations between the works of Gayon, the tradition in which he first studied, and the one he later adopted, but not without reservations. Tracing his intellectual journey, this article explores why he naturally appears as a Canguilhemian, a view against which he defends himself. Attending to the connexions that he established with the philosophers of biology will bring to light the growing tension that inhabits him, from the middle of the 1980s onwards, and which led him to identify some of the limits inherent to the matrix of historical epistemology, without disowning his first philosophical tradition. Finally, in an attempt to reconcile two approaches, I will underline the emergence in the thought of Jean Gayon of a third way seeking to overcome the opposition between a "unitarian" style that seeks to make a clear break with the history of science and its philosophical past, and the "dualist" style which tends to stay above scientific debates, and which accepts a clear distinction between the work of science and the work of philosophy. Reuniting the history and the philosophy of the sciences, privileging the long durée and depths of history, where the problems of biology unfold, while also recognizing the value of analysing the conceptual puzzles which structure the contemporary life sciences, the *oeuvre* of Jean Gayon illustrates the complementary character of these disciplines.

Keywords Historical epistemology · Philosophy of biology · History of science · Richard Burian · Georges Canguilhem · Marjorie Grene · Ernst Mayr

---

Translated by Henry Dicks and modified from Méthot (2018).

P.-O. Méthot (✉)
Université Laval, Quebec City, Canada
e-mail: pierre-olivier.methot@fp.ulaval.ca

P.-O. Méthot (ed.), *Philosophy, History and Biology: Essays in Honour of Jean Gayon*, History, Philosophy and Theory of the Life Sciences 30,
https://doi.org/10.1007/978-3-031-28157-0_2

2.1 Introduction

Seen from this side of the Atlantic Ocean, the academic path of Jean Gayon lies in the wake of the "French style" in epistemology, even though he is also one of the first and most important representatives of philosophy of biology in France.[1] He reminds us in a rare autobiographical text that "it is through philosophy of biology that he became aware of a philosophical reflection on the life sciences that was quite different from that with which he was familiar, especially that of his master, Georges Canguilhem" (Gayon, 2014a, p. 207).[2] "After being slightly sceptical in the 1980s with respect to the 'philosophy of biology'—he writes in *Knowledge of Life Today*, a collection of interviews conducted by Victor Petit—I finally concluded that this type of research was perhaps more interesting than everything I had seen in my own country under the label of 'philosophy of the life sciences'" (Gayon & Petit, 2018, p. 53).

It is in the light of this double philosophical heritage—historical epistemology and philosophy of biology—that I propose in this chapter to re-examine the relations between the works of Jean Gayon, the tradition in which he first studied, and the one he later adopted, but not without reservations. During his career, Gayon went back and forth between historical epistemology and philosophy of biology, between his French masters and his Anglo-Saxon mentors, and between the historical method and conceptual analysis.

To begin, I will come back to several aspects of his intellectual journey, showing why Jean Gayon naturally appeared to many of his colleagues as a Canguilhemian, a view against which he defends himself. Examining this journey from a different angle, I will then document the tight links that he established with the philosophers of biology. This reading will make it possible to bring to light the growing tension that inhabits him, from the middle of the 1980s onwards and which leads him to identify some of the limits inherent to the matrix of historical epistemology. By reconstructing the relation of Jean Gayon to Canguilhem with the help of archival documents, we will then see that these tensions did not, however, lead him to disown his first philosophical tradition but to try to reconcile two approaches the methodological options of which are divergent. Finally, I will underline the emergence in the thought of Jean Gayon of a third way seeking to overcome the opposition between a "unitarian" style that seeks to make a clear break with the history of science and its philosophical past, and the "dualist" style which tends to stay above scientific debates, and which accepts a clear distinction between the work of science and the work of philosophy. Reuniting the history and the philosophy of the sciences, privileging the long durée and depths of history, where the problems of biology unfold, while also recognizing the value of

[1] Guided by the ideal of interdisciplinarity and keeping its distance from the history of science and the epistemology of medicine, philosophy of biology saw a rapid rise in Anglo-Saxon countries from the middle of the 1970s onwards. See Nicholson and Gawne (2014); Méthot (2023). On the "French style" in philosophy of science, see Braunstein (2002).

[2] On the role of Canguilhem in the development of historical epistemology, see Limoges (2015, 2019). On the different meanings of this expression and its recent history, see Méthot (2013, 2020).

analysing the conceptual puzzles which structure the contemporary life sciences, the *oeuvre* of Jean Gayon illustrates the complementary character of these disciplines.

2.2 Jean Gayon's Studies in Philosophy and Biology

After obtaining his "*baccalauréat*" in 1966, the young Jean Gayon chose to study philosophy at the Catholic Institute of Paris, after a brief period at the Lycée Henri-IV in *hypokhâgne* (first year of preparatory class for the Ecole normale supérieure). The following year, at the Sorbonne, he received an education education centered on the history of philosophy and the exegetical analysis of the texts of the Western philosophical tradition. Jean Gayon discovered the philosophy of the sciences thanks to the teaching of Bertrand Saint-Sernin and Claude Tresmontant at the Sorbonne in 1967–68, which introduced him to the French tradition in epistemology (Augustin Cournot, Gaston Bachelard, Jean Cavaillès, Pierre Duhem) and the Anglo-Saxon tradition (Karl Hempel, Ernest Nagel) in philosophy of science, which he read in the English original.

A year after the events of May 1968, at the age of twenty, Gayon began his master's studies and completed a certificate in the history of science in which he took courses from Suzanne Bachelard ("The History of Mechanics" and "Mechanism in the Nineteenth Century") and Canguilhem ("Medical Ideology in the Nineteenth Century"), an experience which he described later as "indescribable astonishment" (Gayon & Petit, 2018, p. 24). At that time, historical epistemology was in full swing and Canguilhem appeared as "the very embodiment of this 'French-style epistemology'" (Limoges, 2012, p. 53). During his last years of study, Gayon was thus nurtured by the teachings of two professors deeply engaged in the development of the philosophical history of science. These teachings had the effect of disturbing the quietude of the historical study of systems of thought to which he had been accustomed. The approach that he then discovered, in fact, differed from the training that he had received earlier, and his conception of philosophy also took an unexpected turn: after the course on ideology, he wrote, "[…] the path was marked out: this approach made sense to me, and counterbalanced the internal and arduous, but so comfortable, exegesis of the great philosophical systems" (Gayon & Petit, 2018, p. 25). Gayon noted that he discovered in the course on medical ideology authors like François Magendie, Claude Bernard, Augustin Cournot, and Sigmund Freud as well as objects of study like vivisection, the historical development of hospitals, and the foundation of psychiatry, which were unknown to him and "contrasted with everything [he] had heard about philosophy and the history of science, not to speak of philosophy at all" (Ibid.).

It was under the supervision of Suzanne Bachelard that Jean Gayon wrote, in 1970, his master's thesis on the relation between formal and transcendental logic in Kant. Back then, the influence of Canguilhem was limited to the advice he gave Gayon at the end of his course: "If you want to the philosophy of sciences, start by going to spend a few years in the faculty of sciences" (Gayon & Petit, 2018, p. 23). In other words, to acquire a solid scientific training before considering a PhD thesis in philosophy of science. Appointed at the Maurepas high school, after having passed his school teaching exam in philosophy ("*agrégation*") in 1972, Jean Gayon signed up for a course in geological and biological sciences, attending courses at the Jussieu campus between 1974 and 1981, while teaching general philosophy at secondary school. Except for temporary work on the history and epistemology of psychology organized by Michael Kail at the University of Créteil, where he met up with Paul Mengal and Françoise Parot, Gayon remained on the margins of university networks during his studies of biology. While at the University of Paris 6 (1974–1978), he studied several biological disciplines: paleontology, biochemistry, botany, zoology, physiology. After completing his elementary studies, he obtained a bachelor's degree followed by a master's degree in plant biology (1981) and deepened his training with certificates in genetics, plant physiology, zoology and botany. Despite this, he was not yet thinking of completing a doctoral dissertation on the long history and epistemology of the Darwinian theory of evolution.

Deeply imbued with the lessons of biology, this long detour of nearly a decade through "the learning of positivity [*positivité*] [...] was almost a one-way ticket" (Gayon, 1984a, p. 1). Following this long detour by biology, he wished however to take up philosophy again, which has initially led him to the sciences. Although he had been outside the academic network for over ten years, he telephoned Canguilhem, who had no recollection of him, to ask for advice. In retirement since 1971, Canguilhem suggested that he contact Michel Foucault. But as Jean Gayon wanted above all to work under the supervision of a philosopher interested in "contemporary scientific knowledge" (Gayon & Petit, 2018, p. 31), Canguilhem recommended him to approach instead François Dagognet, who was mainly known at that time for his writings on Pasteur, natural history and biological philosophy. Their first meeting took place in the courtyard of the Sorbonne, towards the end of the summer of 1981. At that time, Gayon had only met Dagognet once, in 1972, during his oral exam, when Dagognet was president of the jury of the "*agrégation*". However, the list of the philosopher's works on Pasteur, natural history, and biological philosophy "inspired his confidence" (Gayon, 2015, p. 1). Back then, Gayon had not yet turned his back on a career in biology; he would even add to his training a DEA (*Diplôme d'études approfondies*) in evolutionary genetics—and "not without annoying the one who was to become [his] thesis director" (Gayon, 1989a, p. 1). Dagognet must have sensed the ambivalence of his new student, for from their first meeting he warned him against "eternalizing [his] student condition" and made him "respectfully but firmly [...] understand" the incompatibility of his task as a professor with "serious historiographical work" (Gayon's letter to Dagognet, 28 October 1982).[3]

Despite this warning, Gayon enrolled in 1983 in the program at the Paris VII laboratory directed by Claudine Petit, a former student of the biologist Georges Teissier, determined to "learn more" about the problematic of evolution (Gayon, 1998a, p. xiii). His teachers included prominent biologists such as Piotr Slonimski

[3] Gayon's correspondence with Dagognet is in the "Fonds Jean Gayon" at the Pasteur Institute. GNJ.Cor2.

(molecular genetics), Gérard Buttin (formal genetics), Albert Jacquard (population genetics), Daniel Blangy (virology), Armand de Ricqlès (paleontology and evolution), and also Jacques Roger (history of science). It was during this final scientific year that he heard, for the first time, of the epistemological problems raised by the theory of evolution and that he discovered the Modern Synthesis. Developed between 1930 and 1950 by Theodosius Dobzhansky, Julian Huxley, Ernst Mayr , G. Leydard Stebbins and George G. Simpson, the evolutionary synthesis allowed, thanks to the introduction into biology of mathematical formalisms developed by Sewall Wright, R. A. Fisher and J. B. S. Haldane, the reunion of Mendelian genetics and the theory of natural selection (Mayr and Provine, 1981; Smocovitis, 1996).

Having received no training on the theory of evolution during his studies, his philosophical curiosity for this question was awakened when, at the end of his biology course, during an oral exam, an "eminent paleontologist" reproached him for using the term "adaptation", a concept that the latter judged "unscientific and metaphysical" (Gayon, 1998a, p. xiii). Impressed by the predictive models of population genetics and intrigued by the identity of the synthesis and its conceptual core, Gayon admits to having "hesitated about returning to philosophy". Petit offered him the possibility of continuing along this path and to complete a Ph.D. thesis in population genetics under her supervision but Gayon declined. His DEA training, in fact, brought him back to his roots. The conceptual debates that run through the Darwinian theory of evolution in its modern version (the Fisher/Wright debate, the concept of species, the idea of a "synthetic theory" of evolution, etc.) revived his "philosophical flame", which the science of the laboratory had gradually extinguished. Another significant consequence of this year of biology is that it led him to identify a first topic that remained a persistent object of reflection from the beginning to the end of his career, namely the history and epistemology of the Modern Synthesis of evolution (see Gayon & Huneman, 2019). And it was the year spent in Petit's genetics laboratory that oriented him in this direction. In a letter to Dagognet dated October 28, 1982, Gayon summarizes the situation thus:

> A year ago, you agreed to meet with me to develop a thesis project under your direction. I explained to you how my interest had been focused on the singular fate of population genetics in the French scientific institution. [...] This final detour [by the DEA] proved to be beneficial. On the one hand, it definitively convinced me that I was too old to be a student, and that by accumulating knowledge I would end up missing the conceptual adventure (forgive the understatement, the word is yours!) from which my ascetic detour among scientists had proceeded; on the other hand, the content of this DEA proved to be conducive to a maturation of the questions likely to animate my research. The "science" of evolution is in its infancy, and the polemics that make it almost complete have rekindled a philosophical flame that the test tube and quantitifcation [*le nombre*] had smothered over the years.

It was the year spent in Petit's genetics laboratory that oriented him in this direction. Although the epistemology and history of population genetics were henceforth to constitute the backdrop of his philosophical reflection, the problematic of his doctoral research had yet to be determined. Gayon considered three subjects: the concept of phenocopy in Jean Piaget's genetic epistemology, an intellectual biography of J. B. S. Haldane, or the history of evolutionary genetics in France, with a focus on the work of Philippe L'Héritier and Georges Teissier. "Phenocopy, Haldane, L'Héritier," replied Dagognet a few days later. "Obviously, each of them is still only a chapter [...]. You must open the angle as much as possible because a philosopher, in general, does not like to close himself up too much" (Letter from Dagognet to Gayon, 01/11/1982). With words that Gayon would remember forty years later, Dagognet not only led him to pin down his object of research, but he also confronted him with a critical choice between a career in biology and philosophy: "either you sink into biology, even if it means dephilosophering [sic] yourself (that is your right)—or you take a step back and fall on the side of epistemology" (Ibid.).

Although a career in the biological sciences would probably have pleased him greatly, Gayon followed the advice to "open the angle": after having studied the results of the biological sciences for a decade, he turned to philosophy and the history of the life sciences in order not to miss out on what Dagognet called "the conceptual adventure".[4]

In January 1984, while studying for his thesis, Jean Gayon put together an application in the context of the national call for the recruitment of researchers at the CNRS. Drawing on biological knowledge acquired during his studies in evolutionary genetics, this 124-page text, presented to the committee of philosophy, epistemology, and history of the sciences, entitled "History and Epistemology of the Synthetic Theory of Evolution (1930–1950)", includes several themes that he went on to study in more depth over the course of his career: the genesis and identity of the synthetic theory, the concept of species, population genetics, etc. Gayon was not selected but he was ranked 1st on the complementary list. He did not apply again the following year because the University of Burgundy (Dijon) recruited him as an associate professor. If his research in history of biology was developing in several directions, the writing up of his thesis was frequently pushed back, and not only as a result of the time spent preparing classes and lectures, but also because he suspected that "the synthetic theory of evolution" had perhaps been a sociological event rather than a veritable "theory" (Gayon & Petit, 2018, 34). The announcement of a post of full professor at the University of Burgundy (Dijon) in 1988 did force him, however, to start the urgent writing up of his thesis and the preparation of his professorial habilitation. Faced with a short time frame, he began writing his thesis which he ended up calling, following the advice of Dagognet, *Darwin et l'après-Darwin* (Gayon, 1992). The three volumes were completed between the months of October 1988 and May 1989, at a speed of 4 pages a day, and the viva examination took place on July 10, 1989. At the end of the same year, he completed his professorial thesis ("habilitation").

A year later, he was named full professor at the University of Burgundy, where he taught for seven years, before being invited to take up a post at University Paris VII. From 2001, he was professor at University Paris I, and then director of the Institute of the History and Philosophy of Sciences and Techniques (IHPST) from 2010 to 2016. Taking account of the institutional dimension of his journey, as well as the different people that supported him (Canguilhem, Bachelard, Dagognet, etc.), Jean Gayon could be taken for a "pure product of the Canguilhemian school" (Gayon, 2004, p. 5).

[4] The "conceptual adventure" is cited in Jean Gayon's letter to Dagognet (28/10/1982).

Indeed, in 1984, he announced "what [his] work had no intention of being", namely, "an epistemology separated from history" (1984b, p. 46) thus inscribing his research in the framework of a *historical* epistemology of the sort thematized by Canguilhem in certain famous texts.[5] Moreover, Gayon employs in certain strategic places in this text, which relates the genesis of concepts in nineteenth century evolutionary biology, the expression "historical epistemology", calls on the concept of "scientific ideology" introduced by Canguilhem, and renews the criticism of the "myth of the precursor" (Gayon, 1984b, pp. 9, 17, 92). Taking aim at Anglo-Saxon historians who, seeking to write the history of the synthetic theory of evolution, end up dissolving its content in the "ocean of precursors", he recognizes his debt to the evolutionary biologist, Ernst Mayr, while also casting a severe judgment on Mayr, who developed this myth, Gayon writes, "with talent and erudition" (Ibid., pp. 9, 93).

At the time, Jean Gayon was critical of all those who see in the Modern Synthesis only a fact "pertaining to the sociology of the scientific community and not to the history of science" (Lewontin, cited in Gayon, 1984b, p. 93). Thus, convinced that the history of the synthetic theory could not be reduced to a "sociological implosion of the biological community", the protégé of Dagognet aimed to make explicit the "problematic of the synthetic theory" by, on the one hand, grounding it in its "conceptual identity", while also, on the other hand, identifying its "conditions of possibility". It is at the level of conditions of possibility that "external" factors operative in the "epigenesis" of concepts are relevant for the historian of science. Unlike Canguilhem, Gayon expressed greater openness to the role of external factors in the dynamics of scientific change. "In this light", he wrote, "the facts of the sociology of the science start making sense, and even acquire their proper place in a historical epistemology worthy of the name: they become factors in the genesis of concepts" (1984b, p. 63).

It is certainly not an exaggeration to maintain, with Anastasios Brenner (this volume), that Jean Gayon "took" French epistemology to the international scene throughout his entire career, even if his intellectual itinerary cannot be reduced to this. Already in the introduction of the work drawn from his thesis (dedicated to Canguilhem and Dagognet) which appeared in 1992, he situates his method in the wake of an "epistemological history" and presents himself as an "epistemologist-historian" (1992, pp. 6, 412). There followed from the middle of the 1990s several collective volumes, chapters, and research articles that are typically categorized under the rubric of "French" or "historical" epistemology. I include here two books on the philosophy and the reception of Gaston Bachelard (*Bachelard: Le rationalisme appliqué* in 1994; *Bachelard dans le monde*, with J.-J. Wunenburger, in 2000); two books on historical and contemporary developments of French epistemology, *L'Épistémologie française 1830–1970*, with Michel Bitbol, in 2006 (Bitbol & Gayon, 2015 [2006]), and *French Philosophy of Science. Contemporary Research in France*, with Anastasios Brenner, in 2009 (Brenner & Gayon, 2009). To this may be added

[5] One thinks of course of the opening chapter in *Ideology and Rationality in the History of the Life Sciences*, where Canguilhem wrote that "in a sense, epistemology has always been historical" (1988, p. 10 [1977]).

an article on the concept of individuality in the biological philosophy of Canguilhem (Gayon, 1998b) as well as an essay on the relations between philosophy and the history of the science in the work of Dagognet (Gayon, 1998c). Lastly, one also finds a commemorative issue on Canguilhem in the *Revue d'histoire des sciences* (2000 when he was co-editor) as well as a study of "Bachelard and the history of the sciences", which appeared in a collective volume on French epistemology (Gayon, 2003).

This summary of the main institutional, biographical, and bibliographical milestones which mark the career of Jean Gayon from the 1970s to the present confirms the impression that he confided to me in private several years ago, namely, that "viewed from the outside [he is] a good example of a 'Canguilhemian' in action" (email from Gayon to Méthot, 8/4/2009). But, if this roughly sketched trajectory offers a faithful image of the intellectual journey of Jean Gayon, it remains silent about the crisis that animated him on the inside and which led him to take a prudent distance with respect to the French tradition of epistemology, to turn towards philosophy of biology, whose rigour and engagement with problems posed by the contemporary biological sciences he admired, without, however, fully embracing its approach in his own philosophical work.

2.3 The Turn Towards Philosophy of Biology and the Research Visit to Harvard

As he was starting his doctoral studies under the supervision of Dagognet, Jean Gayon accompanied his wife Elisabeth to the United States in 1983 on a study trip to New York, Boston, and Washington as a Fulbright student. At Cambridge (Mass.), he became acquainted with Everett Mendelsohn, a great admirer of the work of Canguilhem and director of the Department of the History of Science at Harvard University. A short while later, on returning to France, he developed the project of coming to Harvard on his own account. The document presented in January 1984 at the CNRS mentioned above fitted directly into this project. Announcing that he contacted Mayr and was waiting for a scholarship, he insisted on the importance of consulting manuscripts in the United States and underlined the "urgency" of meeting "the latest pioneers of the synthetic theory" (1984b, 114). With this research trip in mind, he solicited invitation letters from Mendelsohn, Mayr, Provine, Lewontin and Gould, the generosity of whom made a lasting impression on him (Gayon & Petit, 2018, p. 32). Having arrived, he set out to meet the people who had helped him become a Fulbright scholar and made the acquaintance of Richard Burian who became his best friend and a very close colleague.

If Mayr was able to take a positive view of a young researcher trained in biology who was interested in the history of the synthetic theory, he may possibly have been suspicious of the philosophical background of his interlocutor and especially of his avowed Canguilhemian heritage. From his first letter to Mayr dating from the 21st of January 1984, Gayon in fact claimed to belong to the "epistemological tradition initiated by Canguilhem" while also underlining his recent studies in biology under the supervision of Claudine Petit and the "determining" role of the synthetic theory in his personal intellectual itinerary.[6] But, Mayr had often been critical of the French historians and philosophers of science (Loison, 2020; Méthot, 2018). Further, Canguilhem and Mayr did not entertain particularly warm relations at the time.[7] By contrast, Mayr held Claudine Petit in high esteem, whose laboratory represented in his eyes the "fortress" of population genetics in France (letter from Mayr to Petit, 24/10/1972. Source: Ernst Mayr Paper, box 21, folder 1174), and incidentally had a lively interest in the work of Georges Tessier and Philippe L'Héritier. Thus it was that in March 1984, Claudine Petit announced to Mayr the forthcoming appearance of a series of publications of these authors by the intermediary of Jean Gayon, "a philosopher who took our DEA and [that she] oriented towards the history of genetics in France" (letter from Petit to Mayr, 14/3/1984. Source: Ernst Mayr Papers, Box 32, folder 1384).

A short while after his arrival in Boston, Gayon met Mayr and delved into the archives of the principal figures of the synthetic theory of evolution, whose letters he discovered. After a month of work dedicated to reading exchanges between Julian Huxley, Theodosius Dobzhansky and George G. Simpson, amongst others, he confided to Mayr that these documents "have completely changed his plans" and go "beyond his hopes". Cancelling several planned trips as a result of these discoveries, he shared in writing with the biologist "the intellectual excitement and the emotion" that he felt "when touching these material evidence of an entire chapter of the history of biology that he only knew through books and reports" (letter from Gayon to Mayr, 15/8/1984. Source: Ernst Mayr Papers, box 32, folder 1379). Moreover, he informed his host of his project for an "informative and interpretive" article which would insist on "the strong interaction between institutional factors and conceptual questions in

[6] Here is the letter Gayon sent to several American biologists and historians of biology. "Dear sir, I have read with most interest your main contributions to the evolutionary biology, and a number of your historical studies about the 'evolutionary synthesis'. Both have been determinative, in my personal itinerary, as well as the teaching of Claudine Petit (University of Paris 7). I am indeed working on a thesis under the direction of François Dagognet (Professor at the University of Lyon 3). My research purports to write a comprehensive history of the synthetic theory of Evolution, using as a basis my previous studies in both philosophy and biology. For, while I am 'Professeur agrégé de philosophie' and belong to the epistemological tradition initiated by Georges Canguilhem, I have also received a master's degree in biology and worked one year under the direction of Claudine Petit, in her laboratory of evolutionary genetics" (letter from Gayon to Mayr, 21/1/1984. Source: Ernst Mayr Papers, box 32, folder 1379).

[7] Mayr is said to have expressed a desire to meet Canguilhem at the time of his visit to as a professor at the Collège de France in 1978. Canguilhem apparently slipped away and a reciprocal distrust subsequently arose between the two men (email from Gayon to Méthot, 5/2/2017). Mayr's visit did on the other hand elicit a considerable interest amongst biologists and historians of science. "Various people in Paris have expressed to me their wish to be able to attend the lectures. This includes Claudine Petit (and her assistants), Jean Dorst (and his associates), and various historians of science such as Jacques Roger, Mirko Grmek, and Jean Théodoridès" (letter from Mayr to Jacob, 30/12/1977. Ernst Mayr Papers, box 25, folder 1269).

the rise of the 'synthesis' during the 1940s" (letter from Gayon to Mayr, 3/9/1984. Source: Ernst Mayr Papers, box 32, folder 1379).

Towards the end of his stay in Boston, however, the project of a comprehensive history of the synthetic theory, bringing together "internalism" and "externalism" seemed to him increasingly unachievable. His detour via the archives in fact led him to take seriously the hypothesis that the "synthetic theory" was perhaps less an actual "theory" formulated by Huxley, Mayr, Simpson, etc., with all that that implies from a philosophical point of view, than a phenomenon whose nature was above all sociological. The great "intellectual adventure" and the "heroic history" that he wanted to reconstruct (1998a, p. xiv), as well as the "conceptual identity" that he hoped to reveal on the basis of these unpublished sources, appeared to him over time as the result of pragmatic and concerted decisions, taken by different communities of biologists in the period following the Second World War. Little by little, he thus convinced himself that from the point of view of its epistemological foundations the synthetic theory was perhaps not a "scientific theory" in the classical sense of the term. Before leaving Boston, he revealed his doubts to Lewontin, who confirmed his intuition: "I believe your hypothesis is correct. There is no synthetic *theory*. It is rather a question to be examined from the point of view of the social history of the sciences" (Lewontin, cited in Ibid.). On his return to France, disappointed by the turn of events from a philosophical point of view, he recentred his thesis on the history of the *hypothesis* of natural selection in Darwin and his successors. This new orientation led him gradually to become one of the great international authorities on the history of Darwinism.

2.3.1 The Encounter with Richard Burian and Marjorie Grene

If the meeting with Mayr and the archives of the architects of the Modern Synthesis led him to rethink his thesis project with regard to the notion of heredity, another encounter oriented him towards philosophy of biology and, incidentally, brought him back to the history of science. One morning in August 1984, while he was getting ready to visit William Provine in Ithaca (Cornell University), Jean Gayon was welcomed by "a bearded individual with long curly hair, some years older than himself, who introduced himself as Richard Burian" (Gayon & Petit, 2018, p. 55). On that day, Burian was particularly pleased to welcome Gayon, for he was trying, as best he could, to understand an article by L'Héritier in French on maternal heredity and sensitivity to CO_2 in drosophila (email from Burian to Méthot, 8/2/2017; Burian, this volume). This unforeseen encounter was decisive in many respects for Gayon because Burian "made him aware of the methodological canons specific to history and to the interest of working on archives" (Gayon & Petit, 2018, p. 54; on Gayon's research on archival materials, see Méthot 2022). A long friendship between the two

Fig. 2.1 Jean Gayon and Richard Burian in the Alps in 1997 (Courtesy Violaine Jabbour)

men was rapidly established and gave rise to numerous publications on the history of genetics in France (Fig. 2.1).

In summer 1985, one year after a series of fruitful encounters with American historians and philosophers of science, Gayon finally discovered the new philosophy of biology. Installed in a front row seat, he attended the institutionalisation of this sub-discipline of the philosophy of the sciences, which brought him face to face with an epistemological reflection opposed to that which dominated in his country at the time. Taking part in a conference that proved decisive for the creation of the International Society for the History, Philosophy, and Social Studies of Biology (ISHPSSB) at St-Mary's College in Notre Dame (Indiana), he discovered a mode of philosophical reflection on the life sciences that at first left him doubtful but which impressed him nonetheless. At the time of this conference, where he presented a paper on Piaget (with Paul Mengal), Jean Gayon met David Hull, Robert Richards, and Michael Ruse, the founder, a year later, of *Biology & Philosophy*, a specialist journal which contributed to the standing and the professionalization of philosophy of biology.[8]

Later on in the same year, he attended the Congress of the History of Science Society at Bloomington, where he presented the first article in a long series co-signed with Burian and the biologist Doris Zallen on the history of genetics in France (Burian et al., 1988). It was there that he met the historian Jacques Roger, whose work he

[8] On this journal, see Pradeu (2017) and Gayon (2009a, 2009b). See also Pradeu, this volume.

knew but whom he had never met in person (even though Roger had taught him during in Paris during his DEA). Two years later, Burian introduced to him the American philosopher of science, Marjorie Grene, whom he "had the privilege to know well [...] in the latter part of her life" at her retirement in Blacksburg (email from Gayon to Méthot, 8/4/2009). Marginal, both because of her personal trajectory and her interest for continental philosophy of science (Méthot, 2023), Grene nevertheless had a decisive influence on philosophy of biology as an autonomous discipline within general philosophy of science in the United States (Burian, 2009; Gayon, 2009a, 2009b; Smocovitis, 2009). On the occasion of a meeting at Virginia Tech in 1987, the members of the ISHPSSB decided to found a learned society benefitting from special legal statutes (email from Burian to Méthot, 8/2/2017). It was in 1989, at the following meeting organized by René Roth at the Western University at London, Ontario (Canada) that the members of the ISHPSSB voted in favour of the statutes defined after the meeting of Blacksburg at Virginia Tech. The legal process leading to the foundation of the learned society was brought to completion thanks to the efforts of Jane Maienschein and Richard Burian.

These first contacts with philosophers of biology "wove old and deep ties between [him-]self and this world" (email from Gayon to Méthot, 5/2/2017). During the 1990s, Gayon entertained numerous professional relations with Anglo-American philosophers whom he regularly rubbed shoulders with and with whom he participated in then contemporary debates. Thus, after having written a long critical analysis of *The Nature of Selection* by Elliott Sober, a work little known at the time outside the Anglo-American world, he responded to the call of the biologist, Max K. Hecht, aiming to "evaluate the status of contemporary evolutionary biology" and published a synthesis of the central epistemological issues facing the Modern Synthesis in *Evolutionary Biology*. The atmosphere of "crisis" that beset the biology of evolution constituted a fertile ground for this type of reflective work, he wrote, because it "blurs the distinction between philosophy and science" (1991, p. 42). The acknowledgements show that Gayon henceforth belonged to the philosophy of biology community.[9] By taking an interest in the questions of contemporary science but also in the social and political issues raised by the life sciences and medicine (for example, the biological concept of race, eugenics, and medical enhancement), Gayon sought to make up for the desertion and immobilism that he regretfully observed in the philosophers of science in his own country, who seemed to accord but little interest to these problems.

While working actively on the history of the biological and medical sciences in France, Jean Gayon consolidated his relationships with philosophers of biology in the context of multiple talks, conferences, and research visits that he organized or that he participated in, both in Europe and the United States. In doing this, he contributed to the emergence of philosophy of biology in France and introduced several years later the first taught course of this type in his country. Thus, in the summer of 1992, he invited Hull and Ruse to participate in a conference on the "paradigm of filiation",

---

[9] "This chapter is very much indebted to R. M. Burian, M. Delsol, M. Ghiselin, M. Grene, D. Hull, B. Hecht, M. Hecht, and E. Mayr for their encouragements and fruitful criticisms" (Gayon 1991, p. 44).

which he organized in Dijon and Brussels. He translated for this occasion the text of Hull's paper which appeared in a work that he co-edited (Gayon & Wunenburger, 1995). In 1996, he met up again with Burian and Grene at the time of a visit as "international fellow" at the Centre for the Study of Science in Society at the Virginia Polytechnic Institute and State University. Increasingly interested in the question of biological individuality, he gave a paper on Canguilhem's treatment of this theme at a conference on French philosophy of science at Boston, the text of which appeared in the *Journal of the History of Biology* at the request of Mendelsohn, who attended the event.[10] Unlike Canguilhem, who was not very comfortable speaking English, Gayon mastered this language from childhood, as well as Spanish, a second foreign language that he managed to make use of in the journal of the philosophy of life sciences, *Ludus Vitalis*.[11] This mastery of languages could only facilitate exchanges with foreign researchers. In addition to his almost encyclopaedic knowledge of philosophy, of biological science and the history of the sciences, this linguistic facility, which is today only too rare, constitutes an important factor in the international success of Jean Gayon and in his original academic trajectory within the French intellectual landscape.

His book, *Darwin et l'après-Darwin* (1992) was published in English translation in a revised edition in 1998 by Cambridge University Press in the prestigious series "Cambridge Studies in Philosophy and Biology" under the title *Darwinism's Struggle for Survival. Heredity and the Hypothesis of Natural Selection* (1998a). This publication firmly established Jean Gayon's place within Anglo-Saxon history and philosophy of biology. Beyond the prestige associated with the publishers, its appearance was in large part due to the combined efforts of Michael Ruse and Jonathan Hodge, who advocated its publication with the editors of Cambridge University Press (email from Ruse to Méthot, 23/1/2017).

His later works addressing the role of chance in the Darwinian theory of evolution, the meaning and the history of the expression 'animal models' in biology and medicine, the utility of the notion of function in the life sciences, the 'dissolution' of the concept of the gene, as well as the debates surrounding the definition of the notion of life, are all original contributions he made to the literature on philosophy of biology (Gayon & Petit, 2018, p. 67). That said, Gayon did not practice the philosophy of biology as most philosophers nowadays do or aspire to do, that is, going into a biology lab and working closely with scientists on a conceptual and or methodological issue that arises from within a given scientific field.[12]

[10] On this paper, see Méthot (2022).

[11] See on this point Martínez-Contreras, this volume.

[12] For a different appraisal of Gayon's relation to philosophy of biology, see Loison, this volume.

2.4 Gayon's Critical View Toward Historical Epistemology

On December 18, 1989, in Dijon, before a jury composed of Jean-Claude Beaune, François Dagognet, Michel Delsol, Everett Mendelsohn, Anne-Marie Moulin and Jean-Jacques Wunenburger, Jean Gayon defended his habilitation to direct research, only five months after the defense of his thesis. In the very first pages of his report, he confided the following autobiographical reflection, which illustrates the change that has taken place in his posture as a historian and philosopher of science over the course of his meetings with Anglo-American colleagues:

> Various circumstances, including my thesis subject, a Fulbright grant, and a stay at Harvard University, brought me into contact with the American tradition of the history of science, a tradition that is quite unusual for a young French researcher, since it has been built since George Sarton on the decision to emancipate the discipline from the dual tutelage of philosophers and scientists ["*savants*"]. I was forced to note the extraordinary vitality of this school, and, by comparison, the progressive desertification of scientific historiography in my country. No doubt my generation was dazzled by the work, the power of synthesis, and the academic power of a long tradition which, claiming Auguste Comte's name and illustrated by P. Duhem, G. Bachelard, A. Koyré, G. Canguilhem and many others, was built on the postulate that the history of science is a primordial task of philosophy. Under such patronage, one has sometimes come to forget that beautiful epistemological conclusions do not fall from the sky, and that each generation must, if I may say so, 'return to the coalmine' of history ["*retourner au charbon de l'histoire*"]. It seems quite clear to me today that the philosophy of science, if it really claims to be built on the foundation of established knowledge ["*savoir constitué*"], cannot afford the luxury of compromising on historiographical detail. (Gayon, 1989b, p. 14)

Somewhat paradoxically, the personality, the work and the influence of Duhem, Bachelard and Canguilhem would have had the effect of isolating the philosophy of science in France, and even of weakening it on the international level, notably by encouraging an exodus of philosophers towards the history of science. In particular, the absence of institutional structures (learned societies, congresses, specialized periodicals, university courses, etc.) in France during the 1980s contrasted markedly with the effervescence—observed and experienced from the inside by Gayon—that the history and philosophy of biological sciences experienced in the United States at the time with the creation of the journal *Biology & Philosophy* (1986), then of the International Society for the History, Philosophy, and Social Studies of Biology (1989), and finally of specialized academic courses.

Gayon's testimony at the time of his habilitation, while indicating a certain departure from French epistemology, in no way constitutes a renunciation of a historical type of epistemology. The text of the habilitation nevertheless shows that while refusing to put belief and knowledge on the same level, he considers it fundamental to establish as precisely as possible the historiographical facts whose meaning cannot be presumed in advance, for example by having recourse to written or oral archives. This is the meaning of the injunction to "go back to the coalmine" of history before claiming to draw epistemological consequences from the history of science

Gayon thus claimed a different approach to the history and philosophy of science, without denying Canguilhem's philosophical legacy: "In their distant inspiration, the works presented for habilitation owe much to the example of Georges Canguilhem", he wrote; they "clearly belong to a school in which the history and philosophy of science are indissociable undertakings" (Gayon, 1989b, p. 53). However, Gayon said he is "aware" that this "indissociable" character represents "only one heuristic position among others. The history of sciences without philosophy is not vain" (Ibid.). In other words, the history of sciences does not have to serve as an accessory function to philosophy, contrary to a certain vision promoted by Canguilhem, who made it "a primordial task of philosophy". In the same way, if the history without the philosophy of sciences is not "blind", the philosophy of sciences without the history of sciences is not "empty", in the sense that it does not systematically require a deep anchorage in the history of thought; the philosophical reflection can unfold without being inscribed in a historical horizon. But as Gayon would put it later, the formula "historical epistemology" has precisely the defect "[to affirm] dogmatically that epistemology (the philosophy of sciences) can only be historical" (Gayon & Petit, 2018, p. 95).

Jean Gayon's principal reproach towards the French philosophical tradition is its withdrawal towards the history of the sciences as well as its disengagement from contemporary scientific questions. The "desertification" that he regretfully observed within the philosophy of the life sciences in his country in the 1980s and 1990s is explained in part by the profound influence of Canguilhem on a generation of students and whose retirement left an institutional vacuum that was difficult to fill. Indeed, a large number of Canguilhem's students, like Camille Limoges, Mirko Grmek, Yvette Conry and Christiane Sinding, turned resolutely towards the history of the biological or medical sciences. Although the "philosophical" dimension is never absent from their work, as François Duchesneau noted with respect to Canguilhem and Jacques Roger, they principally illustrate in the minds of their students a "'history of biological science' in a bipolar programme" (Duchesneau, 1997, p. x). On the other hand, the Institute of the History and Philosophy of Science and Technology (IHPST) that Canguilhem directed from 1955 largely focused, after his retirement, on the philosophy of logic and the philosophy of mathematics (Gayon & Petit, 2018, p. 37). With hindsight, this orientation was not favourable to the flourishing and the institutional development of the epistemology of the life sciences in France. Throughout the 1980s, Jean Gayon regretted more generally "the absence of academically structured research in the philosophy of the sciences" (Gayon & Petit, 2018, p. 95). Already in 1989, in the context of an analysis of the concept of natural selection in a work by the philosopher, Elliot Sober, *The Nature of Selection, Evolutionary Theory in Philosophical Focus* (1984) he noted that [he] was welcomed "like many other philosophers and historians" (1989c, p. 221) in Richard Lewontin's laboratory at Harvard. In a footnote, he added the following remark, which bears witness to the crisis he was living through at the time: "One would like such practices to be more common in this country [France]. The immense sadness of French philosophy of the sciences today, if not its desertion or desertification, does it not stem from the fact that too many philosophers have renounced the risk involved in partaking in contemporary debates?" (1989c, p. 221).

In stark contrast to the situation in France at the same time, Anglo-American philosophy of biology benefited from powerful institutional structures which favour interdisciplinary work and which have the remarkable consequence of producing schools of thought. At Harvard, Ernst Mayr, Richard Lewontin, and Stephen J. Gould, for example, welcomed numerous philosophers and historians of science interested in biology, several of whom have become their "protégés".[13] As the historian of philosophy of science Phillip Sloan remarks, "Bachelard, Foucault, and Canguilhem, do not seem to have been able to create anything like the disciplinary following generated by David Hull, Philip Kitcher, Alex Rosenberg, Elliot Sober, Bill Wimsatt, and the larger community of biological scientists who have found these general empiricist perspectives the most congenial to their research agendas" (2002, p. 227).

In the early 2000s, in an interview, Jean Gayon praised the refusal of philosophers of biology to separate in a watertight manner science and philosophy, a move that leads them to immerse themselves in the scientific problems of the time and to elucidate them, if not to resolve them, rather than looking down on them from above (see Hull, 1969, 1982; see Méthot, 2023). This new approach takes the opposite stance to the "position of Canguilhem and of historical epistemology", Gayon wrote, which "is based on the postulate that there is a radical distinction between philosophical activity and scientific activity. The philosopher has a critical exterior perspective on biological issues. He is not aboard the scientific boat" (2004, p. 4). Canguilhem, indeed, argued that without a historical perspective, epistemology would be "a completely superfluous double of the sciences, about which it would claim to speak" (2019, p. 257 [1968]). Echoing the sentiment expressed at the end of the 1980s, Gayon also criticizes the hesitation of philosophers of science to climb aboard the "scientific boat", the effect of which is to distance them from the debates constitutive of contemporary science, leaving them instead "tagging along behind" these problems (2004, p. 4).

Faced with interdisciplinarity, which favours collaboration between philosophers and biologists, the perspective "from above" of the Canguilhemian school appeared to Gayon to be based to a too great extent on the transmission and commemoration of a tradition, and not enough on useful work on concepts. Jean Gayon confided to me some years ago that he no longer recognizes himself in a "militant Canguilhemian school, because [he] believe[s] this attitude to be threatened by immobilism: the focus on commemoration and transmission means one forgets to carry out research in the field" (email from Gayon to Méthot, 8/4/2009). At the occasion of a talk entitled "Is there still a 'French Epistemology'" delivered to the American Philosophical Association in 2008, he explained that it was for "contingent reasons" that he edited three collected works on French epistemology, but that the majority of his publications in the history of biology did not belong to that tradition (2008, p. 3). More recently, he even evoked "the harm that historical epistemology has done to the development of the philosophy of sciences" in France (2014a, p. 207).

[13] As the philosopher of sciences, Alex Rosenberg, remarked to Werner Callebaut, "anyone who comes within their [Mayr, Gould, Lewontin] orbit is likely to become a *protégé*" (in Callebaut, 1994, p. 463).

2.4.1 Rapprochements and Tensions Between Two Philosophical Traditions

So, did Jean Gayon break with his own tradition? If so, then how could his critical remarks be reconciled with the affirmation that "his own style and movement are Canguilhemian" and that the method he employed "have mainly been inspired by the example of Georges Canguilhem" (Gayon & Petit, 2018, p. 53)? Following his frequent intellectual exchanges with American philosophers, amongst whom he discovered new mentors, like Grene, Gayon had to work with two philosophical traditions whose methodological choices and epistemological orientations were a source of great personal tension. Despite his turn towards philosophy of biology, that "space of discussion where philosophers and biologists work together in many ways to clarify major concepts" (Gayon, 2014a, p. 207), his research was still marked by the stamp of French epistemology. The discovery of philosophy of biology, in short, did not signify the pure and simple abandon of the tradition of historical epistemology.

The solidity of the links that tie Gayon to his tradition of origin is particularly visible in the first stage of his vast project on the history of genetics in France that he pursued with Burian and for which he solicited the opinion of Canguilhem, on the ground of his "historian's experience" and his "experience of the period". In a letter of 1987, he summarized how his "philosopher colleague, Richard Burian [...] and [himself] became friends over passionate discussion about the strange singularities of the history of genetics in France" and sets out the hypothesis that he wished to test to explain these "singularities". How can one explain, he writes, that the French scientific context was able to resist "the penetration of Mendelism" while becoming, several decades later, "one of the key sites for the development of molecular genetics"? "In the history that must be reconstructed, it seems to me that one must *at least* consider the hypothesis of a specific philosophical obstacle to the penetration of Mendelism in the French biology community. I have some vague ideas on the subject, but I tend to get somewhat lost", he writes (letter from Gayon to Canguilhem, 22/9/1987. Source: Caphés GC. 39.3.1). To envisage the historical development of a scientific discipline in terms of philosophical "obstacles" to be overcome unmistakably echoes the French tradition of epistemology, illustrated by Bachelard, Koyré, and Canguilhem, to which Gayon continued fully to belong. Later on in the letter, he revealed to Canguilhem the ground of his thinking as regards its overall project: to privilege the (sociological) approach, and to avoid the difficulties, even the possible "absurdity" of a "national" history of the sciences (letter from Gayon to Canguilhem, 22/9/1987. Source: Caphés GC. 39.3.1).

The organization of the conference in honour of Marjorie Grene at the University of Burgundy (which awarded her a *honoris causa* doctorate in 1997) illustrates again the attachment of Jean Gayon to his tradition of origin, as well as his desire to create a space for philosophy of biology in France. Already in 1987, at the time of his first encounter with Grene at Blacksburg, she unmistakably reminded him of

Brunschvicg, Koyré, and Canguilhem[14] and he immediately perceived the affinities between Grene and the style of epistemology practised by these philosophers: the refusal to separate philosophy and the history of science, the critique of positivism, and the clear distinction between science and philosophy. After a conference in the United States in 1993, Gayon and Burian thus decided to organize a conference in Dijon to underline the major contribution of Grene to the philosophy of the sciences. By honouring the work of someone he had only known for a few years and who had straight away taken him under her wing, Gayon also hoped to introduce to a French audience the works of an American philosopher whose theses resonate with his own tradition, and which have, moreover, played a structuring role in the rise of philosophy of biology (Burian, 2009; Somocovitis, 2009).

After having accepted the invitation, Grene sent him a detailed list of the personalities she wished to see present for the occasion. As for Gayon, he invited Canguilhem to attend, writing in a letter of October 1993 that it will be about "epistemology" in the sense that the French tradition has given to this expression, in accordance with Grene's expressed desire: "The conference in honour of Grene will, in keeping with her demands, be about *epistemology in the French sense of the term*, that is, about the theory of science, envisaged in its entire historical and philosophical extension" (letter from Gayon to Canguilhem 6/10/1993. Source: Caphés, GC. 33.9.6; my italics). Canguilhem had refused to attend the conference organized in his honour three years earlier at the Sorbonne. Likewise, declining Gayon's invitation, he confided to Gayon that he would prefer to renounce "any claim to be of interest to a meeting of philosophers and historians of science" (letter from Canguilhem to Gayon, 21/10/1993. Source: Caphés, GC. 33.9.6).

Canguilhem's response gave nothing away regarding his thoughts about Grene or her works, the importance of which for the epistemology of biology he had underlined in "The Question of Normality in the History of Biological Thought" (Canguilhem, 1988, p. 144). This one-off reference occurred after a meeting between Grene and Canguilhem at a conference in Finland in 1973 (see Méthot, 2018, 2023). In any case, Gayon was certainly a little disappointed by the missed encounter between the "doyen" of historical epistemology and the "grandmother" of philosophy of biology. He also wondered why Grene had neglected historical epistemology, given that her style and some of here theses are close to the French tradition in the philosophy of the sciences: "Curiously, Marjorie Grene, who has written so much about French philosophers [...] was not particularly interested in the French tradition of 'historical epistemology'. It is worth noting, however, a late study on Canguilhem where the convergence is established" (Gayon, 2007, p. 18).

This late study resulted from an invitation to a thematic issue of the *Revue d'histoire des sciences*, where he held the role of deputy co-editor in chief. For reasons that are difficult to make out, Grene apparently told him that she did not know the works of Canguilhem and even asked him for more information on the

---

[14] "When I heard Marjorie Grene the first time, I said to myself: 'I've already heard something like this before'. It reminded me of Léon Brunschvicg, Alexandre Koyré, Georges Canguilhem" (Gayon and Petit, 2018, p. 59).

subject of the former director of the Institute of the History of Science. Gayon told me that "it was [he] who, at a certain moment, asked her for an article for a special issue of the *Revue d'histoire des sciences*. […] She evidently noticed the opposition between Canguilhem and analytic philosophy straight away, and his way of tightly articulating history of the sciences, history of philosophy, and epistemology" (email from Gayon to Méthot, 8/4/2009). That Grene had claimed not to know Canguilhem is all the stranger given that she often cited his works approvingly, but could be linked to the tensions within the philosophy of biology itself where continental philosophy was not welcome (Méthot, 2023).

2.4.2 Biological Philosophy Versus Philosophy of Biology

From his first intellectual exchanges with philosophers of biology, Jean Gayon noticed that his American colleagues perceived him as a *historian*, not as a *philosopher* of science, even though he was taught by Suzanne Bachelard, he grew up in the tradition of Georges Canguilhem and François Dagognet, who were protégés of Gaston Bachelard, and he himself had obtained the "*agrégation*" in philosophy. But these personalities are either unknown to Americans (the case of Bachelard and Dagognet) or considered as historians (the case of Canguilhem), and not as philosophers (Méthot, 2020). Gayon said of himself that he was "stunned and hurt" to hear his American colleagues repeat to him "that in France, you don't have any philosopher of the sciences" (2014a, p. 207). According special attention to his professional titles, he took care to present himself as a philosopher to his American colleagues.[15] He also did not hide his astonishment (even his irritation) faced with the way philosophers of biology ignore the existence of an original *philosophical* reflection on the biological and medical sciences in France. With regard to the book, *Philosophy of Biology Today*, by his friend Michael Ruse, for example, Jean Gayon regretted—in the pages of the *Journal of the History of Biology*—that the author "only devotes one chapter to 'other lands', that is to say, other than Canada, the United States, and Great Britain. The only sentence of the book about France says this: "France has little interest on the subject", which is frankly astounding for any French philosopher. The fifty-two pages of bibliography do not mention Canguilhem's name" (1998b, p. 305). Without ceasing to draw on Canguilhem, Gayon thus defends Canguilhem's legacy and speaks in his favour in the company of his Anglo-Saxon colleagues (Méthot, 2022).

Faithful to the historical dimension that characterizes his thought, Gayon published in 1998 a long encyclopaedia entry in which he describes the differences between "biological philosophy" and "philosophy of biology" and in which he also endeavours to explain a "nuance" between these traditions which had "until today gone completely unnoticed" (1998d, p. 2153). Thus, after having rubbed shoulders

[15] For example, in his review of 1991, published in *Evolutionary Biology*, he specifies that he is reacting "as a philosopher" (1991, p. 2).

with American philosophers for around fifteen years he sat down to write the differences between "philosophical biology" and "philosophy of biology". Whereas the former developed during the nineteenth century and the first half of the twentieth, principally in Europe, and consists of philosophical reflections on the nature of life and their implications from a human perspective, the latter presents itself as a philosophical *discipline*, denies any clear-cut frontier between science and philosophy, and concentrates on the clarification and the definition of scientific concepts. Biological philosophy demarcates itself from scientific discourse to the extent that it claims to reflect more generally about "life", whereas, to the contrary, philosophy of biology is interested in the definitional and methodological problems of the contemporary sciences (1998d, p. 2154).

The expression "biological philosophy" is old and its genealogy, retraced by Gayon, shows that it can be retrospectively applied to Aristotle, Descartes, Kant, Bergson, Goldstein, or Canguilhem. On the other hand, the "semantic potential" of philosophy of biology is different and ties in with the post-positivist movement of the second half of the last century. As a specific field it is attached to a tradition of thinking which presupposes the existence of a scientific field called "biology". The definition given by Canguilhem in a course from 1946 to 1947 entitled "Biology and Philosophy" allows one to better evaluate the distance that separate *biological* philosophy from philosophy *of biology*: "By a biological type of philosophy, I mean a philosophy that draws its inspiration and its themes of reflection from a vision of the vital phenomenon understood in its originality" (Canguilhem. Source: Caphés, GC.12.1.8). During the 1960s, when Canguilhem suggested that "contemporary biology [...] is in some way a philosophy of life" (Canguilhem, 2019a, 2019b, p. 746 [1968]), what he saw exactly in this emerging field were specifically philosophical themes of reflection on the nature of the living and of life, a point noticed by Gayon and which allowed him to articulate the difference between the two philosophical styles: "When Georges Canguilhem was asking whether molecular biology brought with it a new 'concept of life', American 'philosophy of biology' was asking if Mendelian genetics was reducible to molecular genetics" (Gayon, 1998d, p. 2154).

Gayon is one of the first to have insisted on methodological differences between these two research traditions. Moreover, before everyone became aware of the place of biological philosophy in the work of Canguilhem—and without however proceeding to its historic-critical study as Camille Limoges did—Gayon had picked up on the commitment of Canguilhem towards this approach, which had for a long time been obscured by the disproportionate attention accorded by commentators to epistemology and the history of the sciences (Limoges, 2015, 2018, 2019). Gayon also pointed out to his American colleagues that such an approach "tightly and rigorously combining philosophy of biology and philosophy of medicine is particularly rare nowadays" (1998b, p. 325), a rarity that is explained, amongst other things, by the desire of philosophers of biology to see a certain style triumph by voluntarily pushing to one side questions relating to medicine and to health (Gayon, 2004, p. 3).

2.5 Conclusion: History and Philosophy of the Sciences: Towards a Third Way?

Let us return to how Jean Gayon managed to reconcile the methodological imperatives of philosophy of biology with his natural inclination towards the history of philosophical and scientific problems. As a practitioner of philosophy of biology not educated in this tradition and working at the frontier between the continental and analytical styles, the philosopher-historian-biologist worked out very early how to adopt an external perspective on philosophy of biology, which was just beginning to take an interest in its own history and whose historiography continued to be dominated by Anglo-Saxon philosophers.[16] But already in 1998, he suspected that philosophy of biology risks losing "the memory of the very particular historical conditions in which it was born" (1998d, p. 2154). In this history, which has yet to be written, it is imperative that these "historical conditions" are recalled in order better to discern the ambitions of the philosophers who, by founding philosophy of biology on the analysis of conceptual enigmas, explicitly or implicitly excluded a whole set of work on the epistemology and the history of biology[17] and medical sciences, distancing themselves both from bioethics and from philosophy of medicine (Méthot, 2023).

During the 2000s, Jean Gayon continued to develop the distinctive traits of the "new" philosophy of biology that he had earlier pinpointed, while at the same time reflecting on its relations with the science and the history of the biological sciences (Gayon, 2009a, 2009b, p. 2014). Thus it is that he came to designate the tradition in which philosophy of biology arose as "unitarian" inasmuch as the difference between philosophy and science is only a difference of *degree* not of *genre*. Conversely, he named "dualist" the tradition in which science and philosophy constitute distinct domains of knowledge. Through the orientation of his research and the originality of his philosophical style, does Jean Gayon belong to the "dualist" French tradition of epistemology and history of the sciences or to that other "unitarian" tradition discovered later on and that he taught in France for twenty years?[18] In which of these categories should he be placed: the historian of the sciences or philosophy of biology?

As we arrive at the end of the present study, it is better not to force the issue unnecessarily: Gayon has worked as a philosopher *and* as a historian of the life sciences, setting this or that problem of epistemology of biology in its proper historical context. Choosing now conceptual analysis and the analytic method, now the

[16] In one of his latest articles, covering the history of philosophy of biology from the nineteenth century to our time, Hull admitted that "German and French philosophers of biology have not had the impact that they perhaps should have had on Anglophone philosophy of biology" (Hull, 2008, p. 11).

[17] The work of Joseph H. Woodger, for example, was rapidly set aside and judged in an excessively negative and critical way by Ruse, Hull and other philosophers of biology. See Nicholson and Gawne (2014).

[18] Further, Jean Gayon frequently encouraged his doctoral students to draw on the best of these different research traditions. On this point, see Giroux, this volume.

path of the genealogical account and the historical method, the original craftsmanship of his work lies most often in the attempt to reconcile these two dimensions, without necessarily always succeeding. Recognizing that he was "torn between two contradictory visions, between "two traditions" that he has "successively embraced", "these questions of belonging", he said, "have become secondary in [his] eyes" (Gayon & Petit, 2018, p. 75). "Am I myself a historian of the science or a philosopher of biology? I myself do not know, and I am not certain that it is of great importance!" (Gayon & Petit, 2018, p. 90) However, belonging to this or that research community was for him an important symbolic issue after his discovery of philosophy of biology in the middle of the 1980s and faced notably with the reactions of his American colleagues.

Let us also underline that it is generally with a view to "using it philosophically", to take up Canguilhem's expression (2011, p. 390 [1949]), that he drew on the history of the (biological) sciences and that he practices the latter professionally. Never renouncing the legacy of Canguilhem, who remained for him his only master, Gayon has shown himself willing to listen to his "philosophical lesson" which consists in "drawing philosophical lessons from the history of the science", while also "being more attentive to the state of the art of scientific knowledge" (Gayon & Petit, 2018, p. 54). In addition, his works in the history of French genetics with Burian and Zallen, and that on the history of population genetics in France with Michel Veuille, was based on classical historical methods (archival research, interviews, etc.) (Méthot, 2022).

Despite his openness, on the one hand, to a philosophy of the science that reduces the traditional gap between philosophy and science, and, on the other, his attachment to a style of philosophy of the sciences which gives a central role to the history of the science and to questions relating to medical sciences, Gayon's position coincides neither with an entirely historicist posture, nor with the canon of philosophy of biology. Rejecting the attitude that he judges "dogmatic", according to which epistemology is only "historical" (Gayon & Petit, 2018, p. 95), he is equally mistrustful, perhaps even more so, of a philosophical reflection on the sciences that is entirely detached from a historical or social perspective. Sceptical with respect to the notion of historical epistemology, "to which he never lays claim" (Gayon & Petit, 2018, p. 76), he called into question the "survival" of philosophy of biology inasmuch as it arises from a total dissociation of the history or the sociology of the sciences (Gayon, 2009a, 2009b, p. 211) and judged excessive the "myth of total innovation" entertained by philosophers of biology with respect to their discipline's past (Gayon, 2014b, p. 23). Neither "dualist", nor "unitarian", Jean Gayon wagers instead on the existence of a third way:

> Between a historical history of the sciences, refusing any philosophical posture, and a purely analytical philosophy of the sciences, there exists a *tertium*. This third way, that I would be reluctant to call historical epistemology, given the heterogenous practices this word covers, consists in making the wager that conceptual depth may go hand in hand with temporal breadth. (Gayon & Petit, 2018, p. 77)

On this path, he rediscovered, perhaps slightly despite himself, the expression of historical epistemology and joins up with the ideal of complementarity proposed by

Thomas Kuhn towards the end of his life: history and philosophy of the sciences have different ambitions and methods, they are autonomous from a disciplinary perspective, but they can shed light on one another and certainly have an "interest in collaborating". If this solution cannot completely appease the tensions between history and philosophy of the sciences, it did however allow Jean Gayon to lay claim to the double philosophical heritage that is his own and to elucidate in a singular way the different facets of science.

Acknowledgements A first version of this chapter was published in 2018. I had the chance to present it to Jean Gayon during the workshop in his honor in Paris a year before. I would like to thank Dick Burian, Élisabeth Gayon, Jon Hodge, and Michael Ruse, who agreed to share with me their memories of Jean Gayon. I also address my thanks to Dick Burian, Jean Gayon, Élodie Giroux, Camille Limoges, Philippe Huneman, Laurent Loison, and Charles Wolfe for their rereading of a first version of this text, as well as Nathalie Queyroux, who oversees the CAPHES (Georges Canguilhem Archives) as well as the archivists at Harvard University (Ernst Mayr Papers) and the Institut Pasteur (Jean Gayon Papers) for their precious assistance with my research.

References

Bitbol, M., & Gayon, J. (Eds.). (2015) [2006]. *L'épistémologie française: 1830–1970*. Éditions Matériologiques.

Brenner, A., & Gayon, J. (Eds.). (2009). *French studies in the philosophy of science. Contemporary studies in France*. Springer.

Braunstein, J.-F. (2002). Bachelard, Canguilhem, Foucault: Le "style français" en épistémologie. In P. Wagner (Ed.), *Les philosophes et la science* (pp. 920–963). Gallimard.

Burian, R. (2009). Marjorie Grene: A remembrance with special attention to her importance for ISHPSSB. *Biological Theory, 4*, 183–187.

Burian, R., Gayon, J., & Zallen, D. (1988). The singular fate of genetics in the history of French biology, 1900–1940. *Journal of the History of Biology, 21*, 357–402.

Callebaut, W. (1994). *Taking the naturalist turn seriously; or how real philosophy of science is done*. Chicago University Press.

Canguilhem, G. (2019) [1968]. L'objet de l'histoire des sciences. In G. Canguilhem (Ed.), *Écrits d'histoire des sciences et d'épistémologie*, t. III. (pp. 253–273). Vrin.

Canguilhem, G. (2019) [1966]. Le concept et la vie. In G. Canguilhem (Ed.), *Écrits d'histoire des sciences et d'épistémologie*, t. III. (pp. 711–746). Vrin.

Canguilhem, G. (1988) [1977]. The question of normality in the history of biological thought. In G. Canguilhem (Ed.), *Ideology and rationality in the history of the life sciences* (Goldhammer, A., Trans.) (pp. 125–145). MIT Press.

Canguilhem, G. (2011). Rôle de l'histoire des sciences dans la philosophie des sciences: l'établissement des faits fondamentaux de la dynamique » [1949]. In G. Canguilhem (Ed.), *Résistance, philosophie biologique et histoire des sciences 1940–1965*, t. IV. (pp. 369–391). Vrin.

Duchesneau, F. (1997). *Philosophie de la biologie*. Presses Universitaires de France.

Gayon, J. (1984a). Audition pour le recrutement à un poste d'assistant, Dijon, 17 septembre 1984. Unpublished manuscript. Source: Fonds Jean Gayon, Institut Pasteur, GNJ.A1.

Gayon, J. (1984b). *Histoire et épistémologie de la théorie synthétique de l'évolution (1930–1950)*. Unpublished manuscript. Fonds Jean Gayon, Institut Pasteur, GNJ.O1.

Gayon, J. (1989a). Exposé de thèse (10 juillet 1989). Unpublished manuscript. Fonds Jean Gayon, Institut Pasteur, GNJ.B3.
Gayon, J. (1989b). Synthèse des travaux, présentée en vue de l'obtention du titre de docteur habilité à diriger des recherches. Unpublished manuscript. Source: Fonds Jean Gayon, Institut Pasteur, GNJ.A1.
Gayon, J. (1989c), Épistémologie du concept de sélection (analyse critique de E. Sober, *The nature of selection*). In P. Jacob (Ed.), *L'Âge de la science*, vol. 2, *Épistémologie* (pp. 201–227). Odile Jacob.
Gayon, J. (1991). Critics and criticisms of the modern synthesis: The viewpoint of a philosopher. *Evolutionary Biology, 24*, 1–49.
Gayon, J. (1992). *Darwin et l'après-Darwin: Une histoire de l'hypothèse de sélection naturelle.* Éditions Kimé.
Gayon, J. (1994). *Gaston Bachelard. Le rationalisme appliqué.* CNED.
Gayon, J. (1998a). *Darwinism's struggle for survival: Heredity and the hypothesis of natural selection.* Cambridge University Press.
Gayon, J. (1998b). The concept of individuality in Canguilhem's philosophy of biology. *Journal of the History of Biology, 31*, 305–325.
Gayon, J. (1998c). Dagognet: quelle philosophie pour quelle histoire des sciences? In R. Damien (Ed.), *François Dagognet, médecin, épistémologue, philosophe. Une philosophie à l'œuvre* (pp. 59–77). Institut Synthélabo pour le progrès de la connaissance.
Gayon, J. (1998d). La philosophie et la biologie. In J.-F. Mattéi (Ed.), *Encyclopédie philosophique universelle* (Vol. IV, pp. 2152–2171). Presses universitaires de France.
Gayon, J. (2003). Bachelard et l'histoire des sciences. In J.-J. Wunenburger (Ed.), *Bachelard et l'épistémologie française* (pp. 51–113). Presses universitaires de France.
Gayon, J. (2004). À quoi servent les philosophes… de la biologie? Interview with Michel Imbert and Jean-Jacques Perrier. *Revue Vivant*, 2.
Gayon, J. (2007). Introduction. In J. Gayon & R. M. Burian (Eds.), *Conceptions de la science. Hier, aujourd'hui et demain. Hommage à Marjorie Grene* (pp. 9–19). Ousia; *à l'œuvre* (pp. 59–77). Institut Synthélabo pour le progrès de la connaissance.
Gayon, J. (2008). Is there still a 'French epistemology'? Unpublished text. Source: Fonds Jean Gayon, Institut Pasteur, GNJ.E19.
Gayon, J. (2009a). Philosophy of biology: An historic-critical characterization. In A. Brenner & J. Gayon (Eds.), *French studies in the philosophy of science: Contemporary research in France* (pp. 201–213). Springer.
Gayon, J. (2009b). Marjorie Grene: Personal memories. *Biological Theory, 42*(2), 188–190.
Gayon, J. (2014a). Jean Gayon par lui-même. *Cités, 2*(2014), 205–209.
Gayon, J. (2014b). Biologie et philosophie de la biologie: Paradigmes. In T. Hoquet & F. Merlin (Eds.), *Précis de philosophie de la biologie* (pp. 11–24). Vuibert.
Gayon, J. (2015). François Dagognet (1924–2015): faits et souvenirs. Unpublished manuscript. Source: Fonds Jean Gayon, Institut Pasteur, GNJ.Cor2.
Gayon, J., & Wunenburger, J.-J. (Eds.). (1995). *Le paradigme de la filiation.* L'Harmattan.
Gayon, J., & Wunenburger, J.-J. (Eds.). (2000). *Bachelard dans le monde.* Presses universitaires de France.
Gayon, J., & Petit, V. (2018). *La connaissance de la vie aujourd'hui.* ISTE Editions Ltd.
Gayon, J., & Huneman, P. (2019). The modern synthesis: Theoretical or institutional event? *Journal of the History of Biology, 52*, 519–535.
Hull, D. L. (1982). Philosophy and biology. In G. Fløistad (Ed.), *Contemporary philosophy. A new survey* (pp. 281–316). Nijhoff.
Hull, D. L. (1969). What philosophy of biology is not. *Journal of the History of Biology, 2*, 241–268.
Hull, D. L. (2008). The history of the philosophy of biology. In M. Ruse (Ed.), *The Oxford handbook of philosophy of biology* (pp. 11–33). Oxford University Press.
Limoges, C. (2012). L'épistémologie historique dans l'itinéraire intellectuel de Georges Canguilhem. In H. Schmidgen, P. Schöttler & J.-F. Braunstein (Eds.), *Epistemology and History.*

From Bachelard and Canguilhem to Today's History of Science (pp. 53–66). Max Planck Institute for the History of Science.

Limoges, C. (2015). Philosophie biologique, histoire des sciences et interventions philosophiques, Georges Canguilhem 1940–1965. *Œuvres complètes de Georges Canguilhem, Tome IV, Résistance, philosophie biologique et histoire des sciences 1940–1965* (pp. 7–48). Vrin.

Limoges, C. (2018). La confirmation de l'historien des sciences et la mise à l'épreuve de sa philosophie biologique: Georges Canguilhem 1966–1995. *Œuvres complètes de Georges Canguilhem, Tome V, Histoire des sciences, épistémologie, commémorations, 1966–1995* (pp. 7–60). Vrin.

Limoges, C. (2019). Épistémologie historique et histoire épistémologique des sciences chez Georges Canguilhem. *Œuvres complètes de Georges Canguilhem, Tome III, Écrits d'histoire des sciences et d'épistémologie* (pp. 7–128). Vrin.

Loison, L. (2020). Enseigner la théorie synthétique au Collège de France. Les cours d'Ernst Mayr dans la chaire de François Jacob (1978). In A. Compagnon & C. Surprenant (eds.), *Darwin au Collège de France, XIXe-XXIe siècles* (pp. 159–171). Collège de France.

Mayr, E. & Provine, W. B. (Eds.) (1981). *The evolutionary synthesis: perspectives on the unification of biology*. Harvard University Press.

Méthot, P.-O. (2013). On the genealogy of concepts and experimental practices: Rethinking Georges Canguilhem's historical epistemology. *Studies in History and Philosophy of Science, 44*, 112–123.

Méthot, P.-O. (2018). De l'épistémologie historique à la philosophie de la biologie: Le double héritage philosophique de Jean Gayon. In F. Merlin & P. Huneman (Eds.), *Philosophie, histoire, biologie, Mélanges offerts à Jean Gayon* (pp. 15–50). Editions Matériologiques (This Volume).

Méthot, P.-O. (2020). Introduction. Georges Canguilhem's critique of biological thought and historical epistemology. In P. O. Méthot (Ed.), with the coll. of J. Sholl, *Vital Norms. Canguilhem's* The Normal and the Pathological *in the Twenty-First Century* (pp. 13–51). Hermann.

Méthot, P.-O. (2022). Jean Gayon face à Georges Canguilhem: Une remontée en trois temps. In M. Arminjon, C. Cherici, & P.-O. Méthot (Eds.), *Le normal et le pathologique: Des catégories périmées? Textes réunis en hommage à Jean Gayon* (pp. 267–306). Matériologiques.

Méthot, P.-O. (2023). Analytic and continental perspectives on biology and philosophy: David Hull and Marjorie Grene on 'What philosophy of biology is not.' In G. Bianco, G. Van de Vijver, & C. T. Wolfe (Eds.), *Canguilhem and continental philosophy of biology* (pp. 13–38). Springer.

Nicholson, D. J., & Gawne, R. (2014). Rethinking Woodger's legacy in the philosophy of biology. *Journal of the History of Biology, 47*, 243–292.

Pradeu, T. (2017). Thirty years of *Biology & Philosophy*: Philosophy of which biology? *Biology & Philosophy, 32*, 149–167.

Sloan, P. R. (2002). Reflections on the species problem: What Marjorie Grene can teach us about a perennial issue. The Philosophy of Marjorie Grene. In R. E. Auxier & L. E. Hahn (Eds.), *The library of living philosophers* (Vol. XXIV, pp. 225–255). Open Court.

Smocovitis, V. B. (1996). *Unifying biology. The evolutionary synthesis and evolutionary biology*. Princeton University Press

Smocovitis, B. S. (2009). Marjorie, matriarchy, and 'wretched reflection': A personal remembrance of Marjorie Grene. *Biological Theory, 4*(2), 191–195.

Chapter 3
Taking Historical Epistemology to the International Scene

Anastasios Brenner

Abstract The aim of this article is to examine the reception of French philosophy of science on the world stage, taking up what was a constant concern of Jean Gayon. Three major features distinguish this tradition. First, the use of the term epistemology to designate philosophy of science, thereby allowing for discontinuities between common knowledge and scientific knowledge. Secondly, the conviction that philosophy of science should be grounded on history of science. Finally, what has come to be called the regional turn: philosophical reflection should be carried out with respect to a specific body of knowledge. These directions of research, which have been cultivated for over a century by French thinkers, emerged belatedly within the Anglo-American tradition. Gayon devoted particular attention to the case of philosophy of biology. I complement his endeavor by an examination of philosophy of chemistry. Such divergences raise the question of the relations between the two traditions. It is only recently, with the advent of a new historical epistemology, that a true dialogue has been established. However, the many misunderstandings that have arisen between French and Anglo-American philosophy of science, show how fragile such a dialogue is.

Keywords French philosophy of science · Historical epistemology · History of reception · Philosophy of biology · Philosophy of chemistry · Regional epistemology

3.1 Introduction

To bring French philosophy of science to the world stage has been a constant concern of Jean Gayon. This concern is reflected in many of his publications and activities. Those who are acquainted with his itinerary know that from the outset he was deeply

Translated and adapted from Brenner (2018).

A. Brenner (✉)
Université Paul Valéry Montpellier 3, Montpellier, France
e-mail: anastasios.brenner@univ-montp3.fr

P.-O. Méthot (ed.), *Philosophy, History and Biology: Essays in Honour of Jean Gayon*, History, Philosophy and Theory of the Life Sciences 30,
https://doi.org/10.1007/978-3-031-28157-0_3

impressed by the teaching of Georges Canguilhem and that he resolved to further the biological philosophy which the latter had pioneered. Gayon pursued his studies under the supervision of François Dagognet, who aligned himself with Bachelard's school. But, unlike his teachers, Gayon sought to confront resolutely this school with others. In this regard, one can point to the work jointly edited with Jean-Jacques Wunenburger, *Bachelard dans le monde* (2000), whose methodological principle was to give the floor to non-French scholars. Two other publications are worth mentioning: *L'épistémologie française 1830–1970* (2015 [2006]) co-edited with Michel Bitbol, and *French Studies in the Philosophy of Science: Contemporary Research in France* (2009), in collaboration with the present author. The first of these volumes, which was originally published in 2006, aims to provide a convenient synthesis on the topic; it was soon translated into Chinese, and an updated edition, including a new preface, was issued in 2015.

These publications raise several questions that I should like to address here. Evidently, French philosophy of science has its specificities; but the exact nature of these distinguishing qualities remains to be determined. Are they linguistic, cultural, institutional or, ultimately, philosophical? Casting an eye on the reception of this tradition on the world stage, one gathers that it has been a mixed one. On the one hand, a reflection on science emerged early in France and undeniably constituted a source of inspiration for other traditions, whether we think of the positivism inaugurated by Auguste Comte, or of the historical epistemology attributed to Gaston Bachelard. Positivism played a major role at the international level up through the logical positivism of the Vienna Circle, and likewise historical epistemology can claim many proponents today, in particular at the Max Planck Institute for the History of Science in Berlin. On the other hand, the divide between historical epistemology and analytic philosophy has long hindered a true dialogue between the two traditions. Finally, the question arises of the evolution of French philosophy over the last forty years and its current trends.

3.2 A Peculiar Path

Let us recall that *L'épistémologie française 1830–1970* was the outcome of a research group that met over several years. In the introduction Bitbol and Gayon go to pains to precisely define the scope of their project. The *terminus ad quo*, 1830, corresponds to the publication of the first volume of Comte's *Cours de philosophie positive* (Comte, 1830–1842). This monumental six-volume undertaking, which provided a state-of-the-art picture of all the sciences recognized as such, led to formulate several questions that are still relevant today. To be sure, Comte did not invent the expression "philosophy of science"; one need only read his advertisement to learn that this expression was already in use (Comte, 1830–1842, vol. 1, p. VI). Rather he was calling for a new stage, which he first named "positive philosophy" in the title of his work and, thereafter, "positivism".[1] By examining the role of hypotheses, the criterion of empirical significance or the interdependence between theory and experience, Comte inaugurated a new way of philosophizing with regard to science. He was of course not alone: one can point to William Whewell and John Stuart Mill in the English-speaking world and Bernard Bolzano and Herman von Helmholtz in the German-speaking world. But French philosophy of science, unlike other traditions, will continue to claim descendance from its founding figure. Within this context, positivism will exert a decisive and enduring influence, even though several thinkers distance themselves in the final analysis from this doctrine. One should not forget that Comte's positivism also had a political dimension, which was perhaps even more powerful. As for the *terminus ad quem* 1970, it corresponds roughly to Canguilhem's retirement; he was indeed another guardian figure of French thought.[2] He developed a biological philosophy as early as 1943, long before the emergence of this specialty in the English-speaking context. These chronological limits make it possible to embrace the French tradition, but with sufficient distance for a dispassionate appraisal. At the same time, Bitbol and Gayon seek to encourage a renewal. The following passage from the preface to the new 2015 edition testifies to this:

> To venture beyond that limit would lead us to a place where the boundaries between intellectual history and philosophical commitment would become blurred. Therefore, with an eye to the younger generations, we prefer to draw attention to the strong, collective commitment of an active and brilliant group of young philosophers of science, participating in a philosophy of science largely open to the international forum. (Bitbol & Gayon, 2015, p. 9)

This open-mindedness, Gayon certainly helped to bring about. But can one speak of a single style in French philosophy of science? In France, as elsewhere, there is a variety of trends, echoing more or less the major philosophical schools of antiquity: Aristotelian, Platonist, Stoic, Epicurean, etc. One encounters, alongside various forms of positivism, advocates of logicism, such as Louis Couturat. As for Émile Meyerson, he defended a realism based on the continuity between common knowledge and scientific knowledge, a position like that of Willard Van Orman Quine. It goes without saying that French thought can boast of a multitude of methods and currents.

Of course, one can point to the French term *épistémologie*, which I shall render temporarily by epistemology. Most French thinkers refer thereby to philosophy of science, unlike their English-speaking counterparts. Let us briefly go back over the history of the word. As early as 1900, Édouard Le Roy employed it, apparently for the first time in the French language, at the first International Congress of Philosophy held in Paris:

---

[1] The term "positivism" occurs in 1835 in the 28th Lesson (Comte, 1975–1998, 1, p. 454). For an overview in English, see Bourdeau et al. (2018).

[2] Canguilhem retired from the chair of philosophy of science as well as the direction of the Institut d'histoire et de philosophie des sciences et des techniques in 1971. He was however active well up until his death in 1995.

> Historically, determinism has increasingly prevailed over the thesis of contingency. This belief assumes that science is an adequate knowledge of reality; whereas, for *modern epistemology* [*épistémologie moderne*], it is aimed only at the action of man on nature. (Le Roy, 1900, pp. 138–153)

Le Roy is alluding here to the path opened by Émile Boutroux in his famous doctoral thesis *De la contingence des lois de la nature* (1874). At the same time, he calls on Henri Bergson's philosophy, which emphasizes the dimension of action. He went on to point out the emergence of a current of thought that he called "new positivism" (Le Roy, 1901, pp. 138–153). Le Roy contrasted this new discourse with the earlier view of science.

The term epistemology spread rapidly. It came to designate the philosophy of science, as this discipline was taking shape at the time, largely nourished by the history of science. It also carried the sense of an attempt to develop a philosophical discourse devoted specifically to scientific knowledge, breaking with the totalizing speculations of the past, as evidenced in the following passage from Abel Rey:

> The recent term *epistemology* can quite well refer to this part of sociology, or this social science, as one would like. It does not yet exist in a rigorously positive form. It will be used for a positive examination of knowledge, specifying the objects, the aim, determining the value and scope [...]. It will be the philosophy of the sciences, until now built on the shifting sands of theories of knowledge. (Rey, 1904, pp. 612–625; emphasis mine)

Bachelard will help to establish the term epistemology through a whole series of expressions: "epistemological obstacle", "epistemological act", "epistemological break". Michel Foucault will go a step further by introducing the notions of "epistemological threshold", "epistomologization", and "episteme". I will end with this passage from Bitbol and Gayon that draws the contours of this tradition:

> French epistemology is the name of a deliberately heterogeneous tradition of thought that has always affirmed, if not the theoretical unity, at least the interrelatedness of problems that other traditions often tend to dissociate: logic, theory of the foundations and limits of knowledge (except for the theory of common knowledge), general philosophy of science, philosophy of particular scientific fields and, to a certain extent, history of science. (Bitbol & Gayon, 2015, p. 12)

Is it necessary to add that Anglo-American philosophers generally employ the term epistemology in a different sense, namely for the theory of knowledge and with the assumption of continuity between ordinary knowledge and scientific knowledge? But this use is under pressure. Hans-Jörg Rheinberger, an advocate of historical epistemology, to whom I shall return shortly, writes:

> My use of the term *epistemology* requires a brief explanation. I do not use it as a synonym for a theory of knowledge (*Erkenntnis*) that enquires into what it is that makes knowledge (*Wissen*) scientific, as was characteristic of the classical tradition, especially in English-speaking countries. Rather, the concept is used here, following the French practice, for reflecting on the historical conditions under which, and the means with which, things are made into objects of knowledge. It focuses thus on the process of generating scientific knowledge and the ways in which it is initiated and maintained. (Rheinberger, 2010, pp. 2–3)

From the outset French philosophy of science took a historical orientation. Indeed, Comte already insisted on the importance of resorting to the history of science. This orientation was seconded by Paul Tannery, Gaston Milhaud, and Pierre Duhem, who bequeathed an abundant historical production. Rey, who drew on these thinkers, then coined the expression historical epistemology. This occurrence has long gone unnoticed, revealing some neglect of the history of the philosophy of science. Rey thus writes:

> Philosophy [...] gave the name of epistemology to transcendent research on the principles and general conditions of science. This word, taken in a positive sense, could designate appropriately the documentary research, the entire set of historical observations necessary for founding an exact overview of the different sciences, a positive science of the sciences. And the historical research concerning the general spirit of each science would then be one of the major divisions of this *historical epistemology*. (Rey, 1907, p. 13; see Braunstein, 2012)

This historical approach, far from being abandoned, will be renewed by Bachelard. What is concerned is a recurrent reading: today's science reveals the significance of past science. Canguilhem and Foucault will build their theories on this basis.

Let us lastly bring up a third specificity of the French tradition. From the outset Comte was careful to distinguish each science in terms of its object and method; he thus separates philosophical discourse according to each fundamental science: mathematical philosophy, astronomical philosophy, physical philosophy, biological philosophy and sociological philosophy. Philosophical considerations must be formulated in relation to specific contexts. It should be noted that Comte developed a reflection not only on mathematics and physics, but also on chemistry and biology. Accordingly, French philosophy of science placed emphasis on special studies. Hence the label regional epistemology.

The contrast could not be more pronounced with an Anglo-American tradition largely committed to the context of justification, the structure of scientific theory and the logical analysis of the language of science.

3.3 A Mixed Reception

In the early 1960s Anglo-American philosophy underwent a deep change, often described as the "historical turn". As we know, the difficulties encountered by logical positivists led several thinkers, such as Thomas Kuhn, Imre Lakatos, Paul Feyerabend and Hilary Putnam, to underline the context of discovery. This reaction arose from the conviction that the doctrine of the Vienna Circle provides a too narrow framework to do justice to genuine scientific activity. The means of investigation need to be broadened to include historical elements. Post-positivism, as it came to be called, opts for the study of science in the making. The focus moves from the structure of scientific theory to theoretical change. Its proponents insist once and again that the philosophy of science must be based on the history of science.

One would expect a convergence with Bachelard's school. Indeed, both these doctrines defend the relevance of the history of science to the philosophy of science. They rehabilitate the context of discovery in relation to the context of justification and reject a linear and cumulative view of scientific development. Kuhn's paradigm thus includes not only symbolic generalizations but also methods, models, exemplars, and values. It would seem that we are moving towards the plurality of facets that Bachelard perceived in scientific practice. The notion of paradigm leads to a periodization: science acquires different forms over time, so many world views. It is hard not to see an analogy with Bachelard's different mindsets or Foucault's epistemes. Scientific revolutions are accompanied by gestalt switches. We find the twofold aspect, general and local, of Bachelard's epistemological break. Thus arises a radical discontinuism, which is both cognitive and historical.

More recently, several thinkers have tried to bring the two traditions closer together. For example, Ian Hacking borrows one of Foucault's expressions for the title of one of his books, *Historical Ontology* (2002). In this work, he amply acknowledges his debt to the French thinker. He certainly shook the North American intellectual establishment by showing what could be gained by taking Foucault seriously. His example reinforces that of Kuhn, in the direction of a greater historization of science. While Foucault remains his main source of inspiration in French thought, Hacking does not ignore his forerunners, Bachelard and Canguilhem. The innovative study of scientific experimentation that Hacking offered follows in the former's steps. But Bachelard was a belated discovery for English-speaking authors, a fact that Hacking laments. The lesson had to be learnt over again.

Let us return to Rheinberger, who proposes to reinstate the program of historical epistemology. Looking back on the history of this tradition in his essay, *On Historicizing Epistemology*, he draws several parallels with post-positivism:

> Feyerabend argues for an understanding of the empirical sciences and their modes of knowledge production that we first encountered not just with Toulmin but already with Fleck, and especially with Bachelard in his non-Cartesian epistemology. (Rheinberger, 2010, p. 63)[3]

One would have expected American and French thinkers to unite under the banner of a historical philosophy of science. This did not happen, while Bachelard's school held sway in France. Post-positivists indeed called on French thinkers such as Poincaré and Duhem, who have had an impact on the Vienna Circle. Lakatos added to the list Le Roy and Milhaud. Moreover, Kuhn pointed to the contributions of Meyerson, Léon Brunschvicg and Hélène Metzger. In comparison, the absence of the next generation is even more conspicuous: a survey of the literature reveals very few references to Bachelard and his school. Only in one of his late works did Putnam evoke Foucault (Putnam, 1981). But if he was ready to acknowledge analogies with Kuhn, he went on to minimize Foucault's legacy, accusing him of obscurity and relativism. His analysis reflects less a serious consideration of Foucault's work than a blatant misunderstanding.

The historical turn was followed by a regional turn: a reflection devoted to specific areas of specialization took shape. Gayon's own contribution was a series of impelling

[3] Concerning Bachelard's non-Cartesian epistemology, see Bachelard (1978, especially Chap. 6).

studies in the philosophy of biology, which did not fail to address the differences between the two traditions (see Gayon, 2009). David Hull's 1969 article, "What Philosophy of Biology Is Not" (Hull, 1969) is often credited with having ushered in this specialty. The latter criticized the way logical positivists had treated biology, namely as an example among others of the standard view of scientific theory and its basic unity. In contrast, Hull brought out the distinctive and irreducible features of biological science.

One cannot help but think of Canguilhem's earlier endeavor, as expounded in his medical thesis *The Normal and the Pathological* of 1943.[4] But Gayon shows that the practices are not the same on both sides of the Atlantic: Continental philosophers speak of "biological philosophy"; North American philosophers speak of "philosophy of biology". The philosophy of biology presents itself as an external and autonomous discourse on science, a meta-biology. It places emphasis on recent research and focuses on some of its directions.

In the North American context, the two turns mentioned are therefore neither simultaneous nor conjoined. Although the two traditions promoted a philosophical reflection adapted to the case of the life sciences, this sparked no dialogue. Gayon further highlights the contrast between the favorable reception of Canguilhem's studies in the history of science and the lack of interest in his work in biological philosophy. A comparison of the concepts of individuality in Canguilhem and in his North American counterparts illustrates the differences in views (see Gayon, 1998, 2015).

Why did the dialogue between the Anglo-American tradition and the French or Continental tradition fail? Several arguments can be put forth. Among them, what gives a certain unity to Anglo-American "post-positivism" or the "historical school" is its polemical stance with regard to the conception arising from the Vienna Circle.[5] Indeed, this school developed by opposing point-by-point the theses of logical positivism: the stability of the empirical base, the axiomatic model, the strictly logical method, etc. In other words, post-positivists found themselves in a discursive context entirely different from their French counterparts. It must be acknowledged that throughout the duration of post-positivism there was no convergence either regarding historical method or regional epistemologies.

3.4 The Sciences that Philosophy Neglected

Has the situation now changed? Over the past twenty years a school of thought has emerged under the name of *historical epistemology,* defended by thinkers such as Lorraine Daston, Ian Hacking and Hans-Jörg Rheinberger. Hacking taught for many

[4] See Canguilhem (1991 [1966]), translated into English only in 1978, long after the major works of his student Michel Foucault.

[5] On this issue, see Donovan, Laudan and Laudan (1988, p. 3); Laudan (1986, pp. 142, 146). For an appraisal, see Brenner (2006).

years at the University of Toronto and Daston at the University of Chicago. They both promoted this philosophical orientation. Daston then joined in 1995 the Max Planck Institute for the History of Science, where she worked with Rheinberger; Hacking was elected to the Collège de France in 2000. What is all the more interesting is that this historical epistemology appears no longer to be linked to a particular national tradition. It has spread to several countries—Germany, Austria, Canada, the United States—to form a truly international program. It is no longer a matter of defending the "French style" in epistemology, as Canguilhem did, but of renewing philosophy of science.

Among the defenders of historical epistemology, some seek to reconcile logical analysis and historical study, while others endeavor to fashion an altogether different approach, in which history takes precedence over logic. The first stance is illustrated by Hacking. The appropriation of historical epistemology does not mean for him to forego his initial training. Hacking argues that some of the motivations that drive analytic philosophy can be treated from a different angle. Historical ontology provides a means of analysis: conceptualizations are coordinated chronologically with their sites of enunciation and their resulting trajectories are scrutinized.

I mentioned above the regional shift in the philosophy of biology. This shift is progressively reaching other areas. Perhaps the most striking case is the philosophy of chemistry. It was not until the 1990s that this specialty was duly recognized in English-language philosophy. One can point to various institutional events such as the creation of scholarly societies and journals: *Hyle: International Journal for Philosophy of Chemistry* in 1995, International Society for the Philosophy of Chemistry in 1997, *Foundations of Chemistry* in 1999. This movement aims to fill a gap. Given the nature of chemistry, it gives rise to a greater attention to the concrete, practical, or technical aspects of scientific research.

Yet French philosophy of science had made room for chemistry from the very beginning. Comte already devoted five lessons of the *Cours de philosophie positive* to "chemical philosophy". A whole series of thinkers followed the path thus opened: Duhem, Meyerson, Bachelard, Metzger and Dagognet. The result was an imposing body of work, exploring the numerous philosophical implications of chemistry.

If we go back now to the Vienna Circle, we find that this science is conspicuously absent. Their manifesto, *The Scientific Conception of the World: The Vienna Circle* (Carnap et al., 1973), enumerated several disciplines: arithmetic, physics, geometry, biology, psychology, and social sciences, including even history and economics. It was not that chemistry was excluded from the sphere of science, that it was considered a pseudo-science. Chemistry undeniably implements powerful investigative methods, and it can yield results of extreme precision. It is worth adding that Neurath, whose concern for practice is known, was one of the few members of the Circle to provide references to chemistry—evidence, if necessary, of the tacit intention to include this science. The subsequent development of logical positivism did not lead to fill this gap. One looks in vain for some remarks on chemistry in the nineteen volumes of the *International Encyclopedia of Unified Science*, a book series initiated by Neurath and jointly pursued by Rudolf Carnap and Charles Morris (see Stadler, 2001, pp. 607–609). It is odd that self-proclaimed empiricists were not attentive to the practice of chemists, both in terms of observation techniques and empirical results. It is also surprising that materialists did not show more interest regarding a science centered on the transformations of matter. The reason for this omission is probably to be found in the difficulty of submitting chemistry to an axiomatic presentation. However that may be, it reflects an incomplete picture of scientific knowledge.

When the philosophy of chemistry finally emerged at an international level, the context had changed radically. Earlier research and alternate programs were no longer ignored. Indicative of this evolution is the collective volume edited by Hendry Robin, Paul Needham and Andrea Woody, *Philosophy of Chemistry* (2012). This volume includes separate chapters on Pierre Duhem and Gaston Bachelard. Joap Van Brakel, in his overview, "Prehistory of the Philosophy of Chemistry", likewise touches on the contribution of several French thinkers. One finds here perhaps the clearest case of recognition of the legacy of the French tradition. But let us pause to consider the exact wording: "prehistory" means strictly speaking what comes before written or recorded history. Now, philosophy of science had yielded, since the time of Lavoisier's chemical revolution, a number of works whose explicit aim was to provide a reflection on the science of chemistry: beginning with Comte's "Leçons de philosophie chimique", five lengthy lessons published in the third volume of the *Cours de philosophie positive* in 1838 (Comte, 1975–1998) and Duhem's 1902 *Le mixte et la combinaison chimique* (Duhem, 2002) to Bachelard's *Le pluralisme cohérent de la chimie moderne* in 1932. Van Brakel is obviously reading the history of philosophy of science in the light of his own research program. The above-mentioned recognition is then far from complete.[6]

Furthermore, one can gesture to an evolution in the new historical epistemology, marked by a deepening of its approach and a distancing from analytic philosophy. Such is the case of *Histories of Scientific Observation*, edited by Daston and Elizabeth Lunbeck (Daston & Lunbeck, 2011). These two authors claim to provide here the first genuine historical investigation on the topic. Obviously, there are many studies on one or another aspect of observation: observation by means of the telescope, observation in the laboratory, observation as opposed to experiment. But what was lacking was an inquiry into observation as such, that is an inquiry sufficiently broad in scope to include all the relevant factors involved in its historical constitution. One reads:

> Observation is a highly contrived and disciplined form of experience that requires training of the body and mind, material props, techniques of description and visualization, networks of communication and transmission, canons of evidence, and specialized forms of reasoning [...]. Scientific observation has emerged, flourished, and diversified under specific historical conditions. (Daston & Lunbeck, 2011, p. 3)

Daston and Lunbeck call explicitly on Foucault's work as well as on socio-cultural history. They react against the conception, tinged with positivism, forged in the nineteenth century, of a contrast between active experimentation and passive observation.

---

6 For more on those sciences neglected by mainstream philosophy, see Hanne Andersen et al. (2013). It is worth noting that this volume is the result of the European program "The Philosophy of Science in a European Perspective".

The criticism is leveled at the idea of an observation statement or protocol statement, the touchstone of empirical verification according to logical positivists.[7] But, according to the two editors, even the post-positivists were not able to free themselves completely from this heritage. To assert, as they do, that the facts are theory-laden is not enough:

> Yet even those who, like Hanson and Kuhn, conceive of scientific observation as something more than bare retinal impressions still described it as a kind of enriched perception, drawing on the work of Gestalt and cognitive psychologists, rather than as the fusion of perception, judgment, memory, and reasoning that early modern philosophers of observation had theorized. (Daston & Lunbeck, 2011, p. 5)

Let us pay heed to the wording here: observation is a complex process that results from the combined contribution of different skills. Only a broad and long-term approach can do justice to this diversity. I shall not go further on this occasion; suffice it to say that Daston and Lunbeck clearly break with the techniques of analytic philosophy. The perspective is decidedly historical, and draws on various specialties: conceptual history, institutional history, cultural history, etc.

To illustrate this trend, one can also cite Matthias Schemmel's *Historical Epistemology of Space* (2016). The subtitle states: "From Primate Cognition to Space–Time physics". This book is the result of a research program implemented at the Max Planck Institute. In the words of this author: "Its main goal is to show that such a historical epistemology is possible at all: that the different forms of spatial knowledge are indeed related in their development, and that the study of this interrelated development may indeed provide insights into their epistemic status" (Schemmel, 2016, p. V).

The historical perspective adopted is accompanied by a break with analytic philosophy:

> Scientific concepts of space such as Newton's absolute space or Einstein's curved space-time are thus presented in their rootedness in pre-scientific knowledge structures. At the same time it is shown how these concepts are part of broader conceptual systems that are able to integrate expanding corpora of experiential knowledge. (Schemmel, 2016, p. V)

By deploying his analysis of space, Schemmel clearly distances himself from the approach initiated by Carnap in his doctoral thesis on space (Carnap, 1922); he reengages with Mach, Poincaré and Duhem. Schemmel also calls on the legacy of Jean Piaget (Piaget & Inhelder, 1981) and Max Jammer (Jammer, 2012). Finally, he points to Marx Wartofsky's use of historical epistemology in the North American context as early as 1973 (Wartofsky, 1979). His own work builds on the research program initiated in 1994 by Peter Damerow and Jürgen Renn (Damerow, 1994, 2007; Renn, 2004). Without sacrificing in the least standards of rigor and clarity, he opposes the theses of logical positivism. The outcome is a study that combines

[7] Carnap coined the term "protocol sentences" to meet the requirement expressed in the Vienna Circle program: "The meaning of every statement of science must be statable by reduction to a statement about the given" (Carnap, 1973, p. 309; see also Carnap, 1987). However, logical positivists did not submit the concept of observability to a critical analysis (Suppe 1977, p. 46).

research from a wide range of disciplines—cognitive science, history, anthropology, etc.—in a vigorous effort to renew the understanding of our notions of space.

Schemmel emphasizes the entanglement of the ontogenetic and the historical-genetic developments of cognition. And he concludes: "The argument underlying the book was that it is only by studying the history of spatial thinking that the epistemic status of spatial knowledge can be assessed" (Schemmel, 2016, p. 107). He rejects the dichotomy between the context of justification and the context of discovery:

> Here we have followed another way of conceiving of the rational reconstruction of the history of science: not as the task to free a logical skeleton from the irrational flesh of the actual historical-psychological process of research and discovery, but rather as the task to identify the rationality within the historical-psychological process. This approach implies a broader concept of rationality, a concept that cannot be reduced to formal logic. (Schemmel, 2016, p. 112)

3.5 Conclusion

It is time to wrap up our remarks. I do not purport to give conclusions per se, but more modestly reflections on new trends and directions. I mentioned a deep change in French philosophy of science. This change has been facilitated by an institutional transformation: a move from a closed university system to an open one, characterized by increasing mobility.[8] It is reflected in the increasing adoption of exogenous methods and practices. The specificity of French philosophy of science is gradually fading. Should one complain about this? Or should one rather seize the opportunity to encourage new exchanges? Admittedly, French philosophy of science as practiced today tends to become one tradition among others within the framework of European philosophy of science, and, from a global perspective, Europe provides the most appropriate category (Brenner, 2015).

We have also observed a change in American philosophy of science, which dominates international research. In recent years, there has been much discussion of a practical turn, a historical turn, and a regional turn. These three turns are bringing about a transformation of analytic philosophy as originally conceived, based on the powerful but narrow model of first order-logic. American philosophy today offers a wide variety of methods and objects. It includes both proponents of a formal approach and advocates of an integrated history and philosophy of science. We are witnessing a growing interest in other schools, from Foucault's historical ontology to phenomenology. Philosophy of science is taking up questions bordering on ethics and politics, which it had discarded previously. It remains to ascertain whether this development will make possible a fruitful dialogue with French philosophy of science. If we are to judge in the light of past experience, nothing is less certain. For it is also a matter of tempo. When historical epistemology began to be adopted in Berlin or

[8] In 1985 all restrictions for foreigners pursuing university careers in France were lifted. This led to a clear sociological evolution.

Toronto, it was going out of fashion in Paris. What we have is a series of missed opportunities, rather than a real convergence.

Finally, I cannot help but express, in the current situation, my concern about international cooperation. For the task we set ourselves is being called into question, owing to the departure of the United Kingdom from the European Union and a period of closure on the part of the United States of America—those countries with which some French philosophers have been striving to establish a dialogue. Such a situation obliges us to recast our task. It is only appropriate that we reflect once again on the purpose of philosophy. As we know, Socrates inaugurated philosophical thinking in the agora, by questioning citizens on the business of the day and the great problems of life. What came out of this was a diversity of schools and ways of living. Philosophizing was intimately bound up with pluralism and democracy. Later, modern states instituted the teaching of philosophy each in their own way. In France this discipline, after an eclipse during the authoritarian regime of Napoleon III, was re-established by the French Third Republic, as the crowning event of secondary education. A century and a half later, this philosophy course caters to most of the young generation. The French case brings out with clarity the surreptitious links between philosophy and politics: the part played by Enlightenment philosophers in the French Revolution, the role of the teaching of philosophy in the reestablishment of the Republic and, more generally, the importance of philosophical discussion in the political debate. The philosopher came to be considered as the herald of critical thought, freedom, and emancipation. Philosophy of science is called on to relate to these fundamental values.

References

Anderen, H., et al. (Eds.). (2013). *New challenges to philosophy of science*. Springer.

Bachelard, G. (1932). *Le pluralisme cohérent de la chimie moderne*. Vrin.

Bachelard, G. (1978) [1934]. *Le nouvel esprit scientifique*. Presses universitaires de France.

Bitbol, M., & Gayon, J. (Eds.). (2015). *L'épistémologie française: 1830–1970*. Éditions Matériologiques.

Bourdeau, M., Pickering, M., & Schmaus, W. (Eds.). (2018). *Love, order and progress: The science, philosophy, and politics of Auguste Comte*. University of Pittsburgh Press.

Boutroux, É. (1991) [1874]. *De la contingence des lois de la nature*. Presses universitaires de France.

Braunstein, J.-F. (2012). Historical epistemology. Old and new. In H. Schmidgen, J.-F. Braunstein, & P. Schöttler (Eds.), *Epistemology and history: from Bachelard and Canguilhem to today's history of science* (pp. 33–40). Preprint, Max Planck Institute for the History of Science.

Brenner, A. (2018). Porter l'épistémologie française sur la scène internationale. In F. Merlin & P. Huneman (Eds.), *Philosophie, histoire, biologie: Mélanges offerts à Jean Gayon* (pp. 65–80). Éditions Matériologiques.

Brenner, A., & Gayon, J. (Eds.). (2009). *French studies in the philosophy of science: Contemporary research in France*. Springer.

Brenner, A. (2006). Quelle épistémologie historique? Kuhn, Feyerabend, Hacking et l'école bachelardienne. *Revue de métaphysique et de morale, 1*, 113–125.

Brenner, A. (2015). Is there a cultural barrier between historical epistemology and analytic philosophy of science? *International Studies in the Philosophy of Science, 29*, 201–214.

Canguilhem, G. (1991) [1966]. *The normal and the pathological* (Fawcett, C. R., & Cohen, R. S., Trans.). Zone Books.
Carnap, R., Hahn, H., & Neurath, O. (1973) [1929]. The scientific conception of the world: The Vienna Circle. In M. Neurath & R. S. Cohen (Eds.), *Empiricism and sociology* (pp. 299–318). Reidel.
Carnap, R. (1922). *Der Raum: Ein Beitrag zur Wissenschaftslehre*. Reuther & Reichard.
Carnap, R. (1987). On protocol sentences. *Noûs, 21*, 457–470.
Comte, A. (1975–1998). *Cours de philosophie positive* (1830–1842) (2 Vols.). Hermann.
Damerow, P. (1994). Vorüberlegungen zu einer historischen Epistemologie der Zahlbegriffsentwicklung. In G. Dux & U. Wenzel (Eds.), *Der Prozess der Geistesgeschichte* (pp. 248–322). Suhrkamp.
Damerow, P. (2007). The material culture of calculation: A theoretical framework for a historical epistemology of the concept of number. In U. Gellert & E. Jablonka (Eds.), *Mathematization and demathematisation: Social, philosophical and educational ramifications* (pp. 19–56). Sense Publication.
Daston, L., & Lunbeck, E. (Eds.). (2011). *Histories of scientific observation*. Chicago University Press.
Donovan, A., Laudan, L., & Laudan, R. (Eds.). (1988). *Scrutinizing science: Empirical studies of scientific change*. Kluwer.
Duhem, P. (2002) [1902]. *Mixture and chemical combination and related essays* (Needham, P., Trans.). Kluwer.
Gayon, J., & Wunenburger, J.-J. (Eds.). (2000). *Bachelard dans le monde*. Presses Universitaires de France.
Gayon, J. (1998). The concept of individuality in Canguilhem's philosophy of biology. *Journal of the History of Biology, 31*, 305–325.
Gayon, J. (2009). Philosophy of biology: An historic-critical characterization. In A. Brenner & J. Gayon (Eds.), *French studies in the philosophy of science: Contemporary research in France* (pp. 201–213). Springer.
Gayon, J. (2015). Le concept d'individualité dans la philosophie biologique de Georges Canguilhem. In M. Bitbol & J. Gayon (Eds.), *L'épistémologie française: 1830–1970* (pp. 389–419). Éditions Matériologiques.
Hacking, I. (2002). *Historical ontology*. Harvard University Press.
Hendry, R., Needham, P., & Woody, A. (Eds.). (2012). *Philosophy of chemistry*. Elsevier.
Hull, D. (1969). What philosophy of biology is not. *Journal of the History of Biology, 2*, 241–268.
Jammer, M. (2012) [1954]. *Space concept: a history of space theories in physics*. Dover.
Laudan, L., Donovan, A., Laudan, R., et al. (1986). Scientific change: Philosophical models and historical research. *Synthesis, 69*, 141–223.
Le Roy, É. (1900). Compte rendu du Premier congrès international de philosophie. *Revue de métaphysique et de morale, 8*, 503–699.
Le Roy, É. (1901). Un positivisme nouveau. *Revue de métaphysique et de morale, 9*, 138–153.
Piaget, J., & Inhelder, B. (1981) [1948]. *La représentation de l'espace chez l'enfant*. Presses Universitaires de France.
Putnam, H. (1981). *Reason, truth and history*. Cambridge University Press.
Renn, J. (2004). The relativity revolution from the perspective of historical epistemology. *Isis, 95*, 640–648.
Rey, A. (1904). Ce que devient la logique. *Revue philosophique de la France et de l'étranger, 57*, 612–625.
Rey, A. (1907). *La théorie de la physique chez les physiciens contemporains*. Alcan.
Rheinberger, H.-J. (2010). *Historische Epistemologie: Zur Einführung* (2007) (Fernbach, D., Trans.). *On historicizing epistemology: An essay*. Stanford University Press.
Schemmel, M. (2016). *Historical epistemology of space*. Springer.
Stadler, F. (2001). *The Vienna Circle: Studies in the origins, development influence of Logical Empiricism*. Springer.

Suppe, F. (1977). *The structure of scientific theories*. University of Illinois Press.

Van Brakel, J. (2012). Prehistory of the philosophy of chemistry. In A. I Woody, R. Findlay Henry, & P. Needham (eds.), *Handbook of the philosophy of chemistry* (pp. 21–45). Elsevier.

Wartofsky, M. W. (1979) [1973]. Perception, representation, and the forms of action: Towards a historical epistemology. In R. S. Cohen, D. Davidson, G. Nuchelmans, & W. C. Salmon (Eds.), *Models: Representation and the scientific understanding* (pp. 188–210). Reidel.

Chapter 4
Jean Gayon and the Historical Perspective in Philosophy of Biology

François Duchesneau

Abstract Jean Gayon can be defined a philosopher of biology with a historical approach to epistemic issues. Evidence for this statement may come from reconsidering his *Darwinism's Struggle for Survival* (Gayon, *Darwinism's struggle for survival: Heredity and the hypothesis of natural selection*, Cambridge University Press, 1998 [1992]), which he presented as "a scenario of rational reconstruction." He was essentially interested in developmental perspectives on concepts, theories and analytic procedures, but the reconstructions he considered had to undergo the test of fitting with a historical analysis of the antagonistic theories that developed and succeeded that of Darwin. The rigorous and successful methodology Gayon applied in this major work and several others may account for the lasting value of his reconstitution of the transformation phases that the Darwinian hypothesis of evolution by natural selection underwent up to the advent of the evolutionary synthesis.

Keywords Philosophy of biology · Darwinism · Natural selection · Historical analysis · Evolution theory · Biometric · Heredity

4.1 Introduction

In his remarkable postscript to a posthumous collection of Jacques Roger's articles, Jean Gayon analyzed what related to philosophy of science in the works of that historian of the life sciences who would only be defined as an historian (Gayon, 1995). Twice, Jean Gayon referred to expressions that reflected his own philosophical approach. The historian, he points out, used with insistence and in a modern sense such terms as 'biology', 'life sciences', 'heredity', 'transformism'. This fact can be explained as follow: "It was the historian's typical way of obliging the concept to reveal itself in the history of science" (Gayon, 1995, p. 464). And he would add a little

Translated from Duchesneau (2018).

F. Duchesneau (✉)
Université de Montréal, Montreal, Canada
e-mail: francois.duchesneau@umontreal.ca

P.-O. Méthot (ed.), *Philosophy, History and Biology: Essays in Honour of Jean Gayon*, History, Philosophy and Theory of the Life Sciences 30,
https://doi.org/10.1007/978-3-031-28157-0_4

further: “It is essential for an historian to understand the emergence and consolidation of concepts” (Ibid., p. 465). According to Jean Gayon, for the historian Jacques Roger, philosophy of science, in connection with contemporary scientific achievements, formed an integrative part of the interpretation of past scientific developments, notably by allowing those forgone theses to be differentiated from their contemporary connotations.

Contrary to Jacques Roger, who was our common friend, Jean Gayon would not primarily define himself as an historian, but as a philosopher. Nonetheless, an historical approach is unquestionably present in the way he approaches questions pertaining to the philosophy of biological sciences. I felt that the best way to verify that point would consist in reading over *Darwin et l’après-Darwin: Une histoire de l’hypothèse de sélection naturelle* (Gayon, 1992), translated into English as *Darwinism’s Struggle for Survival: Heredity and the Hypothesis of Natural Selection* (Gayon, 1998), and re-examining the methodological orientations of that work. We all recognize it to be a major contribution to philosophy of biology and that it seems to bring together all the other fields Jean Gayon’s research has investigated. It is quite tempting to link Jean Gayon’s achievements in research with the tradition of “historical epistemology” (“*épistémologie historique*”) fostered by some of the scholars who have been his teachers. About this tradition I shall say a word to begin with, since Gayon himself investigated that school of thought historically and critically, and defined it largely in contrast with the philosophy of biology as the latter came to be institutionalized. But, beyond intellectual traditions, I believe that Jean Gayon’s epistemic practice possessed its own characteristics and that it deserves an independent appraisal in the context of philosophy of biology today.

4.2 Three Arguments for a Historical Approach to the Philosophy of Biology

In a survey article to which Thomas Pradeu alludes to in this volume, Jean Gayon drew a contrasted picture of philosophy of biology as it had developed since the 1960s, mainly in, or in close connection with, the Anglo-American academic world, contrary to the epistemology of the life sciences that had been mostly practiced in the European and Latin world, where the historical approach would appear as a dominant feature among others (Gayon, 2009). Let me quote from that study of his:

> Conversely, philosophy of biology has clearly not been concerned by the historical turn. The opposite is the case. The most obvious characteristic of philosophy of biology is the common acceptance of the methodological opposition between an analytical (or conceptual) approach to biology and an historical approach to biology. One should not confuse here ‘philosophy of biology’ and ‘philosophers of biology’. In practice, a significant number of philosophers of biology are both philosophers and historians of biology. For instance, Michael Ruse or David Hull, who were so prominent in the construction of philosophy of biology as a discipline, have each made significant historical contributions. It is even probably true that they have written more in the genre of history of science (or history of philosophy) than in the genre of philosophy of science. But they have imposed a certain style of research based on the

> aforesaid distinction. Those who have explicitly challenged this view, and together claimed to be doing 'philosophy of biology' are quite rare. Marjorie Grene and Richard Burian are probably the best examples. I am also one, though I come from a quite different cultural horizon. The non-historical character of philosophy of biology is in fact an implicit norm of behaviour in the academic context, not a well-articulated doctrine. It constitutes a major difference with other traditions of reflection on biology among philosophers of science. (Gayon, 2009, p. 204)

The principal merit of this statement is to suggest, but implicitly, that the historical practice in philosophy of biology is methodologically legitimate since it has been illustrated by the works of major researchers who have been acknowledged as philosophers of biology. Indeed, Jean Gayon sides with this group, even if he would never have suggested that he has been one of its best representatives, which is indeed the case. Two comments come to my mind about the strict opposition of approaches that Gayon states, based on the enquiry he had undertaken and for which he had collected data about a period that went up to almost twenty years ago. And I take due note of the fact that Thomas Pradeu extended that investigation to a more recent period (Pradeu, 2017). My first remark is that the contents of the articles specifically identified with a philosophy of biology without historical dimension bore in fact on historical components that related to more recent conceptual developments, by contrast with analyses that would have dealt with older theories. I shall take as examples of that category articles that bore on the polysemy of the concept of 'gene' and needed to take into account the shift from neo-Mendelian to molecular genetics, since the analytic categories developed by the former had coexisted, in many instances, with those of the latter and would not have allowed a full argumentative reduction.

My second remark is that the horizon of philosophy of biology, presented as rather monolithic in the description that Jean Gayon gave of it, has fractured, or at least, diversified to a considerable degree. My personal belief is that this is due to the cleavages that have affected the Modern Synthesis because of the renewed interest for questions concerning development and epigenetics. But it would also be in order to point to research issues such as those that Thomas Pradeu himself addresses, which have required to reconsider the very configuration of biological objects. One then tends to turn to an order of 'metaphysical' and 'epistemic' considerations that call for an historian's vision. During the congress of the International Society for the History, Philosophy and Social Studies of Biology, held in Montreal in July 2015, Jean Gayon and I talked about the historical and/or thematic configuration of the various papers presented. We both had the impression that we were witnessing a shift—not such that the historical gap in styles had been really bridged—where the argumentation on both sides of the divide tended to relativize, on the one hand, and that more complex and less clearly circumscribed research agendas called for methods that would be more flexible, and above all, more open to developmental perspectives concerning concepts, theories, and analytic procedures, on the other. From this viewpoint, a revisited historical approach is probably timely for many issues in the philosophy of biology. Jean Gayon, who was not specially inclined to blind optimism, has often presented himself as the representative of a fading epistemological style. And he would use the "we" to include me as well in the same disappearance process, as if we were probably heirs to the Bachelard-Canguilhem-Dagognet trio—a condition for which he may qualify to a given point, but with which I cannot truly identify myself.

I cannot partake in Jean Gayon's pessimism about our disappearing epistemological style nor in his too modest dedication to, and respect for, the "*maîtres*" who had influenced him, for three reasons. The first is the remarkable work performed by the researchers he trained, contributed to train, or simply inspired. I know some of them for having evaluated their works, supported their careers and learned much from their publications. I challenge them to argue that they have not been enticed to follow Jean Gayon's research methodology. The second reason I shall evoke relates to the acknowledged value of Jean Gayon's achievements: his personal researches as well as the major academic projects that he led in the professional milieus of philosophy and the sciences. My conviction is that this success was essentially due to the approach he had developed, namely: a vast, rigorous, historically and conceptually grounded approach that he never ceased to apply in his analyses of issues about the life sciences and beyond.

My third argument is from my personal viewpoint the most consequent. Very few contributions in our disciplines may be acknowledged as "lasting monuments" according to the phrase Locke used at the beginning of the *Essay concerning Human Understanding* to characterize the scientific achievements of Sydenham, Boyle, Huygens and Newton, and this is even more so if one takes into account the contingency that applies to our historical and philosophical works and is the hallmark of all researches of the epistemic kind. It is in fact impossible to decide a priori what will be the actual reach and relevance of the enquiries we hold. But the most important of Jean Gayon's contributions can qualify as accomplishments of lasting value. And, as paradoxical as it may seem, this is due to the fact that he has been able to scientifically trace back the genesis and transformations of concepts that had historically structured contemporary biological knowledge, at least for those episodes he was concerned with and that he investigated in detail.

4.3 A Rational Reconstruction of Scientific Developments

Jean Gayon describes his work in *Darwinism's Struggle for Survival* as a "scenario of rational reconstruction" (Gayon, 1998, p. 3), which conforms to the epistemic profile of several researches in the philosophy of science. From the start, he anticipates that this method will breed objections from the perspective of history of science. In fact, he defines what he considers to be the essence of Darwinian theories of evolution, by a characteristic formula he finds already sketched in Darwin's own works, which would also form the matrix of phases of gradual deepening for a significant theoretical furrow: "'Darwinian' is taken to mean any interpretation of evolution as being the product of the gradual modification of species, predominantly guided by a process of natural selection functioning on a field of intra-populational variation" (Gayon, 1998, p. 4). It is obvious that this formula refers in first instance to a major "hypothesis", presented as such by Darwin, that of natural selection. At the same time, the demonstration of this hypothesis required explanatory principles for the various modes of occurrence and transmission of variations that could present a differential advantage to the individuals in concurrence for survival and reproductive capacity within a population and in a given ecosystemic context. The conceptual basis of the Darwinian hypothesis must entail a reference to an explanatory theory about the mechanisms of heredity—a theory that in Darwin's time was very imperfectly, if not at all, established. The intellectual project of Jean Gayon aimed at showing how biometric, population genetics, and the evolutionary synthesis contributed to the justification of the conceptual core of the original Darwinian hypothesis. In this book, Gayon also established how these transformations have been made possible by the correlative and, at times divergent, developments in neo-Mendelian genetics. And it is a fact that without referring to the structure of the Darwinian model as he analyzed it, one might not understand either today's neo-Darwinian theories of evolution, nor those theoretical perspectives that developed as antithetically oriented alternatives to the latter, such as epigenetics, punctuated equilibrium and developmental systems theories.

The core of Jean Gayon's analysis bears on the crisis that affected somehow internally the Darwinian theory, which tended to get resolved only in the first third of the twentieth century. Indeed, during that period, the philosophy of evolution implied three concurrent branches which could even be thought capable to supplant the Darwinian perspective, but, according to Jean Gayon, what should principally draw our attention are patterns of concepts and practices that revealed tensions in the choice of principles and models as well as unsolved problems. The introduction to *Darwinism's Struggle for Survival* evokes certain methodological requirements, one of which proves to be particularly interesting. It relates somehow to the notion of 'rational reconstruction'. It is notorious that some historians of science who identified themselves as such have often discarded from their research projects any intent to rationally reconstruct science, judging that this is precisely what philosophers of science intend to do and what causes them to betray the real way science came into being through the representation they give of it. I remember the arguments thus produced by two historians who were our friends, but would not accept to be considered as philosophers: Jacques Roger, whom Jean eulogized and granted philosophical merits to, and Mirko Drazen Grmek, whom I treated similarly not very long ago (Duchesneau, 2019). The simplest is probably to let here the author of *Darwinism's Struggle for Survival* speak his own mind:

> The history of science might not necessarily have as its objective a 'rational reconstruction'. From the outset, the method employed here needs to be made clear. Particular attention is paid to the conceptual and empirical difficulties encountered by the Darwinian theory of selection, in terms of an epistemological history. The unique aspect of such an approach to the history of science is that it takes into consideration the difficult question of the validity of different theories. The methodological starting point is that science is, among other things but above all else, a problem-solving activity. This implies that whatever the weight of sociological and ideological determinants of a given field of research, it must be possible to describe the history of a scientific theory as an enterprise within which *some* problems are posed in such a way that they can be solved. Of course, this approach is by definition retrospective,

> because, as has often been pointed out, the historian of science cannot do without a 'frame of reference'. (Gayon, 1998, p. 8)

Jean Gayon however tends to dissociate his concept of rational reconstruction from the one Imre Lakatos had proposed (Lakatos, 1971): while the latter could endorse logically reframed rational reconstructions, the former submits all reconstructions to the norm of historical evidence. This implies that the object of philosophy in this instance could not be a formally ideal genesis of knowledge, but one reconstituted as closely as possible to the available historical data (on Lakatos, see the chapter by Loison, this volume).

Let me recall here what Jean Gayon used to declare in his historiographic works on the comparison of styles and objects in the philosophy of biology. If the presently dominant trend set in from the 1960s on, it is essentially due to philosophers, like Michael Ruse, who would draw the inference that the formal model inherited from logical positivism could not account—one might say a priori—for developments accomplished in contemporary biology: on the contrary, one should develop a posteriori an epistemology that would accord with the effective formation modes of the theories in question. I believe this may be wherefrom the non-negligeable historical motivation and interest among those pioneers of the philosophy of biology originate. It might be the case however that the latent influence of the neo-positivist epistemic model would explain the distance often maintained afterwards by philosophers of biology toward the history of science, while historians of science, in the same period, would have tended to adopt orientations that were more sociological and more foreign to the genetic unveiling of concepts and theories.

Jean Gayon shows in the best of ways how natural selection according to Darwin does not act for some variety of a type, but for the benefit of small hereditary individual variations, which implies a system of determinate relations between variation, heredity and transformation in a context of limitation in supply means. Hence a rational reconstruction of the concept, transposed into a twofold formula:

> Natural selection is not the selection of individuals but of 'variations', that is, of hereditarily transmissible peculiarities of structure or behaviour. The unique criterion for the retention of variations is that of the advantage conferred upon individuals in their struggle for survival and reproduction. In other words, natural selection deals only with individual advantage: it acts only through the 'preservation' (i.e., the survival and reproduction) of individuals, in proportion to the advantageous or disadvantageous variations they possess. (Gayon, 1998, p. 66)

This reconstruction makes one apprehend the immense application field for such a hypothesis, but it correlatively reveals the intrinsic risks of refutation to which it lent itself, in particular, concerning the mechanisms of hereditary transmission of individual variations and the capacity of these mechanisms to yield transformations implying sums of variations. In view of the extreme difficulty of delivering such a reconstruction, one should decide which is the more remarkable: the second part of the work, devoted to the heredity crisis, or the third, on the genetic theory of selection. Faced with this alternative, I shall suspend judgment. Jean Gayon in fact reassesses on the one hand the contributions of the biometricians who succeeded

Francis Galton; he traces back on the other hand the mechanisms of replication of the germinal lineage abstracted from the heredity of acquired traits, as professed by the physiologists and cytologists who followed in the footsteps of August Weismann. He shows, how, in opposition to Darwin's 'pangenesis', processes of statistical analysis were applied to variety transmissible among populations, while, at the same time, a more stringent conception of the causal role of natural selection as an hegemonic principle of evolution, gained credit. Duly justified, this analysis reveals the deficit in adequate and especially empirical arguments for what formed the core of the Darwinian theory. For instance, Jean Gayon writes:

> Weismann and Galton show us the fundamental uncertainty that marked theoretical Darwinism as the lack of an experimental science of heredity made itself increasingly felt. Wishing to preserve Darwin's heritage, Weismann radicalised the principle of selection to the point of subordinating it to the principle of heredity. Without natural selection, heredity could do nothing, not even conserve. Natural selection thus found itself in the epistemologically fragile position of having to explain everything, and of being the sole rampart against universal biological disorganisation. Galton, on the other hand, demanded that the principle of selection conform to his idea of heredity as a conservative force of regression able to oppose the modification of populations. In this view, natural selection was confined to the role of sifting 'races' that it had played no role in creating. To summarize, post-Darwinian evolutionary theory was characterised by a profound confusion. (Gayon, 1998, pp. 176–177)

These original and opposite counterfeits of the Darwinian principles will nevertheless be found capable of transformations in ways that Jean Gayon traces back meticulously, and this in the end will contribute to the unfolding of a research program that conforms to the profile of the original Darwinian theory, but endowed with new ontological and epistemic foundations.

One of the main contributions thus achieved consisted in determining what biometricians, such as Walter F. R. Weldon and Karl Pearson, brought on in terms of factual evidence of the selection process. Their aim was to reduce the causal analysis of facts of heredity to sufficiently confirmed statistical correlations; and these would *ipso facto* become compatible with a variety of different explanatory hypotheses. There would result therefrom a setting aside of all properly biological hypotheses, relating in particular to heredity.

> With biometry, Darwinism paradoxically reached the critical point of its history. Thanks to the biometricians, natural selection appeared clearly as a 'fact'. But because the physiological basis of heredity remained unknown, the biometricians reduced heredity to a mere method of regrouping data, thus threatening the theory of selection with ceasing to be a theory at all. (Gayon, 1998, pp. 238–239)

Surprisingly, though logically, the bracketing of ontological presuppositions concerning the varying entity that would only be represented by means of multiple correlations, made it possible for Pearson to arbitrarily conjoin coefficients of differential survival and fecundity into a synthetic concept of *fitness*: "Pearson had underlined [...] that it was necessary to accept that the 'fittest' was also the 'the most fertile', or, more precisely, whatever constituted the best compromise between individual adaptation and productivity" (Gayon, 1998, p. 248).

Jean Gayon establishes then that Mendelism, whose developments are exactly analyzed, proceeds in an epistemically contrary manner. Once the parental genotypes are presupposed, their effects are statistically inferred in order to be empirically confirmed. This approach is right away causal. But, as the law of segregation of genic elements intervenes in a setting of saltationist evolution theories, such as Hugo De Vries' and Wilhelm Johannsen's, natural selection does not act primarily as a mechanism underpinning individual variations, but as a sanctioning process, once variations have been produced within the type in question. It is later, especially through William E. Castle's works and the setting of the dynamic of evolution under aegis of the Mendelian hypothesis, that the Darwinian model resumed its grounding function for the mechanism of evolution. Jean Gayon draws a remarkable conclusion from this phase of conceptual transformation. Thus one got as a result a bicephalous theory: on the one hand, selection could be interpreted from then on as a dynamic of genic replacement among populations; on the other hand, attention was drawn to a new issue about the "systems of interaction in which genes acquire a selective value" (Gayon, 1998, p. 317). The challenge about this second theoretical component has essentially been that of providing mathematical models capable of assisting in the formulation of laws, which would constitute the framework of population genetics—a framework whose various components Jean Gayon has detailed in an epilogue to the study he had undertaken. This is what he would term an attempt at defining a "transcendental topic of the science of evolution" (Ibid., p. 321), whose extreme poles of achievement he located in the work of Ronald A. Fisher and Sewall Wright, once closely analyzed.

I shall not go further in evoking the significant epistemic merits I acknowledge in that study that Jean Gayon implemented by combining the highest respect for historical exactness and, notwithstanding, extreme philosophical relevance. At the very end of *Darwinism Struggle for Survival*, after having endorsed Elliott Sober's thesis according to which the natural selection principle could probably be rated as a "consequence-law" from a physicist's viewpoint (Sober, 1984), he adds:

> There is no reason to believe that [natural selection] 'acts' at every point of biological space-time in the sense that the action of a fundamental force in physics can be thought of in terms of 'field'. Natural selection does not have a 'field of action'; it is always a *resultant* force – the composition of those molecules. Or, in other words, while it may be correct to say that natural selection ultimately controls all the properties of living beings, it would be naive to think that it *determines* them at each and every moment and in the detail of all their constituent elements. (Gayon, 1998, p. 407)

4.4 Conclusion

Jean Gayon seems to me to have thus addressed the question that I personally hold to be the most fundamental in philosophy of biology today, that of conceiving what epistemic status should be attributed to the functional properties of the manifold entities on which biologists work and to which they tend to link law statements or whatever they consider as such. Can it then come as a surprise that Jean Gayon has been one of the leaders of the research program on functions in which several of us have had the honour to participate? (Gayon & de Ricqlès, 2010).

Judging that confidentiality is not to be kept anymore in this case, I would like, in guise of a conclusion, to communicate an excerpt of the evaluation of Jean Gayon's works that I had forwarded in support of his candidacy for a senior fellowship of the Institut Universitaire de France, on February 26, 2010. The assessment I had then given, has remained timely and has been more than confirmed in view of his more recent and unfortunately last contributions.

> We owe him many excellent analyses on central epistemic themes related with developments in the life sciences. The principal merit of Jean Gayon as a researcher is his having analyzed the Darwinian theory of evolution in its various methodological and epistemic aspects, first in the context of its emergence and initial transformations, then along its contemporary and present modalities through the principles, models and enlarged factual horizon of the synthetic theory of evolution, as well as through the prism of the various contestations it underwent. The remarkable achievement in analysis, conceptual reconstruction and criticism that resulted in the form of his great book *Darwinism's Struggle for Survival*, published at Cambridge University Press in 1998, first published in French under the title *Darwin et l'après-Darwin. Une histoire de l'hypothèse de sélection naturelle*, in 1992, has justly been held by the principal international specialists to be a major and lasting contribution on the historical and thematic foundations of contemporary evolution theory.
>
> One ought to underline the considerable spread of Jean Gayon's research interests: these intersect with various domains of either general philosophy of science or philosophy of biology. His remarkable works have notably illustrated major queries in the now well-institutionalized discipline that the philosophy of the biological sciences constitutes. He has provided the French university community with a just critical appreciation of the conceptions that had developed, especially in the Anglo-American milieu, on problems of biological individuality, species, functional explanation, genes and unit of selection. Moreover, he contributed innovative approaches on such issues that are of intense philosophical interest.
>
> Jean Gayon has also been a prominent historian of genetics, a scientific discipline he knew from the inside since he had been trained in it. His competence in that field has resulted in first rate publications, notably on the development of the theories and practices of genetics in France.
>
> If Jean Gayon set his mark as an original and productive researcher through a long series of high-quality works of his own, his leadership in the philosophy of biology also became manifest through the various collective projects for the execution of which he would assume a major share of responsibility. I also wish to underline his exemplary contribution to the formation of the next generation of scientists, which is well attested by the quality output of his former students and postdocs, whom he taught, supervised and generously inspired to emulate his own achievements. Thus, he succeeded in setting forth dynamic and productive networks, open to international collaborations, as part as the mandate and activities of the Institut d'histoire et de philosophie des sciences et des techniques, Université Paris I.
>
> There is no doubt that Jean Gayon counts among the most reputable philosophers of biology worldwide.

References

Duchesneau, F. (2018). Jean Gayon et l'orientation historique en philosophie de la biologie. In F. Merlin & P. Huneman (Eds.), *Philosophie, histoire, biologie: Mélanges offerts à Jean Gayon* (pp. 51–63). Éditions Matériologiques.

Duchesneau, F. (2019). Grmek et l'invention théorique chez Claude Bernard. In P.-O. Méthot (Ed.), *Médecine, science, histoire: Le legs de Mirko Grmek* (pp. 219–238). Éditions Matériologiques.

Gayon, J. (1992). *Darwin et l'après-Darwin: Une histoire de l'hypothèse de sélection naturelle.* Éditions Kimé.

Gayon, J. (1995). De la philosophie biologique dans l'œuvre de Jacques Roger. In J. Roger (Ed.), *Pour une histoire des sciences à part entière* (pp. 459–471). Albin Michel.

Gayon, J. (1998). *Darwinism's struggle for survival: Heredity and the hypothesis of natural selection.* Cambridge University Press.

Gayon, J. (2009). Philosophy of biology: A historico-critical characterization. In A. Brenner & J. Gayon (Eds.), *French studies in the philosophy of science: Contemporary research in France* (pp. 201–212). Springer.

Gayon, J., & De Ricqlès, A. (Eds.). (2010). *Les fonctions: Des organismes aux artefacts.* Presses Universitaires de France.

Lakatos, I. (1971). History of science and its rational reconstructions. *Boston Studies in the Philosophy of Science, 8*, 91–135.

Pradeu, T. (2017). Thirty years of *Biology & Philosophy*: Philosophy of which biology? *Biology & Philosophy, 32*, 149–167.

Sober, E. (1984). *The nature of selection: Evolutionary theory in philosophical focus.* MIT Press.

Chapter 5
Jean Gayon, History and Philosophy of Biology: A New Synthesis

Thomas Pradeu

Abstract In this contribution, I show that Jean Gayon's work operates an original synthesis between the history of science, the philosophy of science and the life sciences. I propose that the philosophy of biology as it has been constructed since the 1970s could usefully draw inspiration from this synthesis to develop a much richer and more open analysis of current biological sciences. While the philosophy of biology is very strongly dominated by the biology of evolution, the history of biology has opened up to practically all the major fields of contemporary biology (for example, cell biology, molecular biology, or virology, among many others). Perhaps Jean Gayon's work is an invitation to draw inspiration from this thematic diversity to construct a new philosophy of biology and, beyond that, a new synthesis between philosophy, history, and biology.

Keywords Philosophy of biology · Evolutionary theory · Cell biology · Development · Virology · *Biology and Philosophy* · PNAS

5.1 Introduction: Jean Gayon's "School"

It was at the sea that I made my first real encounter with Jean Gayon. In September 2004, Anouk Barberousse (then Research Fellow at the CNRS, now Professor at Paris-Sorbonne University) organized a memorable summer school on the theme of the philosophy of science, in the idyllic environment of Saint-Aygulf (Fig. 5.1). I had just returned from Harvard University, where I had spent an academic year as a Visiting Fellow, making decisive encounters with Richard Lewontin and Peter Godfrey-Smith. I officially began my thesis, under the supervision of Jean Gayon, on September 1, 2004. This week in Saint-Aygulf was not only an opportunity to

Translated from Pradeu (2018).

T. Pradeu (✉)
ImmunoConcept, CNRS, University of Bordeaux, Bordeaux, France
e-mail: thomas.pradeu@u-bordeaux.fr

IHPST, UMR8590, CNRS, Université Paris 1 Panthéon-Sorbonne, Paris, France

P.-O. Méthot (ed.), *Philosophy, History and Biology: Essays in Honour of Jean Gayon*, History, Philosophy and Theory of the Life Sciences 30,
https://doi.org/10.1007/978-3-031-28157-0_5

Fig. 5.1 Summer school of Saint-Aygulf (September 2004). Jean Gayon, in red, is second from the left (back row). Picture taken by T. Pradeu

meet many of the best French and French-speaking philosophers of science, but also to hear lessons from Jean Gayon on the theme of functions in biology and to have my first long exchanges with him.

From that moment on, I was struck by the way Jean was able to "create a school" ("*faire école*"). By this I mean that a whole generation of students of various ages have been trained either directly by Jean (who supervised a very large number of Master theses and PhD dissertations), or in his intellectual and human environment (for example, for those who did not complete a thesis under Jean's direction, his influence was still important via his teachings, research seminars of the Institute of History and Philosophy of Science and Technology (IHPST), the meetings of the International Society for the History, Philosophy, and Social Studies of Biology (ISHPSSB), etc.). From this point of view, Jean reiterated a gesture of Georges Canguilhem: many philosophers, including Jean himself, were very strongly influenced by Canguilhem and even became his intellectual "disciples", without having been his students. In the same way, I see in several extremely promising figures of the "young generation" of philosophers of biology in France who did not work under Jean's direction, such as Lucie Laplane for example (now a Research Fellow at the CNRS), continuators of Jean's approach.

Jean was thus able to do something that is, I believe, quite rare: to teach in a methodological rather than thematic way. Indeed, if some of Jean's students worked on themes close to his own (notably Darwinism, the Modern Synthesis, the history of genetics, heredity, or functions in biology), most of them explored very different issues (from systems biology to the origins of life, from microbiology to epigenetics, from ecology to evolutionary ethics or the notion of health—to give just a few examples here). In my opinion, this shows that the common denominator of what these numerous students came to Jean for is less a thematic expertise (although Jean's scientific and philosophical expertise was very diverse and deep) than a certain conception of the method to be adopted when doing philosophy of science.

In my opinion, Jean's method has three dimensions. The first is intellectual rigor, which is obvious to anyone who heard Jean: rigor of sources, theses, arguments. The second dimension is open-mindedness, which is in line with what has been mentioned above: openness to all the biological and biomedical sciences and to all their themes, whether specific to a discipline or transversal. The third is interdisciplinarity: Jean constantly sought to bring together four poles, in an ever-renewed synthesis: biology, philosophy of biology, history of biology, and general philosophy of science. This may be seen, of course, as a continuation of "historical epistemology", of which Jean was unquestionably an heir and a major actor, and on which he wrote extensively (e.g., Bitbol & Gayon, 2006; Gayon, 2003). However, this synthesis is broader and goes further, since it has come to associate and combine the tradition of historical epistemology itself with the approach of the "philosophy of biology" as it developed from the 1960s and 1970s onwards, mainly in the English-speaking world (Hull, 1969, 1974; Ruse, 1973). Jean was himself undeniably a figure of both "French" historical epistemology and "Anglo-Saxon" philosophy of biology. For example, he was one of the witnesses to the birth of the international society ISHPSSB in the 1980s in the United States, and a member of the editorial board of the journal *Biology & Philosophy*, founded by Michael Ruse in 1986.

It is this triple dimension of Jean's method that I would like to examine here, focusing my attention on a particular lesson that Jean, to my knowledge, did not formulate as such but which his work seems to me to outline, and which can be summed up in a very brief sentence: *it would be desirable for the philosophy of biology to be inspired by the thematic diversity of the history of biology*. The exploration of this lesson, which is simple but, in my view, extremely fruitful, will involve two stages: an analysis of the thematic narrowness of the philosophy of biology, and then an examination of how the history of biology could offer a different model of development for the philosophy of biology.

A final remark before going into the details of this analysis. Jean inspired younger people always by leaving a high degree of autonomy to them. When I was a doctoral student at the IHPST (from 2004 to 2007), surrounded by such fascinating people as Pierre-Alain Braillard, Philippe Huneman, Marie-Claude Lorne, Christophe Malaterre and many others, Jean constantly left us "*carte blanche*". As far as I was concerned, this went through an immediate and unreserved support when I created the Philosophy of Biology seminar at the IHPST ("Philbio"), the "Philosophy and Immunology" seminar also at the IHPST (with Leïla Périé and François Asperti-Boursin, then Marc Daëron), or for all the working groups and colloquia that I have organized or co-organized, whether as a doctoral student (2004 to 2007), as an Associate Professor at Paris-Sorbonne University (2008–2014), or when I later joined the CNRS in Bordeaux (2014). Systematically, Jean listened to me carefully and then told me to see my projects through to the end. This model of "autonomy under control"—as our friends from the theoretical biology group in San Sebastian would say, perhaps—regularly inspires me today.

5.2 The Lack of Thematic Diversity in Philosophy of Biology

In 2009, Jean Gayon published a text entitled "Philosophy of Biology: An Historico-Critical Characterization" (Gayon, 2009), in which he examines the origins, central themes and boundaries of the field of "philosophy of biology" as it developed from the 1960s to the 1970s, mainly in English-speaking countries. To do this, Jean proposed a review and a classification of all the articles published in the journal *Biology & Philosophy* from 1986 (the date of the creation of this journal) to 2002 (which is simply the moment when Jean's investigation stopped, for practical reasons). It is, if you like, a text of "meta-philosophy of science", in the sense that it proposes a subtle and "objective" analysis of a particular field of philosophy of science, in this case the philosophy of biology.

The choice of the journal *Biology & Philosophy* is justified by the fact that it is the only journal devoted fully and exclusively to the philosophy of biology.[1] And by the fact that it is the journal generally recognized as dominant in this field.

In this chapter, Jean proposes to divide the articles published by *Biology and Philosophy* into three main groups, the first grouping brings together articles relating to a particular field of life sciences (48% of articles); the second pertains to general philosophical questions (causality, laws, etc.) (37% of articles); and the third concerns highly historical articles (15% of articles). Jean makes three observations based on this classification. First, 72% of the articles relating to a particular field of life sciences concern evolutionary biology. Second, biological fields as central as physiology, biochemistry and biophysics are totally absent from *Biology & Philosophy*. Third, there is no significant change in the representation of the different fields of biology over time (over a period of sixteen years), which suggests that philosophers of biology are not particularly sensitive to the evolution of biology in their time.

This chapter is a particularly convincing illustration of what might be called the method of "critical taxonomy" favored by Jean Gayon, which consists in using description and classification tools (including statistics) to lead the reader to a bold conclusion. Here, a work that is at first thought to be innocently classificatory leads us

---

[1] Among the other journals that publish work in the philosophy of biology, some are devoted to both historical and philosophical approaches to the life sciences (*Studies in History and Philosophy of Biological and Biomedical Sciences*, for example), while others publish research that is relevant to the philosophy of science as a whole (*Philosophy of Science* or *The British Journal for the Philosophy of Science*, for example).

to the formulation of two radical challenges to the field of the philosophy of biology as it has developed over a little less than twenty years (from 1986 to 2002): the lack of thematic diversity in the philosophy of biology and its relative insensitivity to the evolution of biology itself. Any classification, of course, is the result of positions taken, but it can also, as we just saw, generate new claims.

In a recent work, I tried to extend Jean Gayon's analysis of the journal *Biology & Philosophy* (Pradeu, 2017).[2] This work examines all the articles published in this journal for thirty years, from 1986 (year of birth of this journal, as we have seen) to the end of 2015. My classification, which integrates and restructures that of Jean, is purely thematic, which means that I have classified all the articles published in this journal solely according to the discipline (main or unique) of life sciences to which they relate. From 1986 to 2002 (the period covered by Jean Gayon's analysis), 55% of the articles published in *Biology & Philosophy* dealt, from a thematic point of view, with the evolutionary biology. From 2003 to the end of 2015, 62% of the articles published in *Biology and Philosophy* dealt with evolution, 14% with psychological and cognitive sciences. The other fields were only marginally represented, and often because of a handful of articles or a special issue devoted specifically to this field; these were genetics (5%), developmental biology (4%), neurosciences (3%), ecology (3%), microbiology (3%) and medical sciences (2%) (on all these data, see Fig. 5.2). Some fields are completely absent, including biophysics and computational biology, cell biology, immunology, physiology, and systems biology.

The next step in my reasoning in this article was to take as seriously as possible the wish of the founders of the philosophy of biology—foremost among them David Hull (Hull, 1969, 2008)—that their field reflect as much as possible the questionings, results and practices of biology itself. I have therefore compared the distribution of the fields of biology in *Biology and Philosophy* over the recent period (2003–2015) to the distribution of the fields of biology in a major scientific journal offering a rigorous classification of the articles it publishes, namely the *Proceedings of the National Academy of Sciences of the United States of America* (*PNAS*),[3] during the same period.[4] The distribution in the *PNAS* offers a completely different picture of

---

[2] At this point, I would like to mention two potential conflicts, which may distort my judgment. First, I have been Associate Editor of the journal *Biology and Philosophy* since January 2017 (the mission of the Editor-in-Chief and Associate Editors is to receive submissions, select reviewers according to a "double-blind" procedure, and make the final decision whether to publish the submitted article). There is therefore perhaps a risk that I overestimate the importance of the journal *Biology and Philosophy* relative to other journals. Second, my own research in philosophy of biology has focused on the philosophy of immunology, a field that is not primarily evolutionary. There is therefore perhaps a risk that I overestimate the importance of non-evolutionary approaches in the philosophy of biology.

[3] *The Proceedings of the National Academy of Sciences of the United States of America* (*PNAS*) is the official journal of the National Academy of Sciences of the United States, published since 1915.

[4] More specifically, I have used the *PNAS* thematic classification, which I have applied to articles published in *Biology and Philosophy*. This was the best way for me to produce a satisfactory comparison between the two journals. An obvious consequence of this method is the multiplication of "0%" mentions in the representation of biological fields as it appears in *Biology and Philosophy* (Fig. 5.1).

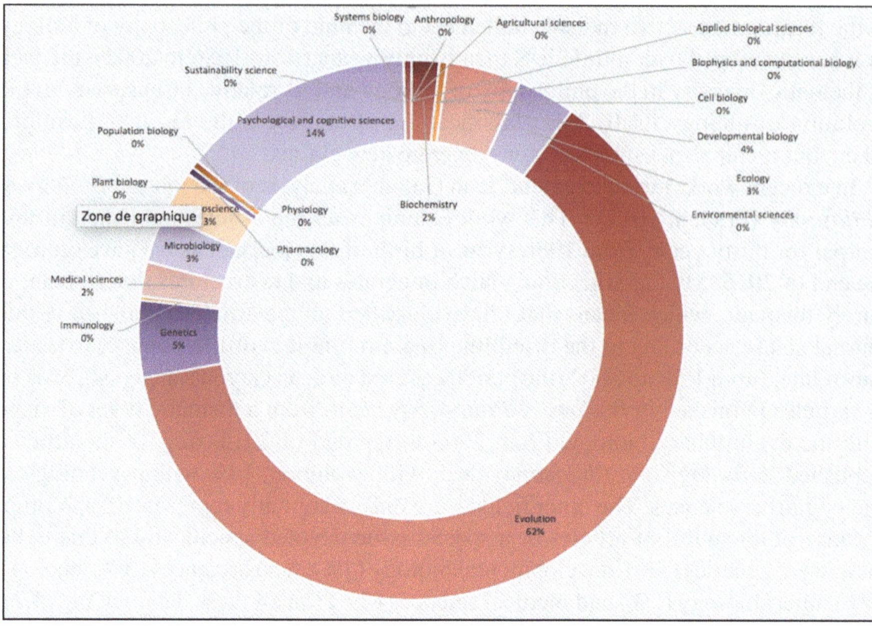

Fig. 5.2 Representation of biological fields in the journal *Biology & Philosophy* from 2003 to 2015. Based on Pradeu (2017)

the biological sciences. The dominant fields are, in order of importance: biochemistry (12%), neuroscience (12%), medical sciences (12%), computational biology (9%), immunology (7%) and microbiology (7%) (on all these data, see Fig. 5.3). When comparing, therefore, the representation of the biological fields in *Biology and Philosophy* and in the *PNAS* from 2003 to 2015, one sees a huge gap, illustrated especially by the fact that evolution represents 62% of the papers in *Biology & Philosophy* but only 5% in the *PNAS*. We can also note that some of the most frequent domains in the *PNAS* are not or hardly represented in *Biology & Philosophy*, notably biochemistry (12% vs. 2%), biophysics and computational biology (9% vs. 0%), cell biology (8% vs. 0%), immunology (7% vs. 0%) and neuroscience (12% vs. 3%). Finally, the representation of domains in *Biology & Philosophy* over time is unaffected by the change in domain representation in the *PNAS*, suggesting that the significant changes in biology from 2003 to 2015 have remained without influence on the themes of biology investigated by the philosophers.

The general conclusion that can be drawn from these analyses proposed by Jean Gayon and then by myself is that the philosophy of biology as it has developed over the last thirty years is "provincial", in the very specific sense that it has been practically interested in only one of the fields of the life sciences, namely evolution.[5]

[5] In a fascinating study, Christophe Malaterre and his colleagues have enriched and challenged Jean Gayon's and my conclusions. Using a different and increasingly popular method, namely text mining, they conclude that although evolution by natural selection has been a much prevalent and

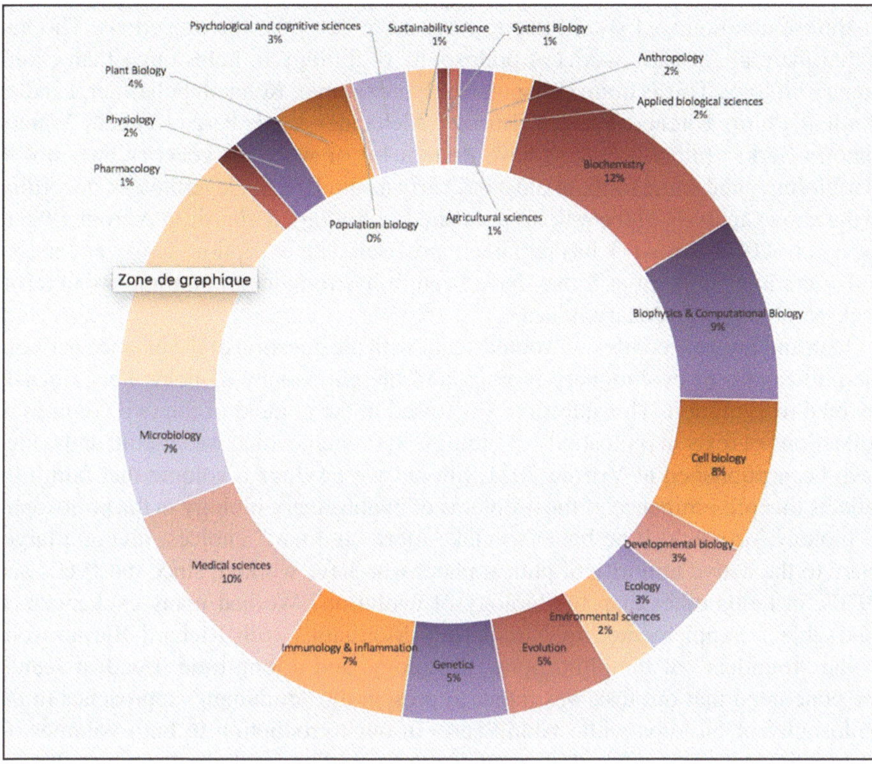

Fig. 5.3 Representation of biological fields in the *Proceedings of the National Academy of Sciences of the United States of America* (*PNAS*) from 2003 to 2015. Based on Pradeu (2017)

The philosophical interest of evolutionary biology is quite undeniable, as generations of biological philosophers have demonstrated (we can think of the likes of David Hull, Michael Ruse, Marjorie Grene, Elliott Sober, Elizabeth Lloyd, Robert Brandon, Peter Godfrey-Smith, or Samir Okasha, among many others). To this should be added the very important observation that most of the biologists who contributed to the birth and institutionalization of the field of philosophy of biology—notably Ernst Mayr, Richard Lewontin, Stephen Jay Gould, Francisco Ayala—were evolutionary biologists.

Nevertheless, the question remains as to whether it would be relevant for the philosophy of biology to open to other areas of the life sciences. In the rest of this chapter, I suggest that this is indeed the case, and that the history of science could constitute, from this point of view, an important source of inspiration for the philosophy of biology.

---

relatively stable topic in *Biology and Philosophy*, a much broader diversity of topics has in fact been addressed in this journal (Malaterre et al., 2020).

Before continuing, I would like to respond to two possible objections. The first is that there are already works of philosophy of biology in fields other than evolutionary biology. This is quite indisputable. For example, Kenneth Schaffner, Lindley Darden, Philip Kitcher, Richard Burian, Alexander Rosenberg, Kenneth Waters, Sahotra Sarkar (among others) have done a lot of work on genetics and molecular biology, and some of them did so as early as the 1970s. Nevertheless, according to the above analysis of the articles published in *Biology & Philosophy* from 1986 to the end of 2015, this work has remained, precisely, the exception: however exciting and sometimes influential it may have been, it has remained in a minority in terms of research on evolutionary issues.

Jean and I were recently confronted again with the question of the balance between the philosophy of evolutionary biology and the philosophy of fields not primarily devoted to evolution. This question was raised in the context of the two volumes of collections of texts in philosophy of biology, in French, which we co-edit, and which have been published at Vrin in 2021. Should we produce a volume that faithfully reflects this pre-eminence of the problems of evolutionary biology in the philosophy of biology, or would it be better to make more "militant" choices, giving a larger share to the active minority of philosophers who have worked, since the 1960s and 1970s, in fields other than the biology of evolution? We had many exchanges on this subject, supplemented by fascinating discussions with Richard Burian—one of the "founders" of the philosophy of biology, and a long-time friend of Jean's. We concluded that our task was rather to present the "dominant" approaches in the philosophy of biology, while taking care, in our introduction to both volumes, to provide important contextual elements suggesting that this field of philosophy has sometimes been (and could be in the future) much more diverse thematically.

The second objection is that certain research themes in the philosophy of biology are published primarily elsewhere than in the journal *Biology & Philosophy*. This objection is probably valid for some fields, such as the philosophy of neuroscience which, probably because of its links with "classical" problems of philosophy of mind and general philosophy, has often found its place in generalist journals of philosophy of science, such as *Philosophy of Science* or *The British Journal for the Philosophy of Science*. Nevertheless, most of the other fields we have mentioned (for example biochemistry, biophysics, cell biology, etc.) are not represented in these generalist journals, which weakens the objection. Moreover, if a field such as the philosophy of neuroscience had taken on a truly major importance in the philosophy of biology, it seems very likely that *Biology & Philosophy* would have been impacted by this (albeit with some delay or with perhaps younger or less prestigious authors).

It therefore seems to me that these two objections, however pertinent they may be, do not invalidate the general observation made by Jean Gayon and then by myself that the philosophy of biology has remained for the most part confined to one field among others of the life sciences.[6]

---

[6] In my 2017 article, I examine and reject several often-advanced explanations for the dominance of evolutionary biology in philosophy of biology, namely the idea of a simple historical contingency,

5.3 History of Science as an Inspiration

Faced with this observation of the thematic narrowness of the philosophy of biology as it has been constructed over the last thirty years, I would like to suggest that the history of science, and especially the history of science as practiced by Jean Gayon, can be a major source of inspiration for the philosophy of biology.

Indeed, without being myself a specialist in the history of biology, it seems to me that this field has been able to avoid the two pitfalls that we have identified with regard to the philosophy of biology: compared to the philosophy of biology, the history of biology is interested in a greater diversity of fields in the life sciences (e.g., cell biology, molecular biology, or virology, among many others), and it is also much more "reactive" to the evolution of biology itself. This seems to me to be one probable cause (though of course not the unique cause) of the fact that biologists are generally more interested in the history of biology than in the philosophy of biology: perhaps they see it as a more faithful reflection of their own intellectual interests and questionings.

Of course, in order to really establish this double observation, it would be necessary to offer a detailed statistical analysis over a long period of time, which will be a job for another time. I will mention here just two clues that tend to corroborate this intuition. The first is a quick analysis of the articles recently published in the *Journal of the History of Biology*. The second consists in a few examples of work in the history of biology—including that of Jean Gayon—showing that themes entirely or almost entirely neglected by philosophers of biology are being explored by historians of biology.

A quick review of articles recently published in one of the major journals in the field of the history of biology, the *Journal of the History of Biology*,[7] offers a very different representation of the life sciences than that offered by *Biology & Philosophy* and *PNAS*. My investigation here only covers the years 2014 and 2015 and is, of course, not statistically significant. The aim is only to have a very general impression, opening research perspectives for the future. Over these two years, evolution is the dominant theme (39%) of the articles published in the *Journal of the History of Biology*, but to a much lesser degree than in *Biology and Philosophy*. Other fields are significantly represented, such as physiology (16%), biochemistry (6%), immunology (6%), anthropology (6%), environmental sciences (6%), agricultural sciences (3%), biophysics (3%), of which six (physiology, immunology, anthropology, environmental sciences, agricultural sciences, and biophysics) were completely absent from *Biology and Philosophy* (see Fig. 5.4). Again, no precise

that of a higher degree of theorization in evolutionary biology, and finally that of a greater intrinsic philosophical interest in evolution. I refer the interested reader to this article.

[7] The *Journal of the History of Biology* was founded by Everett Mendelsohn in 1968. Mendelsohn was Editor-in-chief for thirty-one years, until 1999. Until December 2017, the Editor-in-Chief was Michael Dietrich (Dartmouth College, then Pittsburgh University); since January 2018, the Editors-in-Chief are Karen Rader (Virginia Commonwealth University) and Marsha Richmond (Wayne State University). *The Journal of the History of Biology* is unanimously considered to be one of the major journals, if not the major journal, in the history of biology.

conclusions can be drawn from such a small study. One can only suggest that the themes in a major history of biology journal are different from those of *Biology and Philosophy*, in the sense that they are both more diversified and less neglectful of certain fields of life sciences that are strongly represented in a scientific journal such as *PNAS* (without, however, the distributions of biological fields between these two journals being similar).

The second clue concerns certain works in the history of biology that deal with fields of life sciences entirely or almost entirely neglected by the philosophers of biology. A particularly relevant example is the work carried out by Jean Gayon, Richard Burian, Doris Zallen and, more recently, Laurent Loison on bacteriophages and lysogenesis (Burian & Gayon, 1999; Burian et al., 1988; Gayon & Burian, 2004; Loison et al., 2017). This landmark work spans several biological disciplines (it touches on genetics, molecular biology, cell biology, etc.) and several themes (including French "Lamarckism" (Loison, 2010)), but one of its central dimensions concerns virology. It is striking that virology—one of the most dynamic and conceptually exciting fields in contemporary biology—has been the subject of numerous historical analyses (Bos, 2000; Creager, 2002; Lustig & Levine, 1992; Summers, 2014; van Helvoort, 1994a, 1994b) but has hardly been explored by philosophers

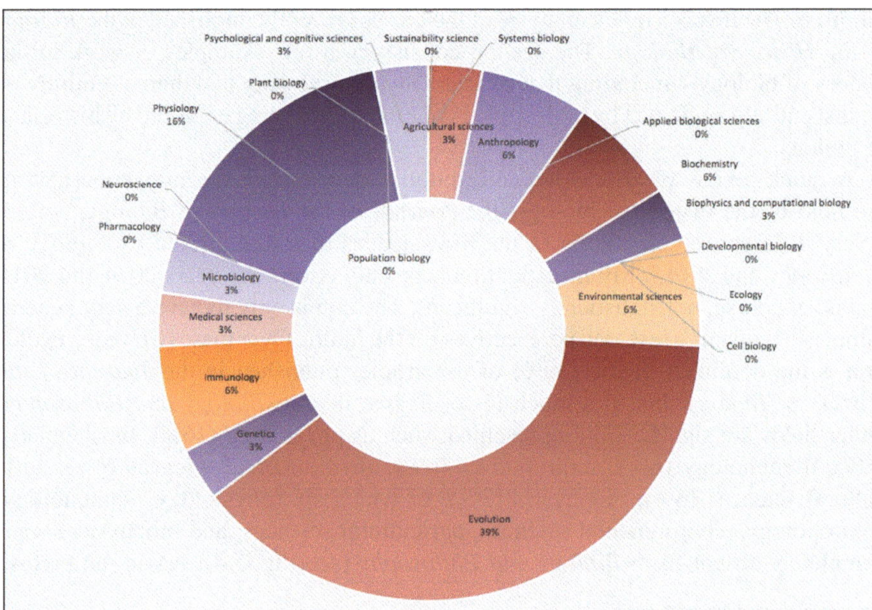

Fig. 5.4 Representation of the biological sciences in the *Journal of the History of Biology* in 2014 and 2015. (This representation has only an example value, the number of articles considered being extremely small)

of biology.[8] The situation may be changing, and recent reflections on giant viruses (Claverie & Abergel, 2010, 2016; Forterre, 2016; Koonin & Dolja, 2013; Raoult & Forterre, 2008) may, among others, inspire philosophers of biology. It is nevertheless striking that the history of biology has long since opened to this central field of present-day biology, while the philosophy of biology, for its part, showed very little interest for it.[9] The recent Covid-19 pandemic showed how few philosophers had worked specifically on viruses. Unsurprisingly, many philosophers showed interest in viruses following the pandemic, which is of course a good thing, although this kind of expertise requires long-term intellectual investments.

Another recent example (and which, this time, is not among Jean Gayon's research objects) is the CRISPR-Cas system, neglected by philosophers of biology but already studied by historians of biology, especially Michel Morange (Morange, 2015a, 2015b). Even more than the thematic diversity of the history of biology, this example shows the sensitivity of this field to the transformations of current biology. The CRISPR-Cas system is generally described as an immune system in archaea and bacteria (Barrangou & Marraffini, 2014; Barrangou et al., 2007; Bondy-Denomy et al., 2013; Horvath & Barrangou, 2010; Koonin et al., 2017; Makarova et al., 2006; Mojica et al., 2005; Pourcel et al., 2005; Pradeu, 2009), which has become one of the most powerful tools for genome manipulation (Charpentier & Doudna, 2013; Morange, 2015b), and its characterization has been long, complex, and controversial (Guttinger, 2018; Lander, 2016; Ledford, 2015, 2016, 2017; Morange, 2015a, 2015b). It is as if history, although turned towards the past of biological disciplines, proved to be much more open than the philosophy of biology to the present and the future of life sciences. Given the conceptual, methodological, and ethical richness of the CRISPR-Case system, it seems almost certain that it will be an important object of study for philosophers of biology in the future. Such philosophical investigations have recently emerged, but the process has been rather slow, and scientists have played an important role in it (Baxter, 2019; Jablonka, 2019; Koonin, 2019a, 2019b; Pradeu, 2019; Pradeu & Moreau, 2019; Veigl, 2019; Wideman et al., 2019; Woolley et al., 2019). The Nobel Prize in Chemistry was awarded in 2020 to Jennifer Doudna and Emmanuelle Charpentier for the discovery of CRISPR-Cas, so hopefully more philosophers of biology will pay attention to it.

---

[8] The most notable exception is the work of Gregory J. Morgan (Morgan 2001, 2006, 2010; Morgan and Pitts 2008).

[9] In 2016, John Dupré, Gladys Kostyrka, and I edited what is, to our knowledge, the first edited volume devoted specifically to the philosophy of virology (Pradeu et al., 2016). It includes the contributions of Gregory J. Morgan (Stevens Institute of Technology), Maureen O'Malley (then at the University of Bordeaux), Thomas Pradeu (CNRS & University of Bordeaux), Jean-Michel Claverie and Chantal Abergel (Université Aix-Marseille), Patrick Forterre (Institut Pasteur & Paris-Sud University), John Dupré and Stephan Guttinger (University of Exeter), Marc van Regenmortel (CNRS & University of Strasbourg), Eugene Koonin (NIH) et Petro Starokadomskyy (UT Southwestern Medical Center), Gladys Kostyrka (IHPST, CNRS & Paris 1 Panthéon-Sorbonne University), and Pierre-Olivier Méthot (Laval University).

Thus, the history of biology—and even more so the approach of "epistemological history" (Gayon, 2003) illustrated by Jean Gayon and Michel Morange[10]—seems to be able to function as a model and a source of inspiration for the philosophy of biology, from the double point of view of its thematic diversity and its sensitivity to current changes in the life sciences.

5.4 Conclusion

The philosophy of biology has unduly limited its object to practically only one field of the life sciences, evolutionary biology. To paraphrase Ernst Mayr, who complained in 1969 that the "philosophy of science" was in reality reduced to the "philosophy of physics" (Mayr, 1969), one could say that the philosophy of biology has practically been reduced to the "philosophy of evolution". Although the philosophical questions raised by evolutionary biology are fascinating, I have tried to show, following Jean Gayon, that the philosophy of biology would be well advised, today, to open up to other fields of the life sciences, which would complement and enrich the problems formulated until now mainly in the light of evolutionary biology.

Through his own work, Jean showed us the path to follow, that of a methodological synthesis between the history of biology, the philosophy of biology and biology itself, which would be in constant dialogue with biologists and would strive to raise problems of a philosophical nature, always bearing in mind the diversity of the life sciences and their rapid and fascinating transformations. It seems to me that Jean's work tells us that the philosophy of biology will probably have to mutate to survive; by diversifying and updating itself, it will be able to prolong the impressive dynamism that has characterized it for almost fifty years now.

Acknowledgements I want to thank Jean Gayon for his constant trust and friendship. He was a model of strength and honor. Being his friend and learning from him has been one of the most important experiences of my life. Thanks to the École normale supérieure, the IHPST and Paris 1 Panthéon-Sorbonne University for allowing me to write my thesis under the direction of Jean Gayon. My research is supported by the European Research Council (ERC; European Union's Horizon 2020 research and innovation programme—Grant agreement n° 637647—IDEM).

References

Barrangou, R., Fremaux, C., Deveau, H., Richards, M., Boyaval, P., Moineau, S., et al. (2007). CRISPR provides acquired resistance against viruses in prokaryotes. *Science, 315*(5819), 1709–1712.

Barrangou, R., & Marraffini, L. A. (2014). CRISPR-Cas systems: Prokaryotes upgrade to adaptive immunity. *Molecular Cell, 54*(2), 234–244.

[10] On Morange, see Loison and Pradeu (2021).

Baxter, J. (2019). CRISPR-Cas changing biology? *Biology & Philosophy, 34*, 15–15.
Bitbol, M., & Gayon, J. (Eds.). (2006). *L'épistémologie française, 1830–1970 (1re éd.)*. Presses Universitaires de France.
Bondy-Denomy, J., Pawluk, A., Maxwell, K. L., & Davidson, A. R. (2013). Bacteriophage genes that inactivate the CRISPR/Cas bacterial immune system. *Nature, 493*(7432), 429–432.
Bos, L. (2000). 100 years of virology: From vitalism via molecular biology to genetic engineering. *Trends in Microbiology, 8*(2), 82–87.
Burian, R. M., & Gayon, J. (1999). The French school of genetics: From physiological and population genetics to regulatory molecular genetics. *Annual Review of Genetics, 33*, 313–349.
Burian, R. M., Gayon, J., & Zallen, D. (1988). The singular fate of genetics in the history of French biology, 1900–1940. *Journal of the History of Biology, 21*(3), 357–402.
Charpentier, E., & Doudna, J. A. (2013). Biotechnology: Rewriting a genome. *Nature, 495*(7439), 50–51.
Claverie, J.-M., & Abergel, C. (2010). Mimivirus: The emerging paradox of quasi-autonomous viruses. *Trends in Genetics: TIG, 26*(10), 431–437.
Claverie, J.-M., & Abergel, C. (2016). Les virus géants - État des connaissances, énigmes, controverses et perspectives. *Médecine/sciences, 32*(12), 1087–1096.
Creager, A. N. H. (2002). *The life of a virus: Tobacco mosaic virus as an experimental model, 1930–1965*. University of Chicago Press.
Forterre, P. (2016). To be or not to be alive: How recent discoveries challenge the traditional definitions of viruses and life. *Studies in History and Philosophy of Science Part c: Studies in History and Philosophy of Biological and Biomedical Sciences, 59*, 100–108.
Gayon, J. (2003). Bachelard et l'histoire des sciences. In J.-J. Wunenburger (Ed.), *Bachelard et l'épistémologie française* (pp. 51–113). Presses Universitaires de France.
Gayon, J. (2009). Philosophy of biology: An historico-critical characterization. In A. Brenner & J. Gayon (Eds.), *French studies in the philosophy of science: Contemporary research in France* (pp. 201–212). Springer.
Gayon, J., & Burian, R. M. (2004). National traditions and the emergence of genetics: The French example. *Nature Reviews Genetics, 5*(2), 150–156.
Guttinger, S. (2018). Riding the wave into a crisper future? *Studies in History and Philosophy of Science Part c: Studies in History and Philosophy of Biological and Biomedical Sciences, 67*, 32–35.
Horvath, P., & Barrangou, R. (2010). CRISPR/Cas, the immune system of bacteria and archaea. *Science, 327*, 167–170.
Hull, D. L. (1969). What philosophy of biology is not. *Synthese, 20*(2), 157–184.
Hull, D. L. (1974). *Philosophy of biological science*. Prentice-Hall.
Hull, D. L. (2008). The history of the philosophy of biology. In M. Ruse (Ed.), *The Oxford handbook of philosophy of biology* (pp. 11–33). Oxford University Press.
Jablonka, E. (2019). Lamarckian realities: The CRISPR-Cas system and beyond. *Biology & Philosophy, 34*(1), 1–5.
Koonin, E. V. (2019a). CRISPR: A new principle of genome engineering linked to conceptual shifts in evolutionary biology. *Biology & Philosophy, 34*(1), 1–30.
Koonin, E. V. (2019b). Lamarckian or not, CRISPR-Cas is an elaborate engine of directed evolution. *Biology & Philosophy, 34*(1), 1–6.
Koonin, E. V., & Dolja, V. V. (2013). A virocentric perspective on the evolution of life. *Current Opinion in Virology, 3*(5), 546–557.
Koonin, E. V., Makarova, K. S., & Zhang, F. (2017). Diversity, classification and evolution of CRISPR-Cas systems. *Current Opinion in Microbiology, 37*, 67–78.
Lander, E. S. (2016). The heroes of CRISPR. *Cell, 164*(1), 18–28.
Ledford, H. (2015). CRISPR, the disruptor. *Nature News, 522*(7554), 20–25.
Ledford, H. (2016). CRISPR: Gene editing is just the beginning. *Nature News, 531*(7593), 156–159.
Ledford, H. (2017). Five big mysteries about CRISPR's origins. *Nature News, 541*(7637).

Loison, L. (2010). *Qu'est-ce que le néolamarckisme? les biologistes français et la question de l'évolution des espèces, 1870–1940*. Vuibert.
Loison, L., Gayon, J., & Burian, R. M. (2017). The contributions—and collapse—of Lamarckian heredity in Pasteurian molecular biology: 1. Lysogeny, 1900–1960. *Journal of the History of Biology, 50*(1), 5–52.
Loison, L., & Pradeu, T. (Eds.). (2021). *La biologie au défi de l'histoire. Mélanges offerts à Michel Morange*. Éditions Matériologiques.
Lustig, A., & Levine, A. J. (1992). One hundred years of virology. *Journal of Virology, 66*(8), 4629–4631.
Makarova, K. S., Grishin, N. V., Shabalina, S. A., Wolf, Y. I., & Koonin, E. V. (2006). A putative RNA-interference-based immune system in prokaryotes: Computational analysis of the predicted enzymatic machinery, functional analogies with eukaryotic RNAi, and hypothetical mechanisms of action. *Biology Direct, 1*(1), 1–26.
Malaterre, C., Pulizzotto, D., & Lareau, F. (2020). Revisiting three decades of Biology and Philosophy: A computational topic-modeling perspective. *Biology & Philosophy, 35*(1), 1–25.
Mayr, E. (1969). Footnotes on the philosophy of biology. *Philosophy of Science, 36*(2), 197–202.
Mojica, F. J. M., Díez-Villaseñor, C., García-Martínez, J., & Soria, E. (2005). Intervening sequences of regularly spaced prokaryotic repeats derive from foreign genetic elements. *Journal of Molecular Evolution, 60*(2), 174–182.
Morange, M. (2015a). What history tells us XXXVII. CRISPR-Cas: The discovery of an immune system in prokaryotes. *Journal of Biosciences, 40*(2), 221–223.
Morange, M. (2015b). What history tells us XXXIX. CRISPR-Cas: From a prokaryotic immune system to a universal genome editing tool. *Journal of Biosciences, 40*(5), 829–832.
Morgan, G. J. (2001). Bacteriophage biology and Kenneth Schaffner's rendition of developmentalism. *Biology and Philosophy, 16*(1), 85–92.
Morgan, G. J. (2006). Why there was a useful plausible analogy between geodesic domes and spherical viruses. *History and Philosophy of the Life Sciences, 28*(2), 215–235.
Morgan, G. J. (2010). Laws of biological design: A reply to John Beatty. *Biology & Philosophy, 25*(3), 379–389.
Morgan, G. J., & Pitts, W. B. (2008). Evolution without species: The case of mosaic bacteriophages. *The British Journal for the Philosophy of Science, 59*(4), 745–765.
Pourcel, C., Salvignol, G., & Vergnaud, G. (2005). CRISPR elements in Yersinia pestis acquire new repeats by preferential uptake of bacteriophage DNA, and provide additional tools for evolutionary studies. *Microbiology (reading, England), 151*(3), 653–663.
Pradeu, T. (2009). *Les limites du soi: Immunologie et identité biologique*. Presses Universitaires de Montréal.
Pradeu, T. (2017). Thirty years of *Biology & Philosophy*: Philosophy of which biology? *Biology & Philosophy, 32*(2), 149–167.
Pradeu, T. (2018). Jean Gayon, l'histoire, la philosophie et la biologie: Une synthèse moderne. In F. Merlin & P. Huneman (Eds.), *Philosophie, histoire, biologie: Mélanges offerts à Jean Gayon* (pp. 89–102). Éditions Matériologiques.
Pradeu, T. (2019). Philosophy of CRISPR-Cas: Introduction to Eugene Koonin's target paper and commentaries. *Biology & Philosophy, 34*(1), 1–5.
Pradeu, T., Kostyrka, G., & Dupré, J. (2016). Understanding viruses: Philosophical investigations. *Studies in History and Philosophy of Science Part c: Studies in History and Philosophy of Biological and Biomedical Sciences, 59*, 57–63.
Pradeu, T., & Moreau, J.-F. (2019). CRISPR-Cas immunity: Beyond nonself and defence. *Biology & Philosophy, 34*(1), 1–6.
Raoult, D., & Forterre, P. (2008). Redefining viruses: Lessons from Mimivirus. Nature reviews. *Microbiology, 6*(4), 315–319.
Ruse, M. (1973). *The philosophy of biology*. Hutchinson.
Summers, W. C. (2014). Inventing viruses. *Annual Review of Virology, 1*(1), 25–35.

van Helvoort, T. (1994a). The construction of bacteriophage as bacterial virus: Linking endogenous and exogenous thought styles. *Journal of the History of Biology, 27*(1), 91–139.

van Helvoort, T. (1994b). History of virus research in the twentieth century: the problem of conceptual continuity. *History of Science; an Annual Review of Literature, Research and Teaching, 32*(2), 185–235.

Veigl, S. J. (2019). A use/disuse paradigm for CRISPR-Cas systems. *Biology & Philosophy, 34*(1), 1–4.

Wideman, J. G., Inkpen, S. A., Doolittle, W. F., & Redfield, R. J. (2019). Mutationism, not Lamarckism, captures the novelty of CRISPR–Cas. *Biology & Philosophy, 34*(1), 1–6.

Woolley, S., Parke, E. C., Kelley, D., Poole, A. M., & Ganley, A. R. D. (2019). Striving for clarity about the "Lamarckian" nature of CRISPR-Cas systems. *Biology & Philosophy, 34*(1), 1–5.

Chapter 6
A Structuralist Method: Or Why Darwin's Pangenesis Remained a Remarkable Blind Spot in Jean Gayon's Writings

Laurent Loison

Abstract That Jean Gayon never paid attention to Darwin's pangenesis might seem like an oddity given that natural selection and biological heredity were his primary focuses for decades. This lack of interest reveals Gayon's specific methodological orientation: he aimed at producing rational reconstructions of the way a scientific hypothesis entered experimentation and subsequently evolved within a specific theoretical pattern. Gayon's most important achievements, *Darwinism's Struggle for Survival* (1998) in the first place, were all based on this "structuralist approach", where philosophy of science provides a strong problematization and even shapes the historical narrative. I argue that Lakatos and Duhem were Gayon's main inspirations, and that his work is thereby substantially different from both philosophy of biology and French historical epistemology.

Keywords Jean Gayon · Rational reconstruction · Natural selection · Heredity · Pangenesis · Duhem · Lakatos · Historical epistemology · Philosophy of biology

6.1 Introduction

Most of Jean Gayon's historical and philosophical work, from the 1980s until his recent death, concerned natural selection and heredity. Yet, it is striking to observe that Darwin's own theory of heredity, namely "pangenesis", was never a topic of interest to him. Out of more than 300 papers, only a short dictionary entry and a few pages in the recent volume *Knowledge of Life Today* are devoted to pangenesis (Gayon, 2013; Gayon & Petit 2019, pp. 177–180). Why was Gayon so uninterested in Darwin's understanding of the concept of heredity even though his "provisional hypothesis" was central to Darwin himself and has given rise to a large body of scholarship in the history of science since the mid-twentieth century?

L. Loison (✉)
Laboratoire SPHERE (UMR 7219), CNRS, Paris, France
e-mail: laurentloison@yahoo.fr

P.-O. Méthot (ed.), *Philosophy, History and Biology: Essays in Honour of Jean Gayon*, History, Philosophy and Theory of the Life Sciences 30,
https://doi.org/10.1007/978-3-031-28157-0_6

What may seem like a merely anecdotal concern at first sight is, I argue, a meaningful path to understand the philosophical ambition of Gayon's historical work, and especially of his *Darwinism's Struggle for Survival: Heredity and the Hypothesis of Natural Selection* (1998), unanimously considered as a major achievement (Méthot, 2018, p. 32). Gayon's disinterest for pangenesis is related to what he sometimes termed, in his last months, his "structuralist approach",[1] i.e., his attempt to understand the history of science within a general abstract scheme. Most of Gayon's major works, beyond *Darwinism's Struggle for Survival*, all follow the same methodology: history is illuminated and reconstructed as unexpected philosophical patterns that shaped theoretical change are made explicit. Gayon's ultimate objective, as he eventually acknowledged, was to "cut up historical problems in such a way that strong structural constraints frame history" (Gayon & Petit, 2018, p. 68, my translation; see also Petit, 2019, p. 357). His 2000 text "From Measurement to Organization: A Philosophical Scheme for the History of the Concept of Heredity" is another textbook example of such a method, where, once again, Darwin's pangenesis is missing.

After a brief characterization of pangenesis, first as Darwin put it, then in twentieth and early twenty-first century historiography, and finally in Gayon's writings, I focus on the issue of the theoretical identity of such a specific and unusual way of doing history of science. In my reading, Gayon's so-called structuralism had two main inspirations, Duhem and Lakatos, and both philosophers are helpful resources in order to precisely understand Gayon's ambition and method. He produced historical reconstructions that made use of some of their conceptual tools, or at least in ways that followed in their footsteps. Both references might initially seem surprising. At the very end of his life, Gayon explicitly acknowledged his debt to Duhem in *Knowledge of Life Today*, one of his final publications, but did not expand on the matter. Lakatos is even more unexpected: nowhere did Gayon endorse Lakatos' legacy and explicit references to Lakatos are rather scant in his writings. Yet, I aim to show that Gayon was remarkably Lakatosian in his historical reconstructions—much more than he probably thought, to such an extent that he was ready to ignore actual history if the philosophical reconstruction required it. This is why he purposefully ignored Darwin's pangenesis.

If my analysis is correct, i.e. if Gayon was in the main a Lakatosian-Duhemian historian interested in producing large-scale rational reconstructions, then it implies that he developed and practiced an original methodology that was at odds both with that of the founders of the philosophy of biology and with the tradition within which he was educated, namely "historical epistemology".[2] His positioning, which he eventually called a "tertium" (Gayon & Petit, 2018, p. 77), represents another alternative that stands *outside* the opposition between normative analytical philosophy and historical epistemology.

[1] This is a phrase he used during the conference in his honor held in Paris in early 2017. He put it in print twice and explained it in few paragraphs: Gayon and Petit (2018, pp. 68–69, 252–253). The term "structuralist approach" is used on page 253.

[2] On the definition and history of historical epistemology in the French tradition, from Bachelard to Foucault, see Braunstein (2002).

6.2 Darwin's Pangenesis

The historical elucidation of Darwin's late theory of heredity, pangenesis, has been a very active field within the "Darwinian Cottage Industry" since 1959 and the anniversary of *On the Origin of Species*. Therefore, the main thrust of Darwin's speculative model is now well known and a reader familiar with the topic could easily skip the next two sections.

6.2.1 An Overview in Darwin's Work (1865–1875)

Before emphasizing the chasm between general scholarly interest (6.2.2.) and Gayon's total lack of interest (6.2.3.) in pangenesis, I shall start with a few words about the content of Darwin's "provisional hypothesis". Darwin added "provisional" because Thomas Huxley harshly criticized the first version that Darwin submitted to his judgement in 1865. He waited another three years before he eventually put it to print, as chapter XXVII of the second volume of his book *The Variation of Animals and Plants under Domestication.* A slightly revised version was published in 1875, in the second and final edition of *Variation*, four years after *The Descent of Man and Selection in Relation to Sex.*[3]

Darwin had been interested in heredity for more than twenty years when he elaborated pangenesis. His model was based on the activity of speculative sub-cellular particles, called "gemmules". In the main, Darwin supported a strongly atomistic account of the organism, "which consists of a multitude of elemental parts, [...] to a great extent independent of each other" (Darwin, 1868, pp. 368–369). Whatever the truth of recent cell theory, it was just a fact "that the body consists of a multitude of 'organic units", each of which possesses its own proper attributes, and is to a certain extent independent of all others" (Darwin, 1868, pp. 370–371). In such a reductionist framework, Darwin was looking for a very general process that would have been at the basis of growth, reproduction (sexual and asexual) and regeneration all together.

Gemmules were supposed to be produced by cells, to "circulate freely throughout the system" in a "dormant" state, and to be able to later develop into new cells, identical to the ones that initially produced them (Darwin, 1868, p. 374). More importantly, because gemmules were collected in the sexual elements, they were thought to be transmitted from the parents to the offspring, and to then be responsible for biological inheritance. Because in that model it was the whole body, at any developmental stage, that directly generates the gemmules (the "pan" of pangenesis), pangenesis was also an assumed explanation of the inheritance of acquired characters.

In *Variation*, Darwin's rhetorical and argumentative strategy was basically the same as in the *Origin of Species.* His hypothesis of pangenesis was presented as

---

[3] In *The Descent of Man*, Darwin did not expand further on the issue of heredity because he believed he had done enough in *Variation*: "I have elsewhere so fully discussed the subject of Inheritance that I need here add hardly anything" (Darwin, 1871, p. 110).

being in accordance with what was known at the time about cells (an aspect strongly contested by his contemporaries, Huxley in the first place), and, more importantly for Darwin, it was able to "connect" many heterogeneous known facts, like hybridization, the characteristics of graft hybrids, inheritance of acquired characters, reversion, etc. (Darwin, 1868, p. 357). In the *Origin of Species*, as is well known, the first part of the book is devoted to understanding why natural selection is the necessary outcome of several processes and phenomena (variation, inheritance, competition), whereas the second part provides an explanation of known facts of natural history within an evolutionary framework. The same was true for pangenesis:

> Before proceeding to show, firstly, how far these assumptions are in themselves probable, and secondly, how far they connect and explain the various groups of facts with which we are concerned, it may be useful to give an illustration of the hypothesis. (Darwin, 1868, p. 376)

Like natural selection and evolution, pangenesis was seen as a natural law by Darwin, and he concluded the pangenesis chapter in a way that evidently echoed *On Origin of Species*' own conclusion:

> We cannot fathom the marvelous complexity of an organic being; but on the hypothesis here advanced this complexity is much increased. Each living creature must be looked at as a microcosm – little universe, formed of a host of self-propagating organisms, inconceivably minute and as numerous as the stars in heaven. (Darwin, 1868, p. 404)[4]

6.2.2 A Significant Part of the "Darwinian Industry"

The first studies of pangenesis from a clearly historical perspective might have been Conway Zirkle's articles "The inheritance of acquired characters and the provisional hypothesis of pangenesis" (1935) and "The early history of the idea of the inheritance of acquired characters and of pangenesis" (1946). As is usually acknowledged, Zirkle convincingly showed that the idea of inheritance of acquired characters dated back to Aristotle (at least) and that many philosophers, physicians and scientists had supported a mechanism close to Darwin's well before 1868. After the Modern Synthesis stabilized, several historians of science started to be interested in pangenesis, in the wake of the anniversary of the *Origin of Species*. In 1963, Robert C. Olby more or less launched this renewal with an article published in the *British Journal for the History of Science* devoted to Darwin's 1865 manuscript, the one he had submitted

---

[4] In *On the Origin of Species* (Darwin, 1859, pp. 489–490): "It is interesting to contemplate an entangled bank, clothed with many plants of many kinds, with birds singing on the bushes, with various insects flitting about, and with worms crawling through the damp earth, and to reflect that these elaborately constructed forms, so different from each other, and dependent on each other in so complex a manner, have all been produced by laws acting around us. [...] Thus, from the war of nature, from famine and death, the most exalted object which we are capable of conceiving, namely, the production of the higher animals, directly follows. There is grandeur in this view of life, with its several powers, having been originally breathed into a few forms or into one; and that, whilst this planet has gone cycling on according to the fixed law of gravity, from so simple a beginning endless forms most beautiful and most wonderful have been, and are being, evolved".

to Huxley (Olby, 1963). The first in-depth study of Darwin's view on heredity was most probably Gerald L. Geison's 1969 "Darwin and Heredity: The Evolution of His Hypothesis of Pangenesis" (Geison, 1969). This 36-page article, partly in response to Olby,[5] discusses most of the pangenesis-related issues that would be bones of contention in the following decades: its origin both in Darwin's trajectory and in the history of science (especially in comparison to Spencer's views, Geison, 1969, pp. 398–403), its conceptual relation to cell theory (Geison, 1969, pp. 391–393), its subsequent eclipse by Mendelian genetics (Geison, 1969, p. 378), its epistemological equivalence with natural selection in Darwin's argumentative strategy.[6]

From then on, pangenesis became a growing part of the so-called Darwinian Industry. It has been regularly reworked from various perspectives, focusing for instance on the way in which pangenesis reflected Darwin's taste for theoretical speculation (Hodge, 1985), its relation to the concept of variation for Darwin (Winther, 2000), its non-integration with natural selection (Hodge, 2010), its reception by Darwin's contemporaries (Holterhoff, 2014), and even its possible relevance for today's biology (Liu, 2008). Dozens and dozens of articles have expanded on pangenesis, which has given rise to a large body of history of science scholarship since the mid-twentieth century.

6.2.3 An Obvious Blind Spot in Jean Gayon's Work

It is thus striking to observe that pangenesis goes almost completely unmentioned in Gayon's work. Of course, Gayon knew in detail the content of *Variation* and the role pangenesis played in the debate regarding biological heredity in the last third of the nineteenth century. It was on purpose that he chose not to engage with Darwin's theoretical hypothesis. Out of more than 300 articles, he only devoted one short dictionary entry specifically to pangenesis (Gayon, 2013); all in all, he merely broached the subject in a few paragraphs,[7] where he simply sums up the main lines of the hypothesis, without any effort to problematize the topic any further.

This is especially true of *Darwinism's Struggle for Survival*, a book devoted to the elucidation of the complex relation between two concepts: natural selection and heredity. Heredity was indeed a major concern for Gayon, including Darwin's ideas about heredity. Yet, Gayon distinguished between hereditary variation by and large

[5] Note that Geison focused on Olby's 1966 book *Origins of Mendelism* rather than on his 1963 article.

[6] "Pangenesis, in short, was not a single-minded attempt to defend against any attack Darwin's commitment to blending inheritance. Neither was it an ad hoc attempt to account solely for the inheritance of acquired characters. It was, rather, an attempt to develop a broad organizing principle, an attempt to connect under one point of view a wide range of previously unconnected and poorly understood phenomena. As such, it had its origin in what Goeffrey West called Darwin's 'naked need for order' – the same need, it might be added, that gave birth to the lasting principle of natural selection" (Geison, 1969, p. 411).

[7] His most extensive account on pangenesis is in Gayon and Petit (2019, pp. 177–180).

in Darwin's work and the more specific model of pangenesis. Darwin never linked pangenesis to natural selection (Hodge, 2010): what he did was to promote a kind of hereditary variation that, at least in his understanding, was supposed to allow the efficiency of natural selection. Gayon directly followed in Darwin's footsteps, in focusing on hereditary variation as the concept was discussed in the *Origin of Species*. In *Darwinism's Struggle for Survival*, which is over 500 pages long, the only significant reference to pangenesis is a short paragraph where Gayon emphasizes Darwin's methodology regarding the relation between theoretical hypothesis and observed facts. For Gayon, and for others before him, as we have seen, it was the very same methodological approach that was at the basis of both natural selection and pangenesis. He then briefly and enigmatically added that "the fact that the same method elicits admiration from the contemporary reader in the first case [natural selection], and astonishment in the second [pangenesis], is another question" (Gayon, 1998a, p. 11).

Why, then, did Gayon simply ignore Darwin's scientific model about heredity, which gave rise to a large body of work that was regularly discussed during Gayon's academic career?

6.3 Rational Reconstruction: A Philosophical Method for Practicing the History of Science

6.3.1 A General Overview

Whether Jean Gayon was a philosopher of science with an interest in history or a historian with a philosophical focus has always been a difficult question to answer, even for himself (Gayon & Petit, 2018, pp. 78–80). What is certain is that during most of his career, his main approach consisted in reconstructing large scale "panoramas"[8] of complex theoretical conceptual issues, with an openly critical perspective. *Darwinism's Struggle for Survival* is no exception: Gayon was looking for a "coherent scenario" (Gayon, 1992, p. 3)[9] that would explain the eclipse of Darwinism during the 1880–1930 period. In contrast to others, such as Peter Bowler, Gayon emphasized the need for an *internal* explanation of the crisis of Darwinism (Gayon, 1998a, p. 6). The fact that alternative evolutionary theories were discussed and taken into account—orthogenesis, neo-Lamarckism, mutationism, etc.—was far from a satisfactory and sufficient explanation of the eclipse of Darwinism. In his understanding, it was the very hypothesis of natural selection itself that was entering a crisis, mostly because its efficiency was strongly dependent on specific assumptions regarding heredity that were in opposition with the current knowledge of the time. In

[8] The word itself was regularly used by Gayon (1998a, p. 3; see also Loison 2018).

[9] The term "scénario cohérent" is absent from the 1998 translation.

short, Gayon strived for an internal and rational reconstruction of the eclipse, rather than a contextual historicization (*à la* Bowler).

For this to be the case, Gayon needed an abstract pattern to elucidate history. He was not ready to content himself with a narrative that would be as erratic and tortuous as history looks like in the first place. Description was not enough: a historical reconstruction must be an ordered one, and this is where philosophy was seen as a fruitful resource: the philosophy of science precisely offers a set of abstract tools that might help making such patterns explicit. Accordingly, he initiated a historical reconstruction that would at the same time be a clarification.

For example, in "From Measurement to Organization: A Philosophical Scheme for the History of the Concept of Heredity", Gayon's aim was to show that "the three major phases in the history of the science of heredity corresponded to three quasi-explicit conceptions of the cognitive status of scientific theories, namely: phenomenalism, instrumentalism and realism" (Gayon, 2000, p. 69). Another example is Gayon's (1998a, 1998b) detailed study of Georges Canguilhem's work.[10] Then, Gayon used the philosophical concept of individuality to make sense of history: it was the "Adriane's thread" (Gayon, 1998b, p. 307) he relied on to organize Canguilhem's "biological philosophy", which "will not be found in a single book, or even in a single paper":

> Individuality played the role of a problematic concept, which happened to drive the three successive major dimensions of Canguilhem's philosophy of biology: the axiological, the ontological, and the gnoseological (theory of knowledge). In other words: three times running, Canguilhem felt it necessary to reassess the concept of individuality in order to develop his conception of life as value, as being, and as knowledge. (Gayon, 1998b, pp. 306–308)

Gayon wanted his historical reconstructions to be ordered, and he sometimes used the term "rational" to emphasize this aspect. What precisely "rational" meant to him remains uncertain. In a strong sense at least, "rational" means that the scientific hypothesis/theory under scrutiny has resisted experimental refutations and is still today understood as scientific. This was the case for natural selection, and this was why a rational reconstruction was grounded. One of Gayon's main concerns was the ways in which a theoretical hypothesis had been able to engage in a fruitful way with experimentation and observation. That was his concern with natural selection (from Darwin up to theoretical population genetics) and with the various models of heredity that were elaborated from the late nineteenth century onwards. He did not deny that in both cases, obviously, ideological and political backgrounds were significant parts of the picture in the way hypotheses were initially framed, yet he always favored the subsequent experimental validation of a hypothesis rather than its original set-up (more on this in section 6.4.1.). In short, he was against the idea that the context of justification could be reduced to a period's ideology:

> The methodology used in the pages that follow is that the history of sciences, like all history, is 'social'. Scientific activity develops in given social and historical conditions, but this does

[10] Pierre-Olivier Méthot, in a recent paper based on an impressive breadth of archives and interviews, has detailed the context in which Gayon studied Canguilhem's work (Méthot, 2022).

> not at all imply that the validity of knowledge is rigidly dependent upon the conditions in which that knowledge was constructed. Or, to put it another way: in order for the history of science not to lose itself in a morass of relativist descriptions of customs and beliefs, in order not to abstract the history of science from everyday human history, we have to allow for a certain retrospectivity and avoid thinking that ignorance of the current state of science in some way helps the historian. Relatively few episodes of the history of what we call science can resist resolution by a strong dose of hindsight. These are the episodes that merit a rational reconstruction of their origins and development. The theory of selection and its evolution from 1859 onwards is such a case. It really is that simple. (Gayon, 1998a, pp. 7–8; see also Gayon, 1992, p. 412)

At least some theories and hypotheses, like natural selection or Mendelian genetics, have shown their rationality in the course of the history of science, and as such "merit a rational reconstruction". Gayon's motivation was in those cases "to explore the rationality of natural selection" (Gayon, 1998a, p. 397). The rationality of natural selection was a historical construct, implying specific restrictions on heredity. Not all theories of heredity were compatible with a creative role for natural selection within the evolutionary theory. In this framework, pangenesis was irrelevant first because it was never integrated with natural selection in a unified theory (Hodge, 2010), and, second, because it was considered almost irrational by Gayon (1998a, p. 397), i.e., unable to fruitfully engage with experimental science.

6.3.2 A Lakatosian Tone

The fact is that the term "rational reconstruction" is widely used in *Darwinism's Struggle for Survival*, and especially in the introduction. Gayon mentions Lakatos only once in the introduction—and in a very few other places in his published writings, in order to stress that the way he used the concept of rational reconstruction differs from Lakatos' in that he was not ready to go beyond actual history (Gayon, 1998a, p. 7).[11] He presents himself as more sensitive to history than Lakatos supposedly was. This is debatable, for two reasons. First, although I will not dwell on this here, Lakatos is usually caricatured as a philosopher who did not consider actual history. His methodology of "scientific research programmes" would only have been speculative philosophy producing historical examples.[12] In my view, Lakatos was a much more subtle and stimulating historian than this standard evaluation usually suggests. Second, and this is the point I am interested in, I think Gayon went remarkably far in a Lakatosian direction; far enough to get rid of actual history, i.e., of pangenesis for instance.

I see at least three characteristics of Gayon's historiographical style that connect him with Lakatos. From the less specific to the more specific, these are:

[11] As far as I know, when Gayon refers to Lakatos, he always cites two sources: "Falsification and the methodology of scientific research programmes" (1970) and "History of science and its rational reconstructions" (1971).

[12] This was for instance Kuhn's main criticism.

The Pre-eminence of Internal History Over External History

In various places, Lakatos expanded on the significance of the demarcation between external and internal history. Not only was this distinction meaningful to him, but he considered that an author's methodology could be best characterized by understanding where he drew the line between those two forms of history (see especially Lakatos, 1971). More importantly, and unsurprisingly, in Lakatos's view there was no symmetry between internal and external history. Internal history was a necessary prerequisite for external history, but not the other way around. Only when and if a rational reconstruction has been produced can sociological contextualization be attempted (Lakatos, 1971, pp. 91–92).

The exact same hierarchy can be found in Gayon's work. Gayon never denied the legitimacy of social history, but in his view, even though social studies on science could sometimes bring new light on old issues, they would never be able to replace internal history. Like Lakatos, he believed that rational reconstructions were prerequisites for a subsequent sociological treatment. This is what he was willing to offer with *Darwinism's Struggle for Survival*:

> Conceptual history is not opposed to the social history of science. But the social history of science attains its goals all the more effectively if it has a pre-existing reconstruction of the intellectual debate involved in that history. The internal history of concepts and methods is no doubt a simplification which sooner or later will be corrected and filled out by the social history of sciences. But it sets out the boundaries, and that is not so bad. (Gayon, 1998a, p. xv)

The Use of Lakatos's Model of a Hard Core and Auxiliary Hypotheses in the Rational Reconstruction of the History of Natural Selection

In order to avoid the limitations of Popper's falsificationism, Lakatos elaborated the extended concept of "research programmes", which was meant to encompass more than just a single theory. A Lakatosian research programme is based on a hard core of theoretical hypothesis that cannot be falsified. Less essential auxiliary hypotheses protect the hard core and can be modified to accommodate experimental results (Lakatos, 1970). In Lakatos's work, there is a distinction between two kinds of theoretical assumptions that do not play the same epistemological role. His model was grounded in an essential asymmetry between those two.

Gayon was thinking of natural selection as something close to a research programme and his most central philosophical interpretation was that "natural selection" referred to two sorts of different epistemological entities, both in Darwin's work and in the subsequent forms of Darwinism: a hypothesis that could be submitted to experimental testing and a more general theory that was liable to reorganize the whole field of natural history (paleontology, embryology, comparative anatomy, ethology, etc.). Gayon explicitly took up Lakatos's model in order to discuss his main distinction. His specific point was that, in contrast to Lakatos, in the case of Darwinism, it was the hard core (i.e., the "hypothesis" of natural selection) that entered a crisis whereas auxiliary assumptions (i.e., the general "theory" of natural selection) were from the outset fully in accordance with known facts:

> These two opening chapters have examined both the implications of [natural selection] and the theoretical context in which it was elaborated. Our next task is to analyse the crisis which rapidly erupted. A particularly knotty epistemological problem emerges at this point: according to the generally accepted idea of the history of science as a 'rational reconstruction', major scientific theories are adopted not because they are initially 'true', but because they give rise to a fertile programme of research. Imre Lakatos has argued that it is essential for such a programme to contain a "negative heuristic" – a range of potentially revisable auxiliary hypotheses that protect the central core of the theory from attempts to refute it. The central propositions of the theory thus have time to take a coherent form and to enlarge their field of application, without being threatened by minor refutations. The Darwinian theory of natural selection appears to have met an exactly opposite fate. The *theory* was immediately applied to a wide range of subjects, extending to the whole of natural history and beyond. It was adopted wholesale by naturalists and also by the general culture of the time. The central hypothesis, however, in the extremely constrained formed outlined by Darwin, was quickly exposed to major internal problems and went into crisis almost as soon as it was born. By rigidly linking the hypothesis of selection to the question of heredity, Darwin did indeed open a long-term research programme that was rich in ideas and potential. But in the half-century that was to follow, the hypothesis was exposed to the permanent threat of refutation. The power of the process of natural selection as conceived of by Darwin did not hold under all conceivable models of heredity. Worse, Darwinian selection was perhaps incompatible with Darwin's own representation of hereditary variation [note that Gayon does not mention pangenesis here]. (Gayon, 1998a, p. 84)

A Rational Reconstruction of the Concept of Heredity that Excludes a Substantial Part of Actual History

Besides *Darwinism's Struggle for Survival* (1998 for the English version[13]), another of Gayon's major achievements is his 2000 text "From Measurement to Organization: A Philosophical Scheme for the History of the Concept of Heredity".[14] This opinion is shared by many colleagues and Gayon himself saw this work as important enough to include it in a recently published anthology devoted to the philosophy of biology (Gayon & Pradeu, 2021, pp. 231–269). Gayon's main argument, as we have seen, is that the history of the concept of heredity had three main phases that can be described using standard philosophical categories: the phenomenalist phase (biometry), the instrumentalist phase (Mendelian genetics) and the realist phase (molecular genetics). What Gayon wanted to offer here was a "philosophical interpretation of the development of the debates" (Gayon, 2000, p. 85), and he was fascinated by the fact that what may at first sight seem to be a contingent history could be reconstructed along such clear-cut philosophical lines:

> Nevertheless, I find it remarkable that the history of this typically modern biological discipline called 'genetics' can be reconstructed along such philosophical lines. I know of no other modern biological discipline the history of which could be reconstructed in terms of such successive epistemological shifts. Of course, the alternative between phenomenalism, operationalism, realism and other philosophical categories concerning the cognitive status of scientific theories are a constant preoccupation for all sciences. But it is not so frequent to

[13] Except for the 'Preface to the English edition' (4 pages), the 1998 English version is very close to the original 1992 one.

[14] The 2000 text is the final version of previous papers: Gayon (1994, 1995a, 1995b).

> observe these philosophical positions as idealtypes shaping massively successive states of a certain area of scientific knowledge. (Gayon, 2000, p. 87)

Nonetheless, Gayon's rational reconstruction came at a cost: it had to exclude from the history of theories of heredity all the speculative models about particulate heredity that flourished during the last third of the nineteenth century, i.e., in the wake of Darwin's own pangenesis (Mayr, 1982, pp. 681–710; Robinson, 1979). Quantitatively speaking, these hypotheses, supported by prominent zoologists and botanists like August Weismann (Churchill, 2015) and Hugo de Vries (Stamhuis et al., 1999), represented a much more significant part of the western biology of the time in comparison to what the biometricians did in the English context.[15] It was on purpose that Gayon excluded these models, in order to promote the purest rational reconstruction possible. In his attempt to justify such a strong methodological decision, he clarified his idea of a *rational* reconstruction:

> One important reservation must be added at this point. The narrative I have proposed here does not take into account the cytological studies which, in the nineteenth and early twentieth century, contributed to shape the modern science of heredity. Cytology, of course, induced a number of biologists, both before and after the rediscovery of Mendel, to adopt, or at least express the wish of a "material" and therefore more realistically orientated view of heredity. Therefore, the periodization I have proposed should not be taken as the only one possible. Nevertheless, the "material" conceptions of heredity that were formulated in the second half of the nineteenth century (e.g., theories of *Germplasm* in the nineteenth century) or Muller's "material conception of the gene" had a strong speculative, not to say a metaphysical taste (see Gayon, 1997; Muller 1926; Weismann). The present paper has been directed primarily at the cognitive status of theories of heredity which had a strong empirical basis in their own time. (Gayon, 2000, p. 87)

As in the case of natural selection, Gayon was interested in theoretical assumptions inasmuch they lent themselves to experimental procedures. "Rational", I argue, primarily referred to such a property of both a scientific hypothesis *and* the historical reconstruction produced. Nineteenth century models of particulate inheritance, and Darwin's pangenesis in the first place, were precisely lacking "a strong empirical basis in their own time". They were only bold speculations, and, in such a view, irrational.

In the end, based on a deliberate methodological decision, Gayon simply excluded from his narrative a substantial part of the history. He did exactly what is the most controversial aspect of Lakatos' approach: he put actual history in the footnotes in order to focus only on the rational reconstruction of the successive theories of heredity produced in accordance with a philosophical scheme.

[15] See for instance Yves Delage's late ninetenth century synthesis: *La structure du protoplasma et les théories sur l'hérédité et les grands problems de la biologie générale* (1895).

6.3.3 Beyond Lakatos: Duhem

At the end of his career, when Gayon was asked which philosopher of science he rated the highest, he answered without hesitation Pierre Duhem (Gayon and Petit 2018, pp. 81–82).[16] Gayon especially valued Duhem's seminal book, *The Aim and Structure of Physical Theory* [*La théorie physique, son objet, sa structure*].[17] He gave it very high praise indeed, claiming that the book had imparted on him "landmarks that [he has] never been able to get rid of afterwards in philosophy of science" (Gayon & Petit, 2018, p. 81, my translation). This late and explicit acknowledgment of Duhem is surprising both because Gayon never mentioned Duhem as a possible mentor or inspiration before and because no colleagues had ever noticed any kind of Duhemian dimension in Gayon's work. Nonetheless, I think that Gayon's final comment on his own path should be taken seriously and that there is, indeed, a sound basis to support a non-trivial Duhemian reading of his positioning.

First, on some occasions at least, Gayon explicitly referred to Duhem and made use of some of his epistemological categories. It was Duhem, for instance, who offered Gayon the philosophical starting point for his 2000 essay on heredity. According to Duhem, as he repeatedly explained in *The Aim and Structure of Physical Theory*, it was the definition of magnitudes that was at the basis of the formation of the physical theory. Gayon wanted to see to what extent this scheme could be applied within the life sciences to the scientific theories of heredity (Gayon, 2000, pp. 69–70). The fact is that heredity itself eventually failed to be treated as a magnitude (like for instance mass in physics), even if measurement did play a key role in the history of biometry and then in Mendelian genetics. Gayon problematized his rational reconstruction following a crystal-clear Duhemian rationale:

> I will show that, having been originally treated as a measurable magnitude, heredity later became a structural property, or the property of an organized object. In fact, while quantitative methods were ever more developed in the experimental science of heredity, the concept of heredity itself became increasingly less like a magnitude. (Gayon, 2000, p. 70)

Second, and more importantly, Duhem's take on the physical theory most probably reflected the way in which Gayon intended to smoothen Lakatos' methodology, to make it more "historical", as he put it in *Darwinism's struggle for Survival*'s introduction. In short, I think that what Gayon wanted to do was to try to write a history that might have been entitled *The Aim and Structure of Evolutionary Theory*. Gayon was especially impressed by Duhem's ability to perform a "historical reconstruction" (Gayon & Petit, 2018, p. 81) within an explicit philosophical framework. *The Aim and Structure of Physical Theory* exemplified a difficult and fragile balance between the history of science and the philosophy of science, between erudite description and normative judgment, a delicate balance that Gayon was always looking for. In my view it was with such a Duhemian model in mind that Gayon wrote *Darwinism's*

[16] Note that this part is absent from Gayon and Petit (2019).

[17] The French version was published in 1906 and already translated into English in 1908.

Struggle for Survival. My claim is not that Gayon's book is reducible to a direct application of Duhem's epistemology in the field of biology. While differences between the physical theory and the evolutionary theory are too substantial to make such a direct use possible, I do think that some of Gayon's major orientations were of Duhemian inspiration. For the sake of concision and clarity, here I limit my argument to only three possible Duhemian dimensions of Gayon's work. Others can most probably be pinpointed and documented.

A Focus on Theories More Than on Concepts

During his entire career, Gayon favored levels of description other than that of scientific concepts (Loison, 2018, p. 85). Concepts are of course always present (the term "concept" is for example used in his 2000 article), but philosophical and historical analysis more usually focus on the theory within which concepts make sense. Natural selection is again a textbook example of this methodological positioning. Gayon's book could perfectly have been about the elucidation of the historical and philosophical path of the concept of natural selection, as historical epistemology would have required (see Sect. 6.4.1.). Nevertheless, Gayon chose not to do this, and instead focused on the distinction between the hypothesis of natural selection and the theory of natural selection, as we have seen. He always favored more comprehensive entities like theories or disciplines instead of notions or concepts: genetics rather than gene, the theory of natural selection rather than the concept of natural selection.

Gayon usually refrained from excessively isolating a single entity, like a concept, within a more comprehensive and articulate framework. This orientation obviously echoes Duhem's own epistemological holism. Duhem's seminal work was on the entire physical theory, not on the concept of force or of gravitation. He was primarily interested in the systemic workings of a theory. To him, the most fundamental issues, such as that of scientific truth, only made sense at the level of the whole theory.

A Strong Interest in the Structure of Theories

In his 1906 opus magnum, Duhem wanted to make the structure of the physical theory explicit. In his reconstruction, a physical theory should be understood as a natural classification of experimental laws. A physical theory starts with the definition of magnitudes, as we have seen, and then comes the experimental production of quantitative laws. The most significant theoretical work consists in the elaboration of abstract principles to encapsulate the diversity of experimental laws in a much more integrated "classification".[18] As such, Duhem's pattern cannot be directly applied to any biological theory, including evolutionary theory. But the spirit of the Duhemian take, so to speak, can be, and this was exactly what Gayon tried to do: he sought, in various works, to elucidate the structure of evolutionary theory, from Darwin onwards. Gayon was particularly impressed by Darwin's own ability to structure his argument in favor of the natural selection hypothesis. In several publications, he tried to make this structure fully explicit and used diagrammatic representations to

[18] Duhem (1906, Chap. 2), "Physical theory and natural classification".

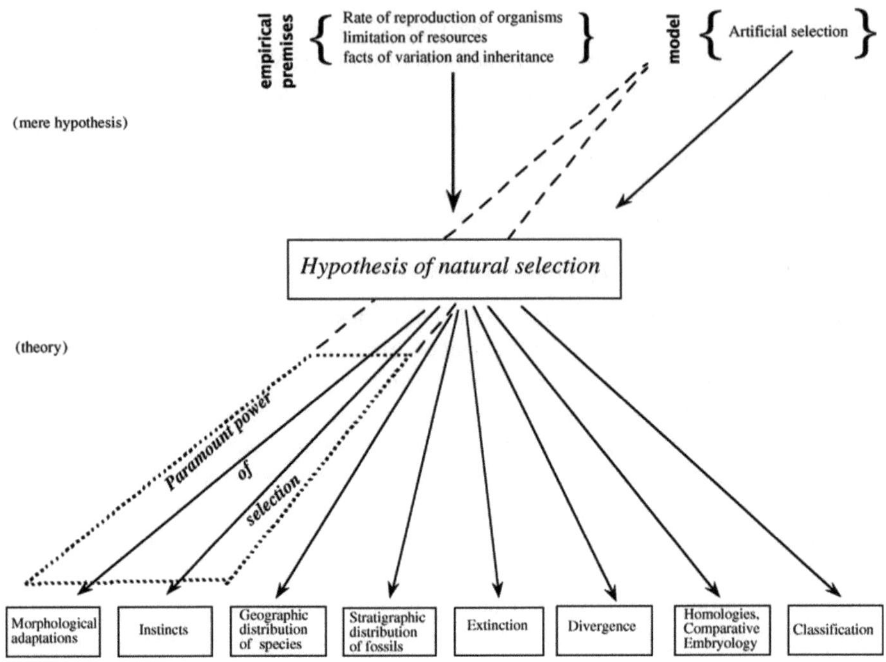

Fig. 6.1 The structure of Darwin's theory of natural selection

make his point. For example, in his 1997 article (Fig. 6.1) "The paramount power of selection: from Darwin to Kauffman", he put it as follows (Gayon, 1997):

This diagram exemplifies the structure of Darwin's *theory* of natural selection. For Gayon, this theory was grounded on two different levels of justification. The first level (upper part) is the different classes of facts (which could be experimentally addressed) that make the existence of natural selection as an adaptive force in populations probable. The second level (lower part) represents most of the facts already known in 1859 in the field of natural history that are better explained within a unified theoretical framework, i.e., the theory of natural selection. For Gayon, natural selection was not just a biological concept among others, but the cornerstone of a new theoretical unification within the life sciences, exactly as the physical theory was for Duhem back in the late nineteenth century.

The Issue of the Empirical Basis: Under-Determination and the Relation Between Theory and Experiments

In direct connection to the issue of theoretical holism, Duhem's work on the under-determination of theory by facts characterizes his established legacy (the so-called Duhem-Quine thesis). Here, again, Gayon did not make a direct use of Duhem, but he often framed his historical investigations with what could be seen as a Duhemian orientation.

In *Darwinism's Struggle for Survival*, Gayon never stopped asking how natural selection, as a causal hypothesis, in relation to heredity, could be submitted to experimental testing. In this book, Gayon's main argument could easily be rephrased as follows: the crisis of Darwinism was the consequence of the considerable empirical under-determination of the hypothesis of natural selection. Given the usual understanding of heredity in the second half of the nineteenth century, natural selection could only be "indirectly corroborated" and had to wait for the advent of theoretical population genetics to become empirically established on firm grounds. In most of his major works from the late 1980s to the early 2000s, Gayon looked at the ways in which a theory (whether natural selection, Mendelian genetics or molecular genetics) could engage with experimental control.

More specifically, Gayon, in the main, seemed to endorse the view that a specific theory historically and progressively produced the conditions of its experimental validation. Experimentation was never at the beginning but must be seen as the final and refined outcome of a theoretical work that could span decades, if eventually it succeeded. This was exactly what happened to the theory of natural selection from Darwin until the 1920s; it was the central aspect of the "coherent scenario" that Gayon hoped to demonstrate. This epistemological positioning perfectly fits Duhem's reflections on the relation between the physical theory and its so-called empirical basis: "The experimental verifications are not the basis of the theory; they are its crowning achievement" (Duhem, 1906, p. 284, my translation). Empirical facts, for Duhem, do not contribute to the construction of a scientific theory, they only come afterwards. The internal crisis of Darwinism was, for Gayon—who was at odds here with other historians like Bowler—a 60-year tortuous theoretical work that ended in population genetics. It was only when the "genetic theory of selection"[19] was ready, i.e., in the interwar period, that this theory was able to produce its "crowning achievement" in the form of an indisputable experimental validation.

6.4 Implications, Corroboration and Extrapolation

6.4.1 Neither Philosophy of Biology nor French Historical Epistemology

If my Lakatosian-Duhemian reading of Gayon is grounded, it should be able to clear up the issue of Gayon's relation to the philosophy of biology on the one hand and to historical epistemology on the other. Ultimately, Gayon sometimes situated himself in a sort of middle ground between these two traditions (Gayon & Petit, 2018, pp. 76–77), as others have also suggested (Huneman, 2018, pp. 312–313; Méthot, 2018; Petit, 2019, pp. 361, 364–365). In my view, neither these labels accurately

[19] This is the title of the third part of *Darwinism's Struggle for Survival.*

reflect what Gayon was doing, if, as I claim, his methodological positioning was in fact more related to Lakatos, Duhem and rational reconstructions.

That Gayon never *practiced* the philosophy of biology (although he taught it a lot), like for instance Michael Ruse or David Hull did, is, I think, quite easy to demonstrate. The philosophy of biology emerged in the 1970s and 1980s in the analytical tradition and aims at clarifying theoretical scientific issues.[20] Its purpose is normative, and the philosophy of biology mostly ignores the history of biology. Philosophers of biology produce critical examinations of standard scientific concepts (like "gene", "fitness", "adaptation", etc.) in order to improve scientific knowledge. Gayon was never at ease with that method; he was not confident enough to teach lessons to scientists. His hesitation is obvious in his only article published in *Biology and Philosophy*, the leading journal in the field. The 1996 text, entitled "The individuality of the Species: A Darwinian Theory", is a puzzling read in the sense that Gayon was not ready to depart from the history of science and thus elaborated an odd position about the concept of "species" that was historically very fragile and that, on the philosophical level, could hardly be seen as a clarification either (Gayon, 1996). In short, this time, his rational reconstruction failed both philosophically and historically and was indeed harshly criticized (Stamos, 1998). In the French version of *Knowledge of Life Today*, Gayon admitted that "if [he] could be faulted for something, it would be that [he] was too often content to 'keep score' in conceptual debates" (Gayon and Petit 2018, p. 79, my translation). One can also directly compare Gayon's *Darwinism's struggle for Survival* with Elliott Sober's *The Nature of Selection* to see how different Gayon's style was from the analytic philosophy of biology.

When it comes to historical epistemology and the French tradition embodied in Bachelard and Canguilhem at the IHPST, things are more complicated. In his last years, Gayon paid his debt towards Canguilhem and while he was reluctant to use the polysemic term "historical epistemology", his own historical style was undeniably more related to the French tradition than to the analytic one. He eventually went as far as to write that "if [he] made use of a method, it had been inspired by Georges Canguilhem's example" (Gayon & Petit, 2018, p. 49, my translation). I think that Gayon's self-characterization is incorrect: the linkage between historical and philosophical analysis is too weak a criterion to attach a work to Canguilhem's legacy or even to historical epistemology. A more substantial characterization is needed to prevent historical epistemology from becoming just a fashionable brand for a mere "conceptual history of science" (Gingras, 2010).

First, it must be highlighted that Gayon never used the classic epistemological tools elaborated by Bachelard and Canguilhem, such as "epistemological break", "epistemological obstacle" or "scientific ideology".[21] These were absolutely not part of his epistemological repertoire. Even the names of Bachelard and Canguilhem are indeed almost entirely missing in Gayon's published work, if one excludes the

---

[20] What is usually considered as the founding paper in the field is Hull (1969).

[21] Pierre-Olivier Méthot showed that Gayon worked first within a Bachelardian perspective and made use of some of these categories. Yet, this Bachelardian early phase never materialized in a published text (Méthot, 2022).

few articles that Gayon specifically devoted to them (mostly Gayon, 1998b; Gayon, 2003). In his day-to-day work, they were not operative references.

Second, when he voiced views about Bachelard or even Canguilhem, Gayon was remarkably critical (Petit, 2019, p. 363). Unsurprisingly, Gayon was critical about Bachelard's approach to the history of science from the beginning. He saw Bachelard's methodology as too presentist, too linear, too committed to an old-fashioned concept of ineluctable scientific progress (Gayon, 2003). Gayon expanded and strengthened his critique in *Knowledge of Life Today* and only credited Bachelard for his literary skills and suggestive and stimulating phrases (Gayon & Petit, 2018, p. 38). He even pointed to Bachelard as one of the reasons why French philosophy of science declined after the War (Gayon & Petit, 2018, p. 39). Gayon's attitude towards Canguilhem was more moderate, and he praised him for taking the history of science more seriously (Gayon & Petit, 2018, p. 47). Nonetheless, when it comes to epistemological concepts, Gayon was no less critical, admitting for instance that he was "not totally convinced of [the] validity" of the concept of "scientific ideology" (Gayon & Petit, 2019, p. 247).[22]

Third, and more importantly, Gayon's approach to the internal history of science was simply not congruent with Canguilhem's. They diverge on several substantial aspects (for a comprehensive characterization of Canguilhem's historical epistemology, see especially Limoges, 2015, 2018, 2019). As I have emphasized, Gayon was interested in theories more than in concepts. Because he thought in terms of research programs, he looked more "forward", i.e., towards the possible progressive experimental deepening of a hypothesis, and less "backwards", in the search for the genealogy of a concept (Méthot, 2013), including within ideology. These differences are obvious when one compares *Darwinism's Struggle for Survival* with Canguilhem's masterpiece, *The Formation of the Concept of Reflex in 17th and 18th Centuries* (1955). Canguilhem's objective was to isolate the concept of reflex from the standard and contemporary mechanistic physiological theory in order to show that this concept was born in a complete different theoretical framework: Willis's seventeenth century vitalism, and not Descartes' mechanism. Canguilhem emphasized the autonomy of scientific concepts, which can circulate through various "theoretical fields".[23] This autonomy makes it possible to complicate the genealogy of concepts by conducting large-scale historical investigations in order to understand their hidden origins.

Gayon, in contrast, was interested in how natural selection was progressively elaborated within commensurable theoretical contexts that were not alien to one another. What he was looking for was the deepening of a theoretical path, the "experimental

[22] In the following paragraph, Gayon specified his criticism: "Canguilhem's theory [scientific ideology] is both too broad and too narrow. It is too broad in that, in my opinion, there is no evidence that all new sciences *must* be preceded by a "scientific ideology". It is too narrow as in most cases the premises of a new theory or new science contain much more than ideology" (Gayon & Petit, 2019, p. 247).

[23] "It is excluded, in any case, that the vigor of a concept in a given theoretical ground can constitute a sufficient presumption to limit to the theoretical grounds of the same composition the research of the places of its birth." (Canguilhem, 1955, p. 6, my translation).

fate", as he liked to call it, of a particular scientific hypothesis. Natural selection has proven to be historically enduring, and as such deserves a rational reconstruction of its experimental path. In short, Gayon favored the analysis of how natural selection produced the experimental basis for its rationality. In very general terms, to use a standard philosophical distinction, it is not incorrect to say that historical epistemology (and Canguilhem) is interested in the historical and epistemological elucidation of the context of discovery, whereas Gayon was interested in the rational reconstruction of the context of justification. Once again, it was a very Duhemian orientation, and Gayon could perfectly have endorsed the idea that "to make the history of a [scientific] principle is, at the same time, to make its logical analysis".

What would have been a Canguilhemian treatment of Gayon's main subject, natural selection? This question can be answered, precisely because that was just what Camille Limoges did in a PhD dissertation supervised by Canguilhem in the late 1960s (for some context on Limoges's stay in Paris as a PhD student, see Méthot, 2009, pp. 49–50). His dissertation came out as a book, published in 1970, *Natural Selection, a Study on the First Constitution of a Concept, 1837–1859* [*La sélection naturelle: Etude sur la première constitution d'un concept (1837–1859)*]. As the title makes clear, Limoges focused on the origin of the concept in Darwin's and Wallace's work, rather than on its subsequent experimental history. Note that "constitution" was a term widely used by Canguilhem that marked out his methodological engagement. Limoges overtly adopted a Canguilhemian style from the outset (Limoges, 1970, p. 5), at the methodological level but also in the writing itself (short, polemical and incisive sentences). What Limoges looked at, as Canguilhem did in 1955 for the concept of reflex, was the theoretical conditions of possibility that made the constitution of natural selection possible before Darwin published *On the Origin of Species*. Limoges's specific thesis was that biogeographical facts and a renewed concept of adaptation that was in opposition with the standard view of "natural economy" were, for both Darwin and Wallace, the theoretical place[24] where natural selection was originally formed. For Limoges, the model of artificial selection came afterwards (Limoges, 1970, pp. 79, 83), and Malthus only stimulated Darwin's elaboration (Limoges, 1970, p. 80). Limoges was also highly interested in showing that Darwin and Wallace had no genuine forerunners—another eminently Canguilhemian concern.[25] He paid special attention to the cases of Matthew, Wells and Lawrence and argued that all of them ignored the ecological dimension that was at the root of natural selection (Limoges, 1970, p. 115). In his book, Limoges not only worked within a Canguilhemian framework, he also used Bachelard as an epistemological resource (Méthot, 2009, p. 50), like for instance when he argued that using artificial selection as a pedagogical device created epistemological problems of its own, a trap that Wallace was able to avoid (Limoges, 1970, pp. 143–150).

If Gayon's *Darwinism's Struggle for Survival* is clearly different from Sobber's *The Nature of Selection*, it is no less different from Limoges's *Natural Selection, a*

[24] Limoges wanted to "situate the place" where natural selection originated as a scientific concept (Limoges 1970, p. 20).

[25] On Canguilhem's famous criticism of this search for forerunners, see Canguilhem (2005 [1968]).

Study on the First Constitution of a Concept. Gayon was not as such interested in the emergence of natural selection in Darwin's or Wallace's work and he discussed in only a few paragraphs some of Limoges's core hypotheses, especially regarding the role (not) plaid by artificial selection in the "constitution" of the concept of natural selection (Gayon, 1998a, pp. 36–37). But even in those paragraphs, he was more concerned with the structure of the final theory than with how it originated: "I have focused on this element [artificial selection] of Darwin's ideas with reference to Darwin's *final theory*, rather than on the genesis of the doctrines involved in that theory" (Gayon, 1998a, p. 37, my emphasis). What Gayon wanted to retrace was the history of the context of justification from Darwin onwards, not the history of its context of discovery, not to mention the possible investigation of its "archeology" as Michel Foucault would have done. Thereby, despite a general but vague similarity, and despite Gayon's final comments, I think that his methodological positioning, at least in his published writings, was never Canguilhemian and never belonged to French historical epistemology if this term is consistently used. It is only if one restricted the available methods to only two opposite possibilities, philosophy of biology versus historical epistemology, that it creates the impression that Gayon was somewhere in between, and maybe closer to the historical pole. My claim is that Gayon was doing something else, which stood outside of this opposition.

6.4.2 Back to Pangenesis

If my hypothesis is correct, it implies that Gayon ignored pangenesis because it cannot be fruitfully reconstructed in accordance with an abstract philosophical scheme. Pangenesis did not deserve a rational reconstruction for two main reasons: its inability to engage with experimentation and its theoretical independence from natural selection. I will come back to the second issue in the next section. Here I would like to reconsider the first, that of the "experimental fate" of pangenesis. This is in fact an almost entirely inexistent fate, except for Galton's brief involvement in disproving pangenesis (note that Darwin's response to Galton's negative results strengthened the metaphysical dimension of his speculative hypothesis). Nonetheless, at least on one occasion, pangenesis gave birth to an actual experimental program within biology. And it is no less remarkable that Gayon made a significant effort to reconstruct its history.

During the late 1910s, new viruses were discovered that were able to attack bacteria. They were called "bacteriophages" and were intensively studied during the interwar period before the era of antibiotics begun. Around 1920, a new phenomenon was suggested by various experimental results and was quickly termed "lysogeny": it appeared that in some cases, bacteriophages were able to be integrated within a bacterium hereditary material and to silently reproduce for dozens of cell generations. The very existence of this latent phase was debated, and even among microbiologists that took lysogeny for granted, several hypotheses were confronted about its precise mechanism. Retrospectively, the most developed research program was Eugène and Elisabeth Wollman's, at the Pasteur Institute in Paris.

Here, I will not go into detail; I am simply pointing out that this highly dynamic and fruitful program, which contributed to the early history of molecular genetics, was first entirely based on Darwin's hypothesis of pangenesis. Eugène Wollman made it clear from the outset that he was reinterpreting Jules Bordet's early results in the terms of the pangenesis model:

> It seemed interesting to us to take up the analogy between these ideas [of Bordet] and those put forward by Darwin in his hypothesis of pangenesis, which, in the mind of the illustrious English biologist, ought to supply a *provisional* schema for the mechanism of heredity in general, and the transmission of acquired variations in particular. [...] Applying the images supplied by the pangenesis hypothesis to the interpretation of the d'Hérelle phenomenon [bacteriophagy] is sufficient to yield the representation proposed by Bordet and Ciuca of the mechanism of this "hereditary nutritive vitiation". Darwin's gemmules become Bordet and Ciuca's "intracellular factors. (Wollman, 1920, pp. 1478–1479, emphasis in the original, my translation)

This time, pangenesis was not just a bold speculation but crystallized into several precise hypotheses that Wollman, his collaborators and his followers (including André Lwoff) submitted to various refined experimental designs. In contrast to Darwin's account of pangenesis, Gayon was highly interested in Wollman's, precisely because this time pangenesis entered an experimental path that allowed Gayon's method to be practicable. For this to be the case, a general pattern had to be worked out in order to organize the historical reconstruction. Such a philosophical scheme was indeed found during a collaborative work engaged in the early 2010s by Gayon, Richard Burian and myself. Our structural hypothesis was that two successive concepts of Lamarckian heredity (a "physiological" account followed by a "particulate" one) framed the general history of the debates regarding the issue of lysogeny. In 2017, in a synthetic article that was intended as the first of two closely linked articles,[26] we wrote in the conclusion: "In our forthcoming article, we will apply the same interpretative *schema* to the history of the concept of enzymatic adaptation in the Pasteur Institute" (Loison et al., 2017, p. 45). This structural hypothesis was supposed to be applied beyond the case of lysogeny. What was at stake was, once again, to understand how a general scheme could inform actual history and then be the basis of a rational reconstruction.

6.4.3 *Back to Darwin's Pangenesis?*

In a Lakatosian-Duhemian reading, pangenesis could be ignored in part, as I have indicated, because it seems to stand outside the rationality of natural selection. Darwin's speculative model of pangenesis was never connected to natural selection, be it by Darwin himself or other nineteenth century biologists. It seems that Darwin

[26] A project not completed yet.

never tried to understand how pangenesis could be responsible for the hereditary variations that could be accumulated by natural selection. In contrast, pangenesis directly offered a refined explanation of Lamarckian inheritance that made natural selection irrelevant to the explanation of adaptive evolution. Thus, it seems perfectly legitimate, from the viewpoint of a rational reconstruction of the hypothesis of natural selection, not to take into account pangenesis, and this is exactly what Gayon did.

I am not disputing the merits of such a choice. Obviously, the success of *Darwinism's struggle for Survival* speaks for itself. More modestly, I would just like to suggest that there is perhaps still some room left for including pangenesis within the rationality of Darwin's research program on natural selection. In other words, what would a "Gayonian" analysis of pangenesis look like if Gayon had eventually decided to stop ignoring the subject? The historical literature devoted to pangenesis depicts it, in the main, as a model designed to account for hereditary variation; since variation might be directed towards adaptation, pangenesis would not leave much to natural selection. Nonetheless, this is not exactly how Darwin termed the issue. In some places at least, Darwin insisted that pangenesis was supposed to explain *variability*, not *variation*. For instance:

> Finally, we can see on the hypothesis of pangenesis that *variability* depends on at least two distinct groups of causes. Firstly, on the deficiency, superabundance, fusion, and transposition of gemmules, and on the redevelopment of those which have long been dormant. In these cases, the gemmules themselves have undergone no modification; but the mutations in the above respects will amply account for much fluctuating *variability*. Secondly, in the cases in which the organisation has been modified by changed conditions, the increased use or disuse of parts, or any other cause, the gemmules cast off from the modified units of the body will be themselves modified, and, when sufficiently multiplied, will be developed into new and changed structures. (Darwin, 1868, pp. 396–397, my emphasis)

Geison rightly pointed out that "natural selection needs a supply of variations on which to work" and that "the problem of how he might account for these variations naturally occurred to Darwin at the same time that his ideas on natural selection were beginning to develop" (Geison, 1969, p. 385). Yet he did not explicitly connect this issue with the rise of pangenesis. Even Winther, who specifically focused on "variation and heredity" did not distinguish between variation and variability and regularly conflated both issues.[27] He concluded that "Darwin theorized about the causes of variation from early on his career" (Winther, 2000, p. 441). Yet it might be the case that Darwin was more concerned with the causes of *variability* (like the impact of external agents on the reproductive system[28]).

---

[27] "The following passage from *The Variation of Animals and Plants Under Domestication* (henceforth *The Variation of Animals and Plants*) epitomizes Darwin's strong externalist position on variation, 'if it were possible to expose all the individuals of a species during many generations to absolutely uniform conditions of life, there would be no *variability*.'" (Wither, 2000, p. 428, my emphasis).

[28] Darwin 1859, chapter V (Laws of Variation). Darwin started the chapter as follows: "But the much greater *variability*, as well as the greater frequency of monstrosities, under domestication or cultivation, than under nature, leads me to believe that deviations of structure are in some way due to the nature of the conditions of life, to which the parents and their remote ancestors have been exposed during several generations" (Darwin, 1859, p. 173, my emphasis).

Therefore, if, from a logical standpoint at least, natural selection requires abundant hereditary variations to be fully creative and if, from a historical viewpoint, pangenesis was motivated by the issue of variability instead of variation (the latter being only a consequence of the former), then, there exists a possible rational connection between pangenesis and natural selection, even if Darwin himself never wrote a word suggesting that (Hodge, 2010, p. 130). To consistently explore this possible link might be a way to renew the historiography on pangenesis; such a reconstruction would follow Gayon's epistemological footsteps.

6.5 Conclusion

Jean Gayon's consistent lack of interest in Darwin's pangenesis can be explained by characterizing his methodological commitments. Whether with heredity, natural selection, or other subjects (the concept of species, Canguihem's work, etc.), he was looking for a philosophical scheme that could retrospectively frame a rational reconstruction aiming at the elucidation of the (experimental) destiny of the hypotheses proposed. Pangenesis, at least at first sight, seems alien to natural selection and never embraced experimental science, two good reasons not to focus on Darwin's "provisional hypothesis". In my reading, most of Gayon's more important work from the late 1980s to the early 2000s at least, i.e., during his most productive years, was performed in accordance with this structuralist approach, beginning with his main achievement, *Darwinism's Struggle for Survival*, and one of his most important articles, "From measurement to organization: a philosophical scheme for the history of the concept of heredity". Gayon was a historian who was interested in patterns, and the philosophy of science was expected to provide fruitful patterns to investigate history.

Nonetheless, my claim is not that Gayon always worked with such a structuralist approach as his main methodological guideline. On other occasions, he favored a more empirical take, with a sharp focus on the institutional context or terminological issues (Gayon, 2006; on this aspect, see Loison, 2018). This was especially the case during his productive collaboration with Richard Burian to elucidate what they called "the singular fate of genetics in the history of French biology". Their founding paper, published in the *Journal of the History of Biology* with Doris Zallen in 1988 was especially representative of this empirically-based approach (Burian et al., 1988). In those cases, the history at stake was about the *reception* of a discipline (Mendelian genetics) and not about the experimental fate of a theory, which requires a patient analysis of scientific journals, university textbooks, teaching programs, institutions' agendas, etc. This does not in any way mean that their work lacked any kind of philosophical perspective.[29] Yet, the methodology it employed was not structuralist in the sense defined in the present paper.

If my reading is grounded, it also follows that Gayon never practiced the philosophy of biology nor historical epistemology, nor even some kind of middle ground between these two traditions. As I see it, Gayon's main inspirations were rather on the side of philosophers of science involved in reconstructing historical panoramas with a strong philosophical focus, like Lakatos and Duhem. It is likely that Gayon would acknowledge the Duhemian inspiration but reject the Lakatosian influence.[30] I hope to have demonstrated why and how both played a significant role in his complex positioning, and in the process highlighted the originality of Gayon's work in the contemporary debate. This also explains, at least to some extent, why Gayon was embarrassed with Canguilhem's legacy and never fully embraced it.

Acknowledgements This paper benefited from several exchanges with Jon Hodge, whom I thank very much for his detailed examination and advice, especially as regards the historiography of pangenesis. I am also indebted to Pierre Olivier Méthot and Camille Limoges for their subsequent attentive reading. Last but not least, Jean-Yves Bart edited the final version.

References

Braunstein, J.-F. (2002). Bachelard, Canguilhem, Foucault. Le 'style français' en épistémologie. In P. Wagner (Ed.), *Les philosophes et la science* (pp. 920–963). Gallimard.

Burian, R., Gayon, J., & Zallen, D. (1988). The singular fate of genetics in the history of French biology, 1900–1940. *Journal of the History of Biology, 21*(3), 357–402.

Canguilhem, G. (1955). *La formation du concept de réflexe aux XVIIe et XVIIIe siècles.* Presses Universitaires de France.

Canguilhem, G. (2005) [1968]. The object of the history of science. In G. Gutting (Ed.), *Continental philosophy of science* (pp. 198–207). Blackwell.

Churchill, F. B. (2015). *August Weismann: Development, heredity and evolution.* Harvard University Press.

Darwin, C. (1859). *On the origin of species by means of natural selection.* John Murray.

Darwin, C. (1868). *The variation of animals and plants under domestication* (Vol. II). John Murray.

Darwin, C. (1871). *The descent of man and selection in relation to sex* (Vol. I). John Murray.

Delage, Y. (1895). *La structure du protoplasma et les théories sur l'hérédité.* C. Reinwald.

Duhem, P. (1906) [2007]. *La théorie physique. Son objet—sa structure.* Vrin.

Gayon, J. (1992). *Darwin et l'après-Darwin: une histoire de l'hypothèse de sélection naturelle.* Kimé.

---

[29] For instance, to explain how molecular genetics became so successful in the French context, they put the emphasis on the positive role played by Pasteurian microbiology. Their perspective was then primarily philosophical: "Formulated in a more conceptual manner, the question becomes: How did transformation of a *problematique of contagion* to a *problematique of heredity* came about in France?" (Burian et al., 1988, p. 383, emphasis in the original).

[30] It should be noted that I never had the opportunity to directly discuss these subjects with Jean. The interpretations proposed here are entirely based on the examination of his published writings.

Gayon, J. (1994). De la mesure à l'ordre; histoire philosophique du concept d'hérédité. In M. Porte (Ed.), *Passion des formes (hommage à René Thom)* (pp. 629–645). ENS, Editions Fontenay St Cloud.
Gayon, J. (1995a). From measure to order: A philosophical scheme for the history of the concept of "heredity". *Workshop "Gene concepts and evolution". Max-Planck-Institut Für Wissenschaftsgeschichte, Preprint, 18*, 53–60.
Gayon, J. (1995b). Entre et force et structure: genèse du concept naturaliste de l'hérédité. In J. Gayon & J.-J. Wunenburger (Eds.), *Le paradigme de la filiation* (pp. 61–75). L'Harmattan.
Gayon, J. (1996). The individuality of the species: A Darwinian theory? From Buffon to Ghiselin, and back to Darwin. *Biology and Philosophy, 11*, 215–244.
Gayon, J. (1997). The paramount power of selection: From Darwin to Kauffman. In M.L. Dalla Chiara, K. Doets, M. Mundici, & J. van Benthem (Eds.), *Structures and norms in science, Volume Two of the tenth international congress of logic, methodology and philosophy of science* (pp. 265–282). Kluwer.
Gayon, J. (1998a). *Darwinism's struggle for survival: Heredity and the hypothesis of natural selection.* Cambridge University Press.
Gayon, J. (1998b). The concept of individuality in Canguilhem's philosophy of biology. *Journal of the History of Biology, 31*, 305–325.
Gayon, J. (2000). From measurement to organization: A philosophical scheme for the history of the concept of heredity. In P. Beurton, R. Falk, & H.-J. Rheinberger (Eds.), *The concept of gene in development and evolution: Historical and epistemological perspectives* (pp. 69–90). Cambridge University Press.
Gayon, J. (2003). Bachelard et l'histoire des sciences. In J.-J. Wunenburger (Ed.), *Bachelard et l'épistémologie française* (pp. 39–85). Presses Universitaires de France.
Gayon, J. (2006). Hérédité des caractères acquis. In P. Corsi, J. Gayon, G. Gohau, & S. Tirard (Eds.), *Lamarck, philosophe de la nature* (pp. 105–163). Presses Universitaires de France.
Gayon, J. (2013). Pangenèse. In P.-A. Taguieff (Ed.), *Dictionnaire historique et critique du racisme* (pp. 1309–1310). Presses Universitaires de France.
Gayon, J & Petit, V. (2018). *La connaissance de la vie aujourd'hui.* Iste Editions.
Gayon, J., & Petit, V. (2019). *Knowledge of life today: Conversations on biology.* Wiley.
Gayon, J., & Pradeu, T. (Eds.). (2021). *Philosophie de la biologie. Explication biologique, hérédité, développement.* Vrin.
Geison, G. L. (1969). Darwin and heredity: The evolution of his hypothesis of pangenesis. *Journal of the History of Medicine, 4*, 375–411.
Gingras, Y. (2010). Naming without necessity. On the genealogy and uses of the label "historical epistemology." *Revue de synthèse, 131*, 439–454.
Hodge, J. (1985). Darwin as a lifelong generation theorist. In D. Kohn (Ed.), *The Darwinian heritage* (pp. 207–244). Princeton University Press.
Hodge, J. (2010). *The Darwin of Pangenesis. Comptes-Rendus Biologies, 333*, 129–133.
Holterhoff. (2014). The history and reception of Charles Darwin's hypothesis of pangenesis. *Journal of the History of Biology, 47*, 661–695.
Hull, D. (1969). What philosophy of biology is not. *Synthese, 20*, 157–184.
Huneman, P. (2018). Jean Gayon (1949–2018). *Revue d'histoire des sciences, 71*, 311–317.
Lakatos, I. (1970). Falsification and the methodology of scientific research programs. In I. Lakatos & A. Musgrave (Eds.), *Criticism and the growth of knowledge, I* (pp. 91–196). Cambridge University Press.
Lakatos, I. (1971). History of science and its rational reconstructions. *Boston Studies in the Philosophy of Science, 8*, 91–135.
Limoges, C. (1970). *La sélection naturelle. Étude sur la première constitution d'un concept (1837–1859).* Presses Universitaires de France.
Limoges, C. (2015). Philosophie biologique, histoire des sciences et interventions philosophiques, Georges Canguilhem 1940–1965. *Œuvres complètes de Georges Canguilhem, Tome IV* (pp. 7–48). Vrin.

Limoges, C. (2018). La confirmation de l'historien des sciences et la mise à l'épreuve de sa philosophie biologique : Georges Canguilhem 1966–1995. *Œuvres complètes de Georges Canguilhem, Tome V, Histoire des sciences, épistémologie, commémorations, 1966–1995* (pp. 7–60). Vrin.

Limoges, C. (2019). Épistémologie historique et histoire épistémologique des sciences chez Georges Canguilhem. *Œuvres complètes de Georges Canguilhem, Tome III, Écrits d'histoire des sciences et d'épistémologie* (pp. 7–128). Vrin.

Liu, Y. (2008). A new perspective on Darwin's pangenesis. *Biological Reviews, 83*, 141–149.

Loison, L. (2018). Ce que je dois à Jean Gayon. Sur quelques caractéristiques du style gayonien en histoire des sciences. In F. Merlin, & P. Huneman (Eds.), *Philosophie, histoire, biologie, Mélanges offerts à Jean Gayon* (pp. 81–88). Editions Matériologiques.

Loison, L., Gayon, J., & Burian, R. (2017). The contribution—and collapse—of Lamarckian heredity in Pastorian microbiology: 1. lysogeny, 1900–1960. *Journal of the History of Biology, 50*, 5–52.

Mayr, E. (1982). *The growth of biological thought: Diversity, evolution and inheritance.* The Belknap Press of Harvard University Press.

Méthot, P.-O. (2009). French epistemology overseas: Analyzing the influence of Georges Canguilhem in Québec. *Humana-Mente, Journal of Philosophical Studies, 9*, 39–58.

Méthot, P.-O. (2013). On the genealogy of concepts and experimental practices: Rethinking Georges Canguilhem's historical epistemology. *Studies in History and Philosophy of Science, A, 44*, 112–123.

Méthot, P.-O. (2018). De l'épistémologie historique à la philosophie de la biologie : Le double héritage philosophique de Jean Gayon. In F. Merlin & P. Huneman (Eds.), *Philosophie, histoire, biologie, Mélanges offerts à Jean Gayon* (pp. 15–50). Editions Matériologiques. This volume.

Méthot, P.-O. (2022). Jean Gayon face à Georges Canguilhem: Une remontée en trois temps. In M. Arminjon, C. Cherici, & P.-O. Méthot (Eds.), *Le normal et le pathologique: des catégories périmées ? Textes réunis en hommage à Jean Gayon* (pp. 267–306). Matériologiques.

Olby, R. C. (1963). Charles Darwin's manuscript of *pangenesis. The British Journal for the History of Science, 3*, 251–263.

Olby, R. C. (1966). *Origins of Mendelism.* Constable.

Petit, V. (2019). Jean Gayon (1949–2018): itinéraire d'un darwinien. *Revue d'histoire des sciences, 72*, 349–380.

Robinson, G. (1979). *A prelude to genetics: Theories of a material substance of heredity, Darwin to Weismann.* Coronado Press.

Stamhuis, I. H., Meijer, O. G., & Zevenhuizen, E. J. A. (1999). Hugo de Vries on heredity, 1889–1903: statistics, Mendelian laws, pangenes, mutations. *Isis, 90*, 238–267.

Stamos, D. (1998). Buffon, Darwin and the non-individuality of species—A reply to Jean Gayon. *Biology and Philosophy, 13*, 443–470.

Winther, R. G. (2000). Darwin on variation and heredity. *Journal of the History of Biology, 33*, 425–455.

Wollman, E. (1920). A propos de la note de MM. Bordet et Ciuca (Phénomène de D'Hérelle, autolyse microbienne transmissible de J. Bordet et M. Ciuca, et hypothèse de la pangenèse de Darwin). *Comptes-Rendus De La Société De Biologie, 83*, 1478–1479.

Zirkle, C. (1935). The inheritance of acquired characters and the provisional hypothesis of pangenesis. *The American Naturalist, 69*, 417–445.

Zirkle, C. (1946). The early history of the idea of the inheritance of acquired characters and of pangenesis. *Transactions of the American Philosophical Society, 35*, 91–151.

Part II
Perspectives on Jean Gayon's Contributions to the History of Evolutionary Theory and French Genetics

Chapter 7
The Modern Evolutionary Synthesis in America and France: The Contribution of Jean Gayon

David Depew

Abstract Jean Gayon's encounters with American evolutionary biologists and philosophers of biology affected his project of justifying the Darwinian research tradition. He challenged entrenched views about evolution in France by showing how the population-genetic theory of natural selection removed conceptual blockages that had thwarted the Darwinian research tradition. He taught his American colleagues how important French physiological genetics was to the Darwinism of the modern evolutionary synthesis and, with Richard Burian, stressed the compatibility of the Synthesis with postwar French molecular genetics. His seamless integration of history and philosophy of biology helped create an international learned society in which historians and philosophers of biology work together (ISSHSPB).

Keywords Richard Burian · Georges Canguilhem · Richard Lewontin · International Society for the History, Philosophy and Social Studies of Biology (ISSHSPB) · Philippe L'Héritier · Population cages · Population genetics · Modern evolutionary synthesis · William Provine · Georges Teissier

7.1 Introduction

In this chapter, I celebrate Jean Gayon as a philosopher-historian who more than anyone helped reunite Francophone and Anglophone approaches to evolution. This achievement benefitted both communities of inquiry but was not easy to achieve. The two traditions had been separated ever since Clemence Royer's unfortunate translation of Darwin's *On the Origin of Species*. I will recount some practical details of how Jean did it. I will then turn to the philosophical insights and historical methods

Adapted from Depew (2018).

D. Depew (✉)
University of Iowa, Iowa City, IA, USA
e-mail: david-depew@uiowa.edu

P.-O. Méthot (ed.), *Philosophy, History and Biology: Essays in Honour of Jean Gayon*, History, Philosophy and Theory of the Life Sciences 30,
https://doi.org/10.1007/978-3-031-28157-0_7

that undergirded his success. Finally, I will recall the admirable character traits that endeared him to students, colleagues, and friends on three continents.

At the beginning of his conversations with Victor Petit, published as *La connaissance de la vie aujourd'hui* not long before his untimely death, Jean recounts how his six years of training in philosophy were followed by nine years studying geology and biology before he attempted to become a university professor (Gayon & Petit, 2018, p. 17).[1] His philosophical education was centered on the great texts of the past. It was only when he took a tutorial under Bertrand Saint-Sernin that he found himself confronting how contemporary philosophers think about knowledge in the making (Gayon & Petit, 2018, p. 22). Saint-Sernin's reading list included not only classics of French history and philosophy of science, but also some seminal English-language papers on scientific explanation by Carl Hempel as well as Ernest Nagel's "marvelous treatise" *The Structure of Science* (1961) (Gayon & Petit, 2018, p. 22). The following year, as a memorable political crisis was unfolding in the streets, Jean earned a certificate in the history of science as part of his master's degree by taking three more epistemologically leavened courses in the history of science: one offered by Suzanne Bachelard on the history of physics, another by Claude Tresmontant on relativity theory, quantum mechanics, the origin of life, and genetics, and a finally a course from Georges Canguilhem on medicine and ideology that he describes as "one of the most stimulating educational experiences of my life" (Gayon & Petit, 2018, pp. 23–24).

Jean's degree secured him a position teaching philosophy in a *lycée.* (It inspires envy that in France philosophy is a high school subject.) He wanted more. In a brief encounter with Canguilhem when he was his student, the latter had assured him that if he wished to go further in his studies, he should not fear the dreaded "*agrégation*" exam in philosophy and gave him tips about how to pass it. Having done so, Jean opted to continue in history and philosophy of science (Gayon & Petit, 2018, p. 27). He also took to heart what both Canguilhem and Tresmontant had told him: To be a philosopher of science, take science courses first and then come back to philosophy. If, like most philosophers, you don't have the requisite background study biology rather than physics (Gayon & Petit, 2018, p. 26). Jean did just that. But it took another nine years learning geology, botany, zoology, and genetics at various campuses of the University of Paris to do it, even as he continued to teach philosophy to high school students and raise a family. "This asceticism," Jean remarks equably, "was interrupted only by military service and later a short episode of depression" (Gayon & Petit, 2018, p. 26).

[1] The English translation of this work, *Knowledge of Life Today: Conversations on Biology* (2019) does not include a biographical chapter that opens the French version. In referring to it and a second untranslated chapter, I cite and translate the 2018 French text. When I quote from the English version, I refer to Gayon and Petit (2019). This chapter expands and corrects Depew (2018). It might be regarded as a counterpart to Méthot (2018, 2022), which examine Jean Gayon's continuity with and departures from French "historical epistemology" in the light of his scholarly adventures in America. My focus is on how Jean's encounter with American historians and philosophers of evolutionary biology led him to forge productive links between the two communities of inquiry and afforded him an opportunity to infuse a bit of historical epistemology into America.

When his studies were approaching an end, Jean called Canguilhem on the telephone, reintroduced himself, and said in effect, I have done what you suggested. What now? Well, he would need an advisor. Having retired, Canguilhem himself was unable to serve. He recommended Michel Foucault. Can you think of anyone who works on contemporary science, Jean asked? Canguilhem mentioned Francois Dagognet (Gayon & Petit, 2018, p. 31). This went well. From the start, Dagognet, who had been one of his "*agrégation*" examiners, was in his corner and helped him develop a research topic (Burian, 2018; Gayon & Petit, 2018, p. 31; Méthot, 2018).

In the course of his biological studies, Jean took courses in population genetics from Claudine Petit and Jean Génermont. He was intrigued by the application, starting in the 1940s, of this abstract set of analytical devices to real-world evolutionary problems that "the French call somewhat grandly 'the synthetic theory'" but in the United Kingdom, the Commonwealth countries, and the United States generally answers to the name the modern evolutionary synthesis (Gayon, 1998, p. xiii).[2] Initially, Jean proposed focusing his thesis on the concept of "genetic load" an issue he believed well suited to bring out the key scientific and ideological issues of population genetical Darwinism. Dagognet thought the topic might be too narrow. Accordingly, Jean broadened it to "The History and Epistemology of the Synthetic Theory of Evolution" (Gayon & Petit, 2018, pp. 31–32).[3]

In the summer of 1983, Jean's wife Elisabeth was scheduled to travel to the United States in connection with her own research. Jean tagged along, taking the opportunity to introduce himself to Everett Mendelson, Director of the History and Philosophy of Science Department at Harvard, and other available historians of biology, including Garland Allen, Joy Harvey, Diane Paul, and Sam Schweber. He also made a preliminary survey of archival resources dealing with the Modern Synthesis. With Dagognet's support, he then applied for and was awarded a Fulbright grant through the Commission Franco-américaine pour la recherche to return to the United States in the summer of 1984. The grant was successful in part because Jean secured letters of support from people he proposed to interview. These included Mendelsohn, whose department would host his visit; the avian systematist and co-founder of the Modern

[2] Jean told Armand de Ricqlès, one of his biology instructors, that he was disappointed to find less evolutionary thinking in his courses than he expected. de Ricqlès, who had undertaken the same rigorous studies, recalled it differently. He found evolution pervading his classes (de Ricqlès, 2018, p. 152). They might not have had quite the same things in mind. Jean appreciated that in applying population genetics to biological problems the makers of the modern evolutionary synthesis turned the study of evolution into a distinct discipline, with its own journals, conferences, and curricula (Gayon & Huneman, 2019).

[3] Genetic load refers to the presumably deleterious persistence of maladapted genotypes in populations. The ideological *frisson* of this phenomenon is that it seemed to some to support eugenical interventions. Those who held this view, notably the accomplished geneticist Hermann Müller, were not counted by its founders as party to the modern evolutionary synthesis. They thought genetic loads are side effects of adaptive mechanisms like the reproductive superiority of heterozygotes (heterosis), and in any case are not as harmful as imagined. In his conversations with Petit, Jean implies that the topic of genetic load could have borne the philosophical weight he wanted to place on it. Dagognet knew little about the issue, but Jean nevertheless deferred to his sense about what would be persuasive to his examiners (Gayon & Petit, 2018, p. 31).

Synthesis Ernst Mayr; the paleontologist Stephen Jay Gould; and the population geneticist Richard Lewontin. All of them were on hand at Harvard. He also secured support from William Provine, professor of history of science at Cornell University in upstate New York and author of the well-regarded *The Origins of Theoretical Population Genetics* (1971). Jean returned to France in Fall, 1984 as the country's most well-informed scholar of contemporary evolutionary theory.

As it happens Jean's summer sojourns in the United States occurred at a time of ferment about the adequacy of the very "theory" he had come to investigate. He found Mayr, Gould, and Lewontin engaged in intense controversies about the explanatory power of the Modern Synthesis. The issues were to some extent empirical, but even more they were methodological, conceptual, and in some cases ideological.

For one thing, Hempel, Nagel and other "logical empiricist" philosophers of science were forced to resort to elaborate and somewhat ad hoc expedients in order to apply their hypothetical-deductive and reductionist ideal of science to the goal-oriented nature of living beings and the functional adaptedness of organic traits. This suggested either that twentieth-century evolutionary biology is not a mature, mathematically tractable empirical science after all or, alternatively, that something is wrong with the logical empiricist approach to scientific inquiry.

"Biological teleology" was not the only anomaly. It had also become apparent that contrary to what logical empiricists demanded the evolutionary biology of the Modern Synthesis does not possess and is unlikely ever to get universally valid laws from which its conclusions can be rigorously deduced. Every object of its inquiries is a product of historical contingencies operating in a particular lineage at a particular time and place. This is true even of Mendel's supposed laws (Beatty, 1995; Gayon & Petit, 2019, pp. 9–10). This inconvenient truth, as well as the fact that even Mendel's laws are full of approximations and exceptions, makes it as reasonable for scientists to doubt the generalizations that bring anomalies to light as it is to throw blame onto inadequate data or methods. In physics, this is not normally an option, although historically it might once have been and is still called for once in a while. In that paragon of a mature mathematized science, laws of nature are so well empirically grounded and exceptionless that they are conditions of the existence of the objects they govern. A violation of the law supports strong doubts whether the theoretical entities in question really exist. Not so for biology. The "law" gets the blame.

When Jean arrived at Harvard a new set of issues was also troubling the waters. It was becoming doubtful whether the Modern Synthesis was as theoretically unified or unifying as advertised. Even though molecular genetics is essential to the confirmation, explanation, and measurement of the heritable variation on which natural selection feeds, population genetics cannot be reduced to or replaced by the genetics of DNA and RNA (Gayon & Petit, 2019, pp. 223–224). At the same time, Gould and his fellow paleontologist Niles Eldredge were adducing good reasons to conclude that macroevolution—evolution above the species level—cannot be treated simply as a long-term effect of microevolution, as the Synthesis stipulated. For one thing, macroevolution is not always gradual (or at least constant), but often occurs in bursts of change between speciation events (Eldredge & Gould, 1972). Otherwise, gene frequencies in populations can remain pretty much the same even though DNA keeps mutating with a regularity that by 1970 was inspiring talk about selectively "neutral evolution" and "molecular clocks."

In the spirit of interdisciplinary collaboration that Jean admired and was soon to foster in his own country, Gould, Lewontin, and Mayr had for some time been inviting a number of young philosophers of science into their labs with a view to working with them to find principled ways of defending the Modern Synthesis in the face of these difficulties. One common theme was that evolutionary biology's objects of inquiry exhibit hierarchical levels—genes, organisms, groups, species, ecological communities—that physics does not. This point puts a higher burden of proof on any purported reduction to its laws of the lowest level.

Mayr, Lewontin, and Gould were not above doing a bit of philosophy of science themselves. In the early 1980s, Mayr, for example, was devoting much of his considerable energy to arguing that biology has its own standards of epistemological success and should not be held hostage to theories of scientific knowledge devised for physics. He took evolutionary biology to be a science that uses interpretive concepts rather than explanatory laws (Mayr, 1985). That was a bit much for the philosophers-in-residence. Their general tendency was to adjust the concept of law to the distinctive characteristics of evolutionary biology.

Jean first met one of these philosophers, Richard (Dick) Burian, in Provine's office at Cornell, which Dick was borrowing at the time in order to work through some articles on population genetics the historian of science stored there. Provine telephoned, asked Dick to have lunch with a visiting French philosopher named Jean Gayon, and invited both of them to dinner at his farm some thirty miles away. In the course of the afternoon, Dick asked Jean's help in interpreting some articles by the French geneticists Georges Tessier and Philippe L'Héritier over which he happened to be puzzling at the time. They hit it off immediately and became life-long friends and collaborators (Burian, 2018, pp. 111–112; Gayon & Petit, 2018, pp. 55–56; Méthot, 2018, pp. 29–30).

At the time, Burian was taking a leading role in establishing the International Society for History, Philosophy and Social Sciences of Biology (ISHSSPB). Jean attended one its formative meetings (in South Bend, Indiana, in 1985) and returned for its summer biennial meetings whenever possible. It was at another of these meetings that he met Marjorie Grene, who has been called "grandmother of the philosophy of biology" in recognition of the catalytic role she played in constituting the field, and other philosophers of biology including John Beatty, Michael Bradie, Robert Brandon, David Hull, Elizabeth Lloyd, Robert Richards, Michael Ruse, Kenneth Schaffner, and Elliott Sober, several of whom had been working with Gould, Mayr, and Lewontin. It was also at meetings of the ISHSSPB (or sometimes the History of Science Society) that he met philosophically informed historians of evolutionary biology such as Joe Cain, Jon Hodge, Phil Sloan, and Betty Smocovitis. It was at least in part through Jean that the ISHSSPB first became truly international. He brought seven students and colleagues with him to the Society's 1987 meeting at Virginia Technical Institute and State University in Blacksburg, Virginia (Burian, private communication). This was the beginning of exchanges between French and

American philosophers and historians of biology that are now more extensive and productive than ever.

7.2 Six Comments and Observations

My first general comment—I will have six—is that Jean's adventures in America fell on fertile ground because he was so well prepared to understand the issues in dispute in American evolutionary biology in the 1980s. He had learned English in his childhood. (He had English cousins with whom he vacationed.) His biological classes equipped him with a "feeling for the organism." "I dissected flowers, lugworms, oysters and mouse embryos," he told Petit, "analyzed extracts of tomato, performed spectrographic analyzes of these extracts … counted populations of snails in the woods … crossed fruit flies, and did endless exercises in mathematical statistics" (Gayon & Petit, 2018, p. 28). In addition, he was as competent as he was eager to contribute his voice to the philosophical disputes that were dogging the Modern Synthesis at the time and, truth be told, still are.

At the theoretical core of the Modern Synthesis are, by my count, five principles or axioms. Evolutionary change begins with heritable genetic variation. Evolution is by definition trans-generational change in gene frequencies in populations. Natural selection operating on (but not causing) heritable genetic variation is the primary force of evolutionary change. Other population-level change-inducing forces—mutation rate, random genetic drift, and inter-populational gene flow in particular—are often in play. Finally, evolution above the species level is assumed to be consistent with, even if not reducible to, changes induced by evolutionary forces.

Are these laws? In physics, laws of nature are both universal and tightly coupled to their empirical instances. That is why there can be a deductive and predictive relationship between the two, as the logical empiricists proclaimed. The evolutionary contingency thesis loosens that bond for evolutionary biology. Jean attributes to two pioneering analytical philosophers of biology, Sober and Brandon, his growing willingness to separate universality from empirical applicability in ways that qualify the former but do not threaten the latter (Gayon & Petit, 2019, p. 9). Sober's *The Nature of Selection* appeared the same year, 1984, in which Jean was doing his research in the United States.[4] He was quite taken by it. Among Sober's contentions was that physics and evolutionary biology both possess (or can discover) laws governing sources of change and others determining its consequences. Sober claims that physics has only a few source laws, but many consequence laws. In evolutionary biology, he says, the opposite is the case. It has many laws governing the sources of evolutionary change but only a few determining their consequences. Contingency enters into the scene of

[4] See Brandon (1978) for a similar but not identical approach to laws. As a graduate student at Harvard, Brandon worked closely with Mayr. In reconstructing Jean's learning curve, whenever I can I cite papers and books by American philosophers of biology that were roughly contemporaneous with his visits in the 1980s.

evolutionary biology with source laws about variation and fitness. Having done so, the consequence laws of population genetics dictate changes in gene frequencies in populations. They are predictive from one generation to the next, and in many cases much further (Gayon, 1998, pp. 212–213; Sober, 1984).

By his own account, it took Jean a lot longer—thirty years, he says (Gayon & Petit, 2019, p. 168)—to fully concur with Lewontin that natural selection is not an empirical generalization but an abstract (and to that extent a priori) model that can in principle be applied at any level of the biological hierarchy from molecules to species so long as populations of these entities exhibit a discriminating form of differential retention of heritable variation (Gayon & Petit, 2019, pp 8–9, 154, 161; Lewontin, 1970; Méthot, personal communication). Because the same goes for other evolutionary factors, Jean told Petit that, "The majority of biological theories take the form of a family of related models" (Gayon & Petit, 2019, p. 7).[5]

I cannot vouch for the adequacy of Jean's analysis or my complete accuracy in reporting it. Rather—this is my second general comment—I mention his approach in order to show how engaged he was with first-generation analytic philosophy of biology and, more importantly, in order to insist that however much he may have changed his mind about this or that point he was always committed to scientific realism and was resolutely optimistic about the empirical validity of the evolutionary biology of the Modern Synthesis.[6]

Jean's epistemological optimism throws into relief a persistent worry that first arose when he read the correspondence between the founders of the Modern Synthesis archived at the American Philosophical Society in Philadelphia. It contains little about theoretical unification and a lot about organizational strategies.

Just before the American entry into World War II, Mayr and another avian systematist, Julian Huxley, worked out a trans-Atlantic proposal for erecting a discipline using the principles of population genetics to solve the problems of systematics, adaptation, and speciation. Huxley's 1942 *Evolution: The Modern Synthesis* is both

[5] Models-based analyses of scientific laws and theories have become increasingly popular since the digital revolution. The most influential versions for population-genetic evolutionary theory were developed by John Beatty (1981) and later by Elizabeth Lloyd (1988/1994). On the so-called "semantic" view of theories, laws are not directly empirical but define the formal rules governing different kinds of systems, such as a Newtonian system. Empirical validity is judged on a case-by-case basis by statistically measured degrees of fit with the model. One of the attractions of the semantic view is that if a model doesn't fit, it isn't falsified but merely declared inapplicable to the case at hand. It might fit other cases well enough. Jean's belated mention of models may suggest that eventually he took up this view, but whether he did is not clear to me.

[6] I use the word 'optimistic' by contrast with the so called "pessimistic meta-induction," which holds that from the fact that every widely accepted scientific theory in the past has been shown to be false in crucial ways we should infer that all future theories will also be. Admitting that the Modern Synthesis was "an episode in the sociology of science" might be interpreted as leaving an opening for the radical solution to this problem proposed by the so-called "strong programme in the sociology of science," according to which social and political factors operating in the background determine choices between rival scientific theories. See for example Steven Shapin and Simon Schaffer's 1985 *The Leviathan and the Air-Pump: Hobbes, Boyle, and the Experimental Life*. That nothing could be further from Jean's thinking explains why he was so concerned to correctly assess the role of social factors in the making of the Modern Synthesis.

a manifesto of the proposal and a summary of its initial results. The forced isolation of the War, however, prompted Mayr, the ringleader, the geneticist Theodosius Dobzhansky, and the paleontologist George Gaylord Simpson to propose a wider Synthesis that prospectively included as many biological disciplines as depend on the field research of naturalists. Jean agreed with the historians that as soon as peace arrived the Americans presented a *fait accompli* to their British, French, German, and Russian colleagues (Cain, 2009; Gayon & Huneman, 2019, p. 26; Gayon & Petit, 2019, p. 147; Smocovitis, 1992). They had already secured funds, formed a society, founded the journal *Evolution*, and were planning incursions into biology departments and curricula. Although Huxley gestured toward the ultimate inclusion of many fields, the ambitious scope of what I will call the American Synthesis raised lasting questions about whether it takes itself to be a complete science of evolution and if so whether its reach exceeds its grasp.

Jean notes that the Synthesis so conceived benefitted from something he was seeing in front of his eyes: the organizational skill and collaborative ways of American academics. "I do not believe that any European country taken in isolation," he told Petit, "would have had the necessary intellectual resources, or that European intellectuals–or [even] Europeans and Americans working together—would have been capable of the level of efficient and effective cooperation shown by American geneticists, systematists, and paleontologists in the 1940s" (Gayon & Petit, 2019, pp. 146–147). If he had wanted to, Jean might have found an odd sort of support for these claims in the fact that two of its three principal architects were not American by birth. In fact, not long before they began working together Dobzhansky was almost deported back to Stalinist Russia and during the war the German-born Mayr's movements were confined because he was listed as an enemy alien. But these enterprising immigrants learned quickly and well how to employ the American academy's flexible scholarly institutions and publicity techniques not just to promote the Synthesis, but also to marginalize rival approaches. Historians are agreed that the Synthesis in both of its incarnations constricted the until-then-wide-open possibility space of viable theories of evolution. Lamarckism, mutationism, inner-driven orthogenesis—all were henceforth ruled out of bounds.

Suspecting for these reasons that the Modern Synthesis was as much an "episode in the sociology of science" as a scientific theory, Jean asked Lewontin about it before returning to France in Fall, 1984. Lewontin was confident that population genetics would keep on generating rigorous knowledge indefinitely, especially since his own experimental work was showing how useful molecular genetics could be in confirming or disconfirming population-genetic hypotheses (Gayon & Petit, 2018, p. 262). But he also confirmed Jean's suspicions that the post-War Synthesis was a sociological event rather than the consolidation of a general theory (Gayon & Petit, 2018, p. 34). It was not a trivial problem. By itself, mathematical population genetics was too formal to be an empirical science, but the integration of biological fields proposed by the makers of the American Synthesis was too diffuse to count as a scientific theory.

Jean's nagging worries about this issue played an important role in the genesis of his masterwork, *Darwin et l'après-Darwin* (1992). Not long after he returned to

France, he became an assistant professor of philosophy of science at the Université de Bourgogne. In 1988, a chance to advance to full professor arose—if he could complete a doctoral thesis and habilitation in time. At first, he intended to expand a 120-page report he had submitted to CNRS in the hope of securing research position. But he soon realized that his discoveries about the formation of the Modern Synthesis would hardly substantiate the claim embedded in the title of his grant application to visit the United States, his report to CNRS, and his doctoral dissertation: "The History and Epistemology of the Synthetic Theory of Evolution." Nor was he likely to persuade his rigor-prizing countrymen "that Darwinism today is any better than the ideology they always thought it to be" (Gayon, 1998, p. xiii). My first scientific love," he plaintively reports, "had disappointed me" (Gayon & Petit, 2018, p. 34).

After a few sleepless nights, a way forward occurred to him. Darwin had distinguished between the hypothesis and the theory of natural selection (Gayon, 1997, 1998, p. xiv; Gayon & Petit, 2019, pp. 92–93). The hypothesis concerns the existence, causal power, and explanatory competence of natural selection. The theory refers to its unifying effect on various biological fields. Darwin explicitly devoted the last part of *Origin of Species* to framing natural selection's unifying effect as a theory, a concern that Jean takes to inform the text even more pervasively (Gayon & Petit, 2019, pp. 93, 96).[7] Challenges to the *hypothesis* of natural selection were so ascendent in the later third of the nineteenth century, however, that Darwin's *theory* came widely to be viewed as either moot or falsified. *Darwin et l'après-Darwin* is the definitive study of how Darwin's hypothesis haltingly re-emerged from its long "eclipse," as Julian Huxley put it, through the development of mathematical tools from statistics and probability theory, many of them devised by people like Francis Galton, Karl Pearson, and Ronald A. Fisher with an explicit view to solving Darwinism's problems (and, it must be said, supporting its putative eugenical implications (Fisher, 1918; Galton, 1892; Gayon & Petit, 2019, pp. 254–259; Pearson, 1910). By the 1930s, these mathematical advances had resulted in a more or less complete account of the genetic forces that can be seen at work when Mendel's combinatory rules are projected onto the level of populations by way of the quadratic "Hardy–Weinberg" equation. It is this story of steadily increasing analytic power that Jean tells in great detail in *Darwin et l'après-Darwin.*

Seen in this light, the Modern Synthesis of the 1940s, while scarcely mentioned in Jean's book, appears waiting in the wings as a more successful attempt than Darwin's at *Unifying Biology*—the title of a book on the subject by Betty Smocovitis (Smocovitis, 1996)—on the basis of epistemological and ontological assumptions different not only from those of Darwin himself, but also from Logical Empiricism, *pace* Smocovitis. In *Darwinism's Struggle for Survival*: *Heredity and the Hypothesis*

[7] This is because Jean takes adaptation of morphological characters in the first half of *Origin* not only as a way for Darwin to explore his hypothesis of natural selection but also as one of the topics that, even though it is discussed early in *Origin*, he proposes in the second half of the book to integrate into a theory by way of what the scientific methodologist William Whewell called a "consilience of inductions" (Gayon & Petit, 2019, p. 93, 96). This isn't quite the way other close students of the text and its history divide up Darwin's book. See Hodge (1977).

of Natural Selection, the aptly titled 1998 English translation and expansion of *Darwin et l'après-Darwin,* Jean summarizes the matter as follows:

> Population genetics is a formal discipline that confronts the naturalist with the implications of various conceivable evolutionary scenarios … In this great formal science … the principle of natural selection was refined and differentiated … As often happens in sciences that are enriched by the intersection of mathematics, evolutionary theory opened itself to experimental testing. (Gayon, 1998, p. 405)

In this respect, Jean's book might also have been titled *A Prolegomenon to the Modern Synthesis*.

My third observation is that *Darwin et l'après-Darwin* is a tale about how conceptual objections that once stood in the way of Darwin's leading ideas were systematically cleared away, revealing an epistemological fecundity in population genetic Darwinism that Jean's mathematically minded colleagues in France were likely to appreciate.

The twentieth century population-genetic revision, even rescue, of nineteenth century Darwinism reveals itself most clearly when we compare how its key concepts are treated in both versions. Jean recognized that analytic philosophers of biology, guided by the practices of working evolutionary biologists, were exceptionally good at reconstructing the logic of genetic Darwinism.

In its nineteenth century form, Darwinian fitness was tied to physical strength, competitiveness, aggression, and sexual prowess, saddling Darwinism with a host of counterexamples—and giving it a bad reputation. The shift to population genetics re-identifies fitness as the expected productive success of the members of one population compared with closely related ones. Expected reproductive success depends on a myriad of context-dependent underlying factors, some of which might be markedly cooperative. Fitness thus defanged, says Gayon, is no more objectionable than the calculations actuaries provide to insurance companies—or any less real or causal for being statistically determined (Gayon, 1998, pp. 208–212; Sober, 1984). Nor is fitness the empty tautology Spencer's phrase "survival of the fittest" suggests (Gayon, 1995).

What about natural selection? One of Jean's main points in *Darwin et l'après-Darwin* is that population-genetic reframing allowed Darwinians to assign quantitative measurements to mutation rate, natural selection, genetic drift, and gene flow and to add and subtract them in a mathematically "homogeneous parametric space in which the effects of very different biological or ecological factors can be expressed *in the same terms*" (Gayon & Petit, 2019, p. 148, my italics). The absence of just such a parametric space was what the astronomer and scientific methodologist John Herschel was complaining about when he gave Darwin grief by dismissing natural selection as a "law of higgledy-piggledy",[8] The solution was made possible by shifting to the dynamics of statistically and probabilistically constituted populations, making it possible rigorously to state when, for example, mutation rate would or wouldn't trump natural selection and bringing in phenomena like genetic drift, which

[8] Darwin to Charles Lyell, December 10, 1859, Darwin Correspondence Project, Letter #2575, http://www.darwinproject.ac.uk.

before then couldn't even be seen, let alone explained or put to use in accounting for speciation. To enforce the shift to populations, Dobzhansky went so far as to *define* evolution as change in gene frequencies in populations. The analytical philosopher David Hull made an equally categorical claim by saying that individual organisms develop, but unlike populations don't evolve (Hull, 1994, and elsewhere). This was the conflation that had held back late nineteenth century and that was now resolved.

Another conceptual consequence of the population-genetic shift was that, in contrast to the one-shot saltationism that undergirded the mutationism of early twentieth century geneticists, the gradualism on which Darwin insisted (but in the absence of mathematics still to be invented was unable to defend) turned out to be right. For one thing, the sudden emergence of a new species in a single generation is statistically highly improbable. For another, the reason for the popularity of saltationist macro-mutationism was thereby discredited. The physicist Fleeming Jenkin had told Darwin that natural selection working on heritable variations of small effect cannot get a foothold because successive generations will tend to regress to a less than adapted norm, swamping any new and potentially useful differences. Darwin's half-cousin and champion Francis Galton assumed the same thing even as he tried to find ways around it. From the perspective of population genetics, however, this worry revealed itself to be unfounded. The Hardy–Weinberg equilibrium formula shows that gene frequencies recur unchanged in successive generations until and unless they are disturbed by the computable genetic forces of mutation, selection, genetic drift, and inter-populational gene flow. There is no inherent tendency to regress. On the contrary, mutations of small effect can spread relatively easily through populations.

More differences between Darwinism's old and new dispensations follow. In population-genetic Darwinism, for example, natural selection is not necessarily driven by Malthusian scarcity, as Darwin held. "The principle of natural selection can be applied to populations in a situation of unlimited expansion," Jean remarks, "at least for a certain time" (Gayon, 2015, p. 415; Gayon & Petit, 2019, p. 160). All selection requires is a gradient between available genetic variation and environmental opportunity. Moreover, natural selection is more than a sieve through which the unfit pass. As Darwin himself had maintained, it is the "creative" factor in evolution, a point Mayr, Dobzhansky, and the other makers of the Modern Synthesis stressed.

Things improved for biological teleology too. Darwin and his contemporaries had trouble distinguishing adaptation from the intentional design. By distinguishing between "selection of" and "selection for," Sober pointed out that the new Darwinism supports the claim that natural selection can result over trans-generational time in traits—adaptations—whose fixation in populations is not a matter either of design or chance but rather of concerted natural "selection for" that can truthfully and literally be said to spread through a population because it performs a function and is to that extent biologically teleological (so-called "consequence etiology"; see Brandon, 1981; Gayon & Petit, 2019, pp. 36–41; Wright, 1958, 1967).

Finally, Sober's distinctions proved useful in explaining neutral mutations in DNA sequences as selection of, but also in showing that selection for can occur at the molecular level (genic selection). The latter can disrupt proper functioning at the organismic level, as the protections of the immune system and relics of old battles in the genome itself show. Contrary to genic reductionists like Richard Dawkins, however, genic selection does not usually explain adaptations at the organismic level. By distinguishing between "a herd of fleet deer" and "a fleet herd of deer," Sober also countenanced selection for at the level of groups if and when biologists identify selection pressure for a trait that is not reducible to an organismic-level one.

At one point or another in his writings and his conversations with Petit, Jean endorses most if not all of these conceptual clarifications.[9] This leads me to my fourth general comment. I hypothesize that *Darwin et l'aprè-Darwin* could get a hearing from his French colleagues because it exhibits the spirit of one of the leading themes of the French school of history and philosophy of science to which Jean was heir: Scientific hypotheses, theories, and fields cannot find their footing until conceptual blockages that stand in their way have been removed (Gayon & Petit, 2018, p. 47; Méthot, 2018, p. 29). French history of physics is full of accounts of how the lingering assumptions of medieval mechanics were removed one by one until Newton's synthesis of celestial and terrestrial physics and subsequently its full articulation in nineteenth century French mechanics is shown in all its glory. Jean's tale about Darwinian evolutionary biology's evolution is scarcely less glorious, and equally exemplifies what Canguilhem called "historical epistemology."[10]

To British and American ears, the word 'epistemology' in Jean's title, "History and epistemology of the synthetic theory of evolution (1930–1950)," can sound a bit odd. For them epistemology is a philosophical field whose claims about knowledge are imposed on the sciences, not inferred from them—and even less from their history. With the specter of skepticism always lurking around the corner, sensory awareness the only foundational given, and the distinction between knowledge and belief its foci, Anglophone epistemology has been largely empiricist in tenor. Jean is referring to Logical Empiricism when he says that twentieth century ideas about knowledge that had previously rested on inspecting psychological data about what is given to sensory consciousness came to be "founded on the logical analysis of language" (Gayon & Petit, 2018, pp. 43, 88–89). This empiricist penchant, I suspect, explains why Anglophone philosophers of science only belatedly accepted what the history of science inescapably tells them: Conceptual issues are internal to science and their resolution is at least as important as finding confirming or disconfirming empirical evidence (Gayon & Petit, 2018, p. 88).[11] By contrast, in French philosophy of science the high road to studying what counts as knowledge, and hence epistemology, is preeminently found in the development of the sciences, and pride of place often goes to

---

[9] When he was Visiting Professor at the University of Notre Dame in 1999, Jean taught many of the seminal papers in which American philosophers of biology set out these conceptual analyses. Thanks to Phil Sloan for providing me with a copy of Jean's syllabus.

[10] Gayon & Petit (2018, pp. 38–49), contains praise for Canguilhem's "injunction not to separate history and philosophy of science" and a critique of Canguilhem's predecessor Gaston Bachelard's more poetic approach. Nonetheless, Jean says that "the concept of an epistemological obstacle, that is to say, an internal blockage to the progress of scientific knowledge, is assuredly a major contribution of Bachelard" (p. 47).

[11] Laudan (1977) is a good example of the post-positivist stress on conceptual problems in formulating and answering scientific questions.

spotting conceptual issues that hinder or advance them.[12] It is characteristic of the French approach that Jean's teacher Saint-Sernin translated Nagel's *Philosophy of Natural Science* as *Éléments d'épistémologie* (Gayon & Petit, 2018, pp. 22–23).

Jean saw conceptual clarification to be essential to the formation and justification of the Modern Synthesis. *Darwin et l'après-Darwin* showed historical epistemology at work in the slow but steady clarification of Darwinism's key concepts. He also saw this process in the collaboration between evolutionary biologists and analytic philosophers of biology that he witnessed first-hand in America. He told Petit, "As early as the mid-1980s, I was paying close attention to the very young 'philosophy of biology' then developing in the English language, and I tried to reconcile the two traditions" (Gayon & Petit, 2018, p. 49).

My fifth comment is that melding the two traditions in this way enabled Jean's colleagues in France to see, often for the first time, the epistemic strength of the Darwinian tradition in general and of the Modern Synthesis in particular. Giroux (2018) and Méthot (2018) testify to that. But I can also testify that it wasn't a one-way one street. The appearance of Darwinism's *Struggle for Survival* in 1998 taught Anglophone philosophers and historians of biology, including myself, to aim at seamlessly integrating biology, history, and philosophy if they wish to appreciate and defend the epistemic depth and fecundity of contemporary evolutionary theory.[13]

From this epistemologically reassuring perspective, the fact that the advent of the Modern Synthesis was an episode in the sociology of science turns out to be a positive. It is of course trivially true that every new or reformed science needs to be institutionalized. Beyond that, however, the fact that the Synthesis is not a theory in the strict sense reduces the temptation to reify it and encourages its adherents to develop new hypotheses as it incorporates more fields and by doing so is changed by them (Gould, 1983). The challenges the Modern Synthesis has from the outset faced are better read as healthy expressions of what Gould called its "expansion" than as threats of looming falsification (Ibid.).

Nowhere is this truer than in an area to which Jean contributed a great deal: the evolutionary implications of molecular genetics, to which I now turn.

In Mayr & Provine's country-by-country survey of contributions to the formation of *The Evolutionary Synthesis,* published few years before Jean first arrived in America, Mayr asserted that France contributed little or nothing (Mayr, 1980). It is true that Mayr's view of the Synthesis was idiosyncratic. According to him it didn't really gel until a conference he himself organized in Princeton in 1949. In saying this, Mayr was implying that while population genetics supports the Synthesis it is

---

[12] I leave aside how German epistemologists approach this topic. Kant, the fountainhead of German epistemology, drew on the development of physics for evidence, especially in his *Prolegomenon to Any Future Metaphysics*. His example was followed by the neo-Kantians, some of whom sought to isolate the conceptual foundations of other sciences, including history and the other social sciences. Until recently British interpretations of the *Critique of Pure Reason* have tended to focus on the cognitive status of sense data, as if Kant had to defeat Hume on Hume's own ground of inspecting the contents of sensory consciousness without looking at the rise of modern science.

[13] This ideal gives added luster, significance, and persuasiveness to Provine's *The Origin of Mathematical Population Genetics* (1971) and *Sewall Wright and Evolutionary Biology* (1986).

not foundational to it. He even counted some people as parties to it who knew little or no population genetics—and excommunicated others, such as Goldschmidt and Muller, who know a lot of it.[14] Accordingly, it was not just for reasons of national pride that Jean pointed out that population genetics is not only the foundation of the Modern Synthesis, but that some of the most crucial experimental findings on which its application to populations in the wild rested, and with them the success of the Modern Synthesis, were made in France.

These findings were facilitated by an innovative experimental device: L'Héritier and Teissier's "population cages" or "*démomètres*." In 1932, L'Héritier attended the Sixth International Congress of Genetics at Cornell University in New York (and in 1938 the important 7th Congress in Edinburgh in 1938, the last before the war). He also did a stint in a genetics laboratory at the University of Iowa and visited Thomas Hunt Morgan's lab in Pasadena, returning to France with valuable strains of *Drosophila.* Teissier and his students—among them the population geneticist Maxime Lamotte, Charles Boquet, Jean's teacher Claudine Petit, and Dobzhansky's last co-author Ernest Boesiger (Veuille, 2018)—used demometers to experimentally validate discoveries that went directly into the Modern Synthesis: the context dependency of the selective value of a mutation; the phenomenon of frequency dependent selection; and the outsized selective value of heterozygotes (Gayon & Veuille, 2001). Sewell Wright, who along with Fisher, Haldane, and the Russian Chetverikoff, was a founding father of population genetics visited Teissier's lab and used his findings to support his claim that speciation begins in small, isolated populations. Wright's long distance collaborator Dobzhansky, who at the time was resident in Morgan's lab in Pasadena as a sort of glorified post-doctoral student, used population cages to discover the wide extent of polymorphism in natural, not just laboratory, populations; the unexpected number of heterozygotes these populations contain; the idea of balancing selection, which selectively favors heterozygotes ("heterosis"); and the role of small isolated populations in igniting if not fully evolving speciation (Gayon & Veuille, 2001; Veuille, 2018, p. 139). Since the Modern Synthesis profoundly depends on these discoveries, Jean rightly remarks that, "It is truly astonishing that L'Héritier and Teissier's work was met with general indifference" both at home and abroad (Burian, 2018; Gayon, 1998, p. 365; Gayon & Veuille, 2001).

It is commonly supposed that indifference in France to the work of L'Héritier, Teissier, and Gustave Malécot, who inspired Motoo Kimura's selectively neutral theory of molecular evolution, springs from a deep-seated cultural preference for Lamarckism and concomitant resistance to Mendelism. Michel Veuille notes a certain incredulity among early twentieth century French anti-Mendelians about whether the holism we see in organisms could possibly come from particulate units of heredity (Veuille, 2018, p. 138). It is presumed that these factors kept genetics (and *a fortiori* its expansion into population genetics) out of French universities and confined to

[14] One of these was the University of Pennsylvania physical anthropologist Carleton Coon, who conspired with his cousin Carleton Putnam to give scientific cover to Southern states that were resisting the Supreme Court's 1954 *Brown v. Board of Education* decision, which declared racially segregated schools unconstitutional. See Jackson & Depew (2018, pp. 194–195).

research institutes, albeit important ones like the Institut Pasteur, until after World War II. A good example of this resistance is afforded by Émile Guyénot's attempts between 1909 and 1917 to falsify Morgan's Mendelian theory of inheritance. Instead, Guyénot had the misfortune of confirming it and in consequence having his career in France thwarted (Burian, Gayon & Zallen, 1988, p. 375; Veuille, 2018, pp. 149–150).

In a series of thorough historical studies, Jean, Dick Burian, and several collaborators affirm the long-standing national preference for Lamarckism. Cytoplastic inheritance and environmental induction of adaptive variation were often presented as providing evidence for other, perhaps deeper, forms of inheritance and the possibility of laboratory contamination and metabolic errors caused by harmful microbes were regularly invoked to cast doubt on even strong Mendelian experimental results, as they were in the case of the unfortunate Guyénot. Still, Gayon and his collaborators showed that the sciences of life in France were also tied to Claude Bernard and Louis Pasteur's strict experimental standards and their focus on physiology (Burian & Gayon, 1999; Burian, Gayon & Zallen, 1988; Gayon & Burian, 2000; Loison et al., 2017). Bernard's distinction between normal and the pathological, for example, dominated the sciences of life, pre-eminently medicine, up to and including Canguilhem, whose interpretation of Darwinian adaptation as normal functioning astonished the American philosopher Ken Schaffner and provoked hostility from another, David Hull, who insisted that adaptive natural selection requires rejecting the very idea of the normal and seeing potentially adaptive differences in what biological "essentialists" treat as pathological defects (Méthot, 2018, p. 44; Hull, 1994).

But that's not the end of the story. Gayon and Burian go on to argue that the rigorous experimental standards and conceptual proclivities of the French sciences of life were amply met after World War II by the physiological approach to molecular genetics of Boris Ephrussi, André Lwoff, Francois Jacob, and Jacques Monod. While post-Watson and Crick molecular geneticists in America and Britain were still trying to nail down the genetic code for producing structural gene products, Jacob and Monod were already paving the way for the transformative effect that the study of regulatory genetics has been exerting on evolutionary biology for some time (Burian, 2018; Burian & Gayon, 1999).

These reflections prompt my sixth comment. Roughly speaking, we might say that in Anglophone countries Darwinian natural selection preceded and took up population genetics and population genetics preceded and took up the molecular genetics of structural gene products—only to be met by persistent challenges coming from regulatory and developmental genetics. Something like the opposite was true in France. Beginning with Monod and Jacob's work on the *lac operon*, regulatory molecular genetics helped population genetics get a belated hearing, as Jean's education in the subject in the 1970s suggests. In turn, population genetics opened a path for an even more belated reception of the Darwinian tradition in evolutionary biology. Taking the last step was to a considerable degree the work of Jean Gayon.

I come finally to the topic of ideological distortions of scientific inquiry. As I noted at the outset, Canguilhem introduced Jean to the subject. Still, he told Petit that he was dissatisfied with Canguilhem's distinction between the "ideology of a scientist" and a "scientific ideology" (Gayon & Petit, 2019, p. 245). The latter is a somewhat self-contradictory concept. It puts together things that should be kept apart. Canguilhem mentions Spencerism as an example—Jean complains it is his only example—of a "scientific ideology." It coopted the name Darwinism for a long time. But the fact that Spencerism did not result in new scientific discoveries (in part, I suggest, because it didn't have a statistical bone in its body) told Jean that it "forms no part of scientific knowledge" and thus cannot be a "*scientific* ideology" (Gayon & Petit, 2019, p. 245, my italics). Jean recognized "eugenics as an ideology in the classic Marxist sense of the term: discourse with a defensive function motivated by interests based on class and race" (Gayon & Petit, 2019, p. 249). Eugenics certainly influenced, even drove, Darwinians like Galton and Fisher. Indeed, Galton invented the term. But that makes it an ideology of scientists, not a "scientific ideology."

Jean's papers on the repudiation of biological racism in official Statements by UNESCO in 1950 and 1951 show why. The repudiation of biological racism owes more to the population genetics of the Modern Synthesis than to any ideological shift in the wind (Gayon, 2002, 2003, 2008). The fact that genetic variation within conventional races far exceeds variation between them deprived evolutionists associated with the Modern Synthesis of whatever lingering traces of racial essentialism and eugenic sympathies they may have had, especially after Lewontin confirmed it by molecular genetics (Gayon & Petit, 2019, p. 249; Jackson & Depew, 2017; Lewontin, 1972). That is the work of science, not of "scientific ideology." I don't know if Jean's is the last word on the subject. It depends on what you mean by ideology. I would like to draw your attention once more, however, to his overarching insistence that science *is* knowledge–and that one can see this most perspicuously in the light of its history (Burian, 2018). If Jean is the opposite of a biological reductionist, he is even more the opposite of a social constructionist.

7.3 Conclusion

Jean's interest in the epistemic status of the Modern Synthesis remained high throughout his life. In the light of its growing influence, he was especially keen to gauge the effect of the molecular revolution on the Modern Synthesis of the 1940s and 1950s. He did methodologically innovative work in computer-assisted searches into the ups and downs of terms like "evolutionary synthesis," "modern synthesis," "modern evolutionary synthesis," "neo-Darwinian theory," and "synthetic theory" in scientific journals in order to frame this question and hopefully answer it objectively. Under the auspices of the IHPST, whose care he inherited from Bachelard and Canguilhem, he initiated an international interdisciplinary research effort that would review how new biological fields were successively brought into the Synthesis, as its founders hoped and expected, and how in turn the Synthesis was affected by them. His last paper, brought to press by Philippe Huneman, took a last look his old, generative question of whether "The Modern Synthesis" [was a] Theoretical or Institutional Event?" It, was, he concluded

> not so much a well-defined theory as a discipline able to attract people interested in evolution who come from particular fields of expertise with their own norms and doctrines into a new discipline, with its proper norms and organization but also possessing a minimal characterization of what methods and concepts would be admissible in this new professional area of scientific knowledge. (Gayon & Huneman, 2019, p. 525)

That new developments, most recently in developmental genetics, will modify what were once taken as canonical formulations of its core principles is only to be expected.

My final, seventh comment is about the man himself. Jean Gayon combined intelligence, clarity, learning, integrity, humanity, modesty, generosity, and a talent for friendship. It is not surprising that he left a significant mark on scholarship on both sides of the Atlantic. Nor is it surprising that his former students uniformly revere him as a collaborator who held them to high standards and empowered them to make their own marks on the learned world, as many of them have (see Giroux, 2018 and Longy, 2018, for example). That so many of those of us in the United States, Canada, Mexico, and other countries who learned with and from him became his friends reveals both his warmth and his openness to the contributions of others. He was one of the most honorable, insightful, and congenial people I have ever had the honor of knowing. I will always cherish his memory.[15]

References

Beatty, J. (1995). The evolutionary contingency thesis. In G. Wolters & J. G. Lennox (Eds.), *Concepts, theories, and rationality in the biological sciences: The second Pittsburgh-Konstanz colloquium in the philosophy of science* (pp. 45–81). University of Pittsburgh Press.

Beatty, J. (1981). What's wrong with the received view of evolutionary theory? In R. Asquit, & R. Giere (eds.), *PSA 1980, Vol. 2* (pp. 341–55). East Lansing: Philosophy of Science Association.

Brandon, R. (1978). Adaptation and evolutionary theory. *Studies in the History and Philosophy of Science, 9*, 181–206.

Brandon, R. (1981). Biological teleology: Questions and answers. *Studies in the History and Philosophy of Science, 12*, 91–105.

Burian, R. (2018). Jean Gayon on the history of French genetics. In F. Merlin & P. Huneman (Eds.), *Philosophie, histoire, biologie: Mélanges offerts à Jean Gayon* (pp. 111–113). Éditions Matériologiques. This volume.

Burian, R., & Gayon, J. (1999). The French school of genetics: From physiological and population genetics to regulatory molecular genetics. *Annual Review of Genetics, 33*, 313–349.

Burian, R., Gayon, J., & Zallen, D. (1988). The singular fate of genetics in the history of French biology, 1900–1940. *Journal of the History of Biology, 21*, 357–402.

Cain, J. (2009). Rethinking the synthesis period in evolutionary studies. *Journal of the History Biology, 42*, 621–648.

Depew, D. (2018). Jean Gayon: An appreciation. In F. Merlin & P. Huneman (Eds.), *Philosophie, histoire, biologie: Mélanges offerts à Jean Gayon* (pp. 105–109). Éditions Matériologiques.

[15] This paper has been improved by the helpful comments of Dick Burian, Pierre-Olivier Méthot, and Phil Sloan. Any factual errors and interpretive misfires are mine.

de Ricqles, A. (2018). Brève histoire de l'enseignement de l'évolution en France: l'expérience Jean Gayon. In F. Merlin & P. Huneman (Eds.), *Philosophie, histoire, biologie: Mélanges offerts à Jean Gayon* (pp. 151–161). Éditions Matériologiques.

Eldredge, N., & Gould, S. J. (1972). Punctuated equilibria: An alternative to phyletic gradualism. In T. J. M. Schopf (Ed.), *Models in paleobiology* (pp. 82–115). Freeman, Cooper.

Fisher, R. A. (1918). The correlation between relations on the supposition of Mendelian inheritance. *Transactions of the Royal Society of Edinburgh, 52*, 399–433.

Galton, F. (1892). *Hereditary genius* (2nd ed.). Macmillan.

Gayon, J. (1992). *Darwin et l'après-Darwin: Une histoire de l'hypothèse de sélection naturelle.* Kimé.

Gayon, J. (1995). Sélection naturelle ou survie des plus aptes? Éléments pour une histoire du concept de fitness dans la théorie évolutionniste. In C. Blanckaert, J.-L. Fischer, & R. Rey (Eds.), *Nature, histoire, société: Essais en hommage à Jacques Roger* (pp. 263–287). Klincksieck.

Gayon J. (1997). The 'paramount power' of selection: from Darwin to Kauffman. In M.L. Dalla Chiara, K. Doets, D. Mundici & J.van. Benthem (eds.), *Structure and norms in science* (pp. 265–282). Kluwer Academic Publisher.

Gayon, J. (1998). *Darwinism's struggle for survival: Heredity and the hypothesis of natural selection.* Cambridge University Press.

Gayon, J. (2002). Faut-il proscrire l'expression 'race humaines'? UNESCO, 1950–1951. *L'Aventure humaine 12/2001: La societé et ses 'races.'*. Paris: Presses Universitaire de France.

Gayon, J. (2003). Do biologists need the expression 'human race'? UNESCO 1950–1951. In J. Rozenberg (Ed.), *Bioethical and ethical issues surrounding the trials and Code of Nuremberg* (pp. 23–48). Mellon Press.

Gayon J. (2008). Is there a biological concept of race? In S. Muller-Wille & H-J. Rheinberger (eds.), *Race and genomics: old wine in new bottles. NTM Zeitschrift fur Geschichte der Wissenschafte, Tecknik and Medizen, 16*(8), 365–370.

Gayon, J. (2015). What future for Darwinism? In P. Sloan, G. McKenny, & K. Eggleston (Eds.), *Darwinism in the twenty-first century* (pp. 404–423). Notre Dame University Press.

Gayon, J. & Burian R. (2000). France in the era of Mendelism. *Comptes-rendus de l'Academie des sciences. Série des sciences de la vie 323*(12), 1097–1106.

Gayon, J., & Petit, V. (2018). *La connaissance de la vie aujourd'hui.* ISTE Editions Ltd.

Gayon, J & Petit, V. (2019). *Knowledge of life today: conversations on biology.* London: ISTE Ltd. & Hoboken. NJ: John Wiley & Sons.

Gayon, J., & Huneman, P. (2019). The modern synthesis: Theoretical or institutional event? *Journal of the History of Biology, 52*, 519–535.

Gayon, J., & Veuille, M. (2001). The genetics of experimental populations: L'Héritier and Teissier's populations cages. In C. Singh, D. Krimbas, D. Paul, & J. Beatty (Eds.), *Thinking about evolution: Historical, philosophical, and political perspectives* (pp. 77–102). Cambridge University Press.

Giroux, E. (2018). Épistémologie historique et philosophie analytique: la voie d'une synthese singulaire. Témoignage d'une ancienne doctorante. In F. Merlin & P. Huneman (eds.), *Philosophie, histoire, biologie: mélanges offerts à Jean Gayon* (pp. 267–273). Paris: Éditions Matériologiques. This volume.

Gould, S. J. (1983). The hardening of the modern synthesis. In M. Grene (Ed.), *Dimensions of Darwinism* (pp. 71–93). Cambridge University Press.

Gould, S. J. (1982). Darwinism and the expansion of evolutionary theory. *Science, 216*, 380–387.

Hodge, M. J. S. (1977). The structure and strategy of Darwin's 'long argument.' *British Journal for the History of Science, 10*, 237–246.

Hull, D. (1994). Species, races, and genders: Differences are not deviations. In R. Weir, S. Lawrence, & E. Fales (Eds.), *Genes and human self-knowledge* (pp. 207–231). University of Iowa Press.

Jackson, J., & Depew, D. (2017). *Darwinism, democracy and race.* London & New York: Routledge.

Laudan, L. (1977). *Progress and its problems*. Berkeley & Los Angeles: University of California Press.

Lewontin, R. (1970). The units of selection. *Annual Review of Ecology and Systematics, 1*, 1–18.

Lewontin, R. (1972). The apportionment of human diversity. *Evolutionary Biology, 6*, 381–398.

Lloyd, E. (1988/1994). *The structure and confirmation of evolutionary theory*. Princeton University Press.

Loison, L., Gayon, J., & Burian, R. (2017). Contributions – and collapse – of Lamarckian heredity in Pasteurian molecular biology: 1. Lysogeny, 1900–1960. *Journal of the History of Biology, 50*, 5–52.

Longy, F. (2018). Le pilote de ACI 'fonction.' In F. Merlin & P. Huneman (Eds.), *Philosophie, histoire, biologie: Mélanges offerts à Jean Gayon* (pp. 275–281). Éditions Matériologiques.

Mayr, E. (1985). How does biology differ from the physical sciences? In D. J. Depew & B. H. Weber (Eds.), *Evolution at a crossroad: The new biology and the new philosophy of science* (pp. 43–63). MIT Press.

Mayr, E (1980). Introduction. In E. Mayr &. W. Provine (eds.), *The evolutionary synthesis: perspectives on the unification of biology.* Cambridge, MA: Harvard University Press.

Méthot, P.-O. (2018). De l'épistémologie historique à la philosophie de la biologie: Le double héritage philosophique de Jean Gayon. In F. Merlin & P. Huneman (Eds.), *Philosophie, histoire, biologie: Mélanges offerts à Jean Gayon* (pp. 15–50). Éditions Matériologiques. This volume.

Méthot, P.-O. (2022). Jean Gayon face à Georges Canguilhem: Une remontée en trois temps. In C. Cherici, M. Arminjon, & P.-O. Méthot (Eds.), *Le normal et le pathologique: Des catégories périmées? Textes réunis en hommage à Jean Gayon* (pp. 267–306). Éditions Matériologiques.

Pearson, K. (1910). *Nature and nurture: The problem of the future*. Eugenics Laboratory.

Smocovitis, V. B. (1992). Unifying biology: The evolutionary synthesis and evolutionary biology. *Journal of the History of Biology, 25*(1), 1–65.

Smocovitis, V. B. (1996). *Unifying biology: The evolutionary synthesis and evolutionary biology*. Princeton University Press.

Sober, E. (1984). *The nature of selection: Evolutionary theory in philosophical focus*. MIT Press.

Veuille, M. (2018). Jean Gayon et l'école française de génétique des populations. In F. Merlin & P. Huneman (Eds.), *Philosophie, histoire, biologie: Mélanges offerts à Jean Gayon* (pp. 135–150). Éditions Matériologiques. This volume.

Wright, L. (1958). Functions. *Philosophical Review, 82*, 139–168.

Wright, L. (1967). *Teleological explanations*. Berkeley & Los Angeles: University of California Press.

Chapter 8
Local Traditions and the Death of Diffusion: Jean Gayon's Contribution to Contemporary Narratives of 'The Molecular Vision of Life'

Edna Suárez-Díaz

Abstract Recent historiographies have put a focus on *place* and *movement*, with important implications for the history of science and technology outside the traditional locus of Western Europe and the United States. In this chapter, I argue that some of the central problems and concepts advanced in Jean Gayon's works, together with contributions in other areas of the history of science, set the ground for a thorough critique of the *diffusionist* view of knowledge, famously associated with George Basalla. Though the themes and tools in contemporary history of science are nowadays coming from afar the industrialized western countries, the movement of twentieth century life sciences into French traditions and institutions captured some of the same basic questions almost thirty years ago. In particular, Gayon contributed to our current narrative of how Mendelism, and later the molecular approach to heredity, made their arrival to France to be accommodated within existing disciplinary fields, traditions, and institutions. In doing so, he and his colleagues introduced a perspective that challenged a merely "rationalist" or even a "spreading" view of the displacement of science, providing an analytical tool kit that emphasized the diversity and fruitfulness of local research traditions.

Keywords Movement · Diffusion · Traditions · Lamarckism · Mendelism · French molecular biology

8.1 Introduction: The Movement of Science

In the history of science, the idea of *place* has been with us for a while. *The Power of Place*, the second volume of Janet Browne's biography of Charles Darwin was published in (2002), and David Livingstone's *Putting Science in its Place*, appeared the next year (2003). Browne argued for the impressive gravitational field that

Adapted from Suárez-Díaz (2018).

E. Suárez-Díaz (✉)
Universidad Nacional Autónoma de México, Mexico City, Mexico
e-mail: ednasuarez@ciencias.unam.mx

P.-O. Méthot (ed.), *Philosophy, History and Biology: Essays in Honour of Jean Gayon*, History, Philosophy and Theory of the Life Sciences 30,
https://doi.org/10.1007/978-3-031-28157-0_8

attracted natural history to the British Empire during the mid-nineteenth century, while Livingstone made a more general argument about the fundamental importance of geography in the production and consumption of knowledge and introduced—in his last chapter—the idea of *movement*: the circulation of knowledge. Nevertheless, historians of science were relatively late newcomers to the interrogation of movement and the dynamics of space. Cultural historians, and American Studies' scholars, to name just a couple of research fields, had approached travels, hybridization, and translation since at least two decades ago, and in the late 1990s international and diplomatic historians took the lead to reflect on transnational history (in Akira Iriye's formulation, 2004; Fisher-Fishkin, 2005).

James Secord's influential article "Knowledge in Transit" (2004), extended the reflection on movement and space into the history of science, and made the "circulation" of knowledge a central topic—and a catch word—for researchers in the field. However, while Secord made reference to *movement*, *communication*, and *translation* as requirements for knowledge in transit, the bulk of his essay was narrowly focused on "communication" and the communication practices of scientists and mediators. As I have argued before, this perspective—paradoxically—is oblivious to the materiality and the contingencies of displacement, or the experience of *travel* (Mateos & Suárez-Díaz, 2019; Suárez-Díaz, 2015).

Not surprisingly, the recent focus on place and movement has important connotations for new approaches and answers for the history of science and technology outside the traditional locus of Western Europe and the United States. *Movement*, and *displacements*, however, can be—and have been—interpreted in different ways. Be it local traditions, scientific practices, or even those socio-professional mammoths called "scientific disciplines", a major question for historians of science has been and still remains that of *how is possible, for highly contextualized knowledge productions, to make their way into completely different contexts*: a different academic field, another institution, or an entirely new geographical location, like a new country or even a new continent. Though the themes and tools in contemporary history of science are coming more and more from afar the industrialized western countries, the movement of twentieth century life sciences into French traditions and institutions captured some of the same basic questions and concepts almost thirty years ago.

In a series of now classical contributions to the history of twentieth century biology, Jean Gayon, together with Richard Burian, Doris Zallen and other colleagues (Burian et al., 1988, 1991) contributed to our current narrative of how Mendelism, and later the molecular approach to heredity, made their arrival to France to be accommodated within existing disciplinary fields and local traditions and institutions. In doing so, Gayon and his colleagues introduced a perspective that challenged a merely "rationalist" or even a "diffusionist" view of the displacement of science.[1] From the introduction of Mendelism to France, to the recognition of rich local traditions that

[1] Burian (1988) do use the reference to "the diffusion of Mendelism in France". However, as I will argue, their reconstruction is far from the view associated with Basalla's model of the spread of science from Western to non-Western countries (Basalla, 1967).

nurtured a peculiar branch of molecular biology in this country, Gayon's historical trajectories informed the contemporary historical accounts of how scientific knowledge and practices make their way into different settings. Later authors refined, and even challenged the emphasis in national contexts, emphasizing the importance of laboratory cultures, local traditions and informal collaboration networks (Gaudillière, 1993), while others called attention to fields, namely agriculture, where Mendelism made its way through alternative itineraries (Bonneuil, 2006). Nevertheless, many of the questions and concepts raised in Gayon's work regarding the movement of science opened new venues to interrogate the past that have remained with us.

Those interrogations, among others made by a generation of historians and philosophers of science three decades ago, set the ground for thick reconstructions that challenged the *diffusionist* view of displacements and, eventually, the idea of the movement of knowledge "from the center to the periphery". Very briefly, this view assumed the coherence of "Western science", and a dynamic dominated by its "spread" to different latitude and cultures. In this essay, I will examine some of the central problems and concepts advanced in some of Gayon's works that, together with contributions in other areas of the history of science, have set the ground for a thorough critique of this view, famously associated with George Basalla (1967). The classic and fiercely criticized essay by Basalla advanced a conceptual framework for the movement of science and technology from western to non-western countries, in successive stages of exploration, colonization and appropriation, that assumed the universality of scientific knowledge and the superiority (colonial, imperial) of Western science which was predominant in the philosophy and history of science during the Cold War period (Anderson, 2009). While there is a deluge of recent historical and postcolonial literature that challenges this model as an adequate depiction of the dynamics of knowledge between different nations, regions and continents, the question of how theories and practices go from one place to another, even in more restricted spatial or disciplinary coordinates, remains.[2] Basalla was obviously thinking of non-Western latitudes and long travels when he set up his model: places like South East Asia, Africa or Latin America. However, his view of the spread of science assumed a rationality of incoming ("arriving") knowledge and practices, and a void to be filled and later appropriated by the "receiving" end. In this essay, I take the liberty to use Basalla's view of the *spread* of science as a background to highlight some of the most enduring contributions of Jean Gayon and his colleagues to the history of the life sciences in the twentieth century, by focusing on their analysis of the movement of science to a place far more familiar to Western histories of science: France.[3]

---

[2] For contributions in this realm see the volume edited by Suman Seth (2009).

[3] Interestingly, the recruitment of France as the "receiving" end of developments in Anglo-Saxon and Northern European developments in genetics and molecular genetics, had parallels with the qualifications of Basalla of the majority of non-civilized nations beyond the "small circle of Western European nations" (p. 611). France's biological tradition was seen at the time as *lagging behind* England in their historic commitment to Lamarckism and anti-Darwinism.

8.2 Resistance

Once upon a time, resistance to new scientific theories and practices was largely viewed as a failure to assimilate new evidence, and/or the result of a fraught understanding of theoretical and experimental developments. It had to do either with lack of proximity, or worse, prejudice; "the forces of light against those of darkness", in Gayon and his colleagues' fit words (Burian et al., 1988, p. 360). The idea that alternative explanations and innovative areas of research had to carve their way within previously established relevant areas of research and traditions was already present in Ludwick Fleck's classic monograph, and most famously in Thomas S. Kuhn's study of scientific revolutions. In Fleck and Kuhn's views, which preceded Basalla's essay, there is no exploratory recognition of an unknown land, nor empirical facts to accumulate in order to consolidate new forms of knowledge.

In a paper co-authored by Richard Burian, Jean Gayon and Doris Zallen, published in 1988, this idea was given tremendous content in the case study of the arrival of genetics to France. The analytic tool kit of that paper is still worthy of a careful reading.

Gayon and his colleagues offered a multi-varied and symmetrical account that challenged the then prevailing metaphor of "the spread" of knowledge as the asymmetric arrival of a "superior" form of knowledge into a primitive land or, in the case of French biology, a land dominated by an "obsolete"—or backward—perspective. Instead of appealing to the long-standing tradition of French Lamarckism, the authors of this path-breaking essay chose to recognize the importance of "non-Mendelian research bearing on the problem of heredity" (Burian et al., 1988, p. 358). The *resistance to Mendelism*, thus, could "also be understood as the result of complex interactions among the diverse and fruitful research traditions that played an important role in the [future] development of molecular biology" in France. The words "diverse" and "fruitful" bear all the importance here. It was not an *old-fashioned Lamarckism* that prevented French researchers and academia to embrace the new transmission heredity, but other competing approaches to the same subject, notably development (causal embryology), physiology and microbiology. In the words of the authors, "the country of Lamarck was also the country of Claude Bernard and Louis Pasteur" (Ibid., p. 358).

Thus, while it was true that the lack of genetics in the institutional curriculum until after the Second World War resulted in a "whole generation deprived of learning and information to carry out research", and that "institutional obstacles deriving from the structure of the French university system" played also a role (Ibid., p. 369), it is the complex disciplinary ecology of France's life sciences which played the major role in Burian, Gayon and Zallen's perspective. Mendelian genetics, in brief, had little to offer to physiology, development and microbiology, an appreciation which—in retrospect and somehow anachronistically—was basically correct.

Nevertheless, a few researchers rapidly incorporated the new Mendelian approach into their previously existing research programs. This is the case of Lucien Cuénot, who famously framed advances in genetics within his physiological investigations of aspects of coat color (Burian et al., 1988, p. 373). He advanced a research program on multi-allelic interactions in animal genetics, but still found it difficult to find a position at a French university with this type of project, and eventually returned to evolution and physiology. In Cuénot's case, thus, the institutional setting also became a locus of resistance to genetics, leaving him with no intellectual descendants (Ibid., p. 376). And so, genetics was researched outside the French universities, at the École normale supérieure, the Institut Pasteur, and the Institut de Biologie physico-chimique, oftentimes with the intervention of the Rockefeller Foundation funds and the help of the collaborative network created around the French Resistance.

More illuminating than the mere question of the "arrival" of a foreign disciplinary approach is the question of how new perspectives carve their way into previously existing disciplinary ecologies. But posing the question like this fails to recognize that a fruitful research program cannot be simply "colonized" or invaded by a new perspective. Doing this would mean that there is an intrinsic superiority in the arriving field, which was the thesis rejected by Burian, Gayon and Zallen in the first place. In their view, the existing French traditions of research in the life sciences were *fruitful* and *relevant*. Nowhere was this situation more obvious than in the case of microbiology, "the main glory of French biology" (Burian et al., 1988, p. 383). Genetics and microbiology have to be put on an equal foot, and the question became not one of agency—on the part of genetics- and receptivity—in the case of microbiology. It was, to put it in other terms, a question of symmetry, of *how genetics was made relevant for microbiology*. And in this, the agency rests on the supposedly receiving side: the translation of microbiology's medical problems into problems of genetics, or "the transformation of a *problematique of contagion to a problematique of heredity*" (Ibid., p. 383; emphasis in original). Though this is not the place to repeat the article's main argument, it is worth to mention the detailed narrative that explains the slow transformation of viruses and bacteria into tools for the study of heredity, most clearly in André Lwoff's research of growth factors in microorganisms, Eugène Wollman's research on lysogeny and phage infection, and the embryologically informed investigations of Boris Ephrussi (Burian et al., 1988, pp. 400–402, Burian et al., 1991).

An alternative interpretation to the origins of genetics and molecular biology in France was given by Jean Paul Gaudillière (1993). In his view, it was the local culture of a small number of laboratories within a few institutions, and their connections with an international network of practitioners that explained the framing of medical problems within the new molecular biology, that is, the transformation from problems of contagion to problems of heredity within microbiology. Gaudillière's detailed and extensive account involved the role of structural biochemistry, but at the same time the impact of favorable institutions and collaborations supported by increased financial support of a few but influential groups. His most important contribution to the debate of the "arrival" of the new molecular biology to France, however, was the subversion of the "national" framework as a legitimate analytic category; in Gaudillière's terms the "national" is a result, an a posteriori artifact produced by the legitimation of influential *international* networks of scientists, and the end-product—not the cause—of the subsequent domination of a set of scientific practices and problems which accommodated within the existing alternative genealogies of practices. The *national*, then, becomes the question to be answered by historical research, and not the analytical category that helps us to understand the development of "French genetics" or French molecular biology. Which takes us to a different subject.

8.3 National Borders and Collaborative Networks

Gaudillière (1993) challenged Gayon and his colleagues' emphasis on national contexts and traditions as useful categories to understand the movement of knowledge and practices in the biological sciences. In Gaudillière's terms, the "nation" is the *explanandum*, not the *explanans* in histories of science; in contrast, he argued for a detailed analysis and a thorough genealogical account of laboratory cultures, experimental problems and tools, and local traditions. But this extremely focal interest on *local cultures* produced, paradoxically, a narrative about the privileged role played by international networks and collaborations, and an emphasis on the contingencies of contacts, decisions and institutional frameworks, even if they took place across continents. Where Gayon and his colleagues had seen a *national* science, Gaudillière saw an *international* network. In today's historiographical terms, these networks—most of them organized around the Rockefeller Foundation—may be better described as *transnational*, given the high mobility across borders of the people and practices associated with the new molecular genetics. By means of material exchanges, postdoctoral students and sabbatical stays, attendance to the same conferences and grants from the same sources, two laboratories across the Atlantic Ocean could be closer in goals and means than two research institutions across the Seine. And indeed, one of the most paradoxical consequences of the *local* turn in the history of science has been the recent *transnational* turn.[4]

The problem of the *reception* of Mendelism in France, or of the creation of a French school of molecular biology within French scientific institutions, was interpreted, in Gayon and his colleagues' reconstruction, as one of resistance and accommodation within previously existing—and productive—national cultures, traditions and institutions. In this nuanced interpretation, they challenged the idea of a "spread", and of the overarching rationality of one universal science which is the hallmark

[4] The most obvious feature of the transnational turn is its abandonment of the idea that national states, or even national traditions, provide the relevant context for the history of science. In this, it shares common problems with other perspectives, such as the cross-national perspective, which focus on material and human travels across borders. Like its close relative, international history, the cross-national perspective assumes the integrity of national states, but interrogates the conditions and obstacles standing between them: infrastructures and bureaucracies, visa systems and security clearances of people, technologies and knowledge (Cohen & O'Connor, 2004, see Krige; 2006; Miller, 2006). The *transnational* turn, in contrast, seeks to transcend the strict limits and interests of national states, to look beyond, into the international context, to account for local developments and laboratory cultures.

of Basalla's model. Nevertheless, this was an interpretation still rooted in localized productive research programs, which assumed the national context and its institutions as given. In Gaudillière's account, by contrast, the extremely localized research groups and traditions that benefitted from transnational networks of collaboration accommodated these developments, reinterpreting them as a *French* tradition when, in fact, many other research groups in the United States and the UK shared the same tools and problems. The transnational umbrella to these events included the still existing inter-war networks created by the Rockefeller Foundation, and the extended collaborations between European and United States' laboratories. The most important developments, however, included a postwar belief in the need to reconstruct the scientific institutions in Europe—and in France in particular–, and the availability of exchanges and interactions between the United States and its allies (without forgetting the role of visas and state power to overrule those interactions, as it is visible when visas were denied for political reasons as in Hermann Muller and Linus Pauling's well-known cases). At a more extended level, the role of governmental national agencies, like the US Public Health Service, and international agencies such as the United Nations' World Health Organization, became instrumental in the promotion of scientific exchanges and training of huge numbers of scientists and technicians around the world, beyond the European continent.

The global character of research programs in twentieth century life sciences has provided a number of exemplary case studies of transnational interactions, and a window to explore the specific features of contemporary science. There are many examples, but I will only mention the impact of some of these movements in the biomedical understanding of disease and its relation to human variation—a theme that has evolved into one of the major fields of research today: pharmacogenomics.

It is well known that during the first part of the century, the Rockefeller Foundation and other private philanthropic funds and international agencies, such as the League of Nations Health Organization, and the Pan American Health Organization, played a large role in promoting important programs in public health (examples include Birn, 2012; Cueto, 2007; Cueto & Palmer, 2014; Packard, 2007). These programs also included the training of experts in the United States and Europe, and an important number of human population studies in nutrition and metabolism, and later on genetic (serological) diversity. Such programs had an everlasting impact in a few Latin American research groups (Pohl-Valero, 2014; Vargas-Domínguez, 2017), and eventually played a role in the establishment of international nutrition standards, and in the recognition of unexpected amounts of human diversity after the Second World War—with an impact on population genetics, evolutionary theory and biomedicine among other biological fields (Suárez-Díaz, 2019). These developments were not minor. Notable examples include the discovery of the Diego blood group by Miguel Layrisse in Venezuela, and Tulio Arendt in Colombia; or the discovery of new hemoglobin variants in the work of Rubén Lisker in Mexico (Suárez-Diaz, 2017, 2019).

In all these cases, the familiarity with postwar experimental cultures, and the use of new technologies in hematology, medical genetics and molecular biology—saliently, gel electrophoresis—were the result of a close collaboration and participation in transnational networks. Lisker, for instance, had studied hematology at the Michael Reed Hospital in Chicago in the mid-1950s, where he met Arno Motulsky, an early promoter of pharmaco-genetics. Lisker was trained as a hematologist, and he became acquainted with the chemical and physiological methods that allowed the identification of hemoglobin and other blood serum protein variants. At the beginning of the 1960s, however, Lisker's career would shift after the discovery of a G6PD variant in a Jewish boy (glucose-6-phospho dehydrogenase, an enzyme of the glucose metabolism), and the launching of a research project to survey blood serum protein variants in indigenous populations over the country. Though Liker never abandoned the use of hematology's tools, his interactions with Motulsky resulted in an eighteen-month research stay at the Department of Genetics at the University of Washington in Seattle, between 1965 and 1966. Lisker learned the new methods of gel electrophoresis, which he brought to Mexico to use on his population surveys, and soon after, thanks to his connections at Seattle, he participated in the amino acid sequencing of two hemoglobin variants found among Mexican indigenous populations, *hemoglobin Chiapas* and *hemoglobin Mexico* (Suárez-Díaz, 2014). These contributions translated into the creation of a new Genetics Department, leaded by Lisker, at the National Institute for Nutrition, a Mexican institution which has become a leading site for nutrition studies and biomedicine, and whose researchers often made the transition as experts in United Nations' agencies, like the FAO and WHO. This brief recount exemplifies the relevance of transnational actors in advancing common scientific agendas and experimental cultures across countries, with the possibility of shaping local scientific cultures, but also of contributing to the new biomedical approach to disease, that relied on molecular techniques. The legitimate question of the therapeutic benefits of the biomedical approach is off the limits of this paper. It is worth to mention, however, that Lisker's research on G6PD variants in indigenous population became part of the *Campaña Nacional para la Erradicación del Paludismo*, as the use of quinine and its derivatives might have unexpected haemolytic effects in carriers of a specific type of genetic deficiency (Suárez-Díaz, 2017).

This is not to deny the power of nation states in the history of science. John Krige, one of the main authors currently exploring the relations between diplomatic and international history, and the history of science and technology, sees the *circulation of knowledge* across national borders as one of the main subjects for a transnational history of science (Krige, 2006). Other authors seem to concur with this definition, including Turchetti et al. (2012), who equate transnational history with the *flow* of people, goods and ideas across national borders. The problem with metaphors of "flow" and "circulation", however, is the same as the use of metaphors of "spread" and diffusion, namely, they make abstraction of the materiality of movement. Warwick Anderson, appropriately has labeled this as an "hydraulic" approach, in its unproblematic appropriation of still another physical metaphor (Anderson, 2014). Knowledge, practices, organisms, techniques and instruments, however, are moved by people, with the help and obstacles of institutions and infrastructures.

Moving the methods and artifacts of science cost money, common interest and effort, including an active side on the "receiving" end. This is obvious in cases where across-borders exchanges took place between countries with reciprocal ties, like the United States, France and the United Kingdom during the postwar years. Indeed, as I mentioned, the transnational scientific collaboration between allied countries was encouraged both as a political instrument (Creager, 2002, 2009), and also as a double-sided policy of internationalism and secrecy (Krige, 2006; Miller, 2006) in this period. The reciprocal exchange between allied countries, however, took place along an "asymmetrical field of power" (Krige, 2012).

As I illustrated with the case of Rubén Lisker's research, non-reciprocal and asymmetrical movements of science and technology also became a major force in the expansion of an impressive number of fields within the life sciences in the twentieth century. Public health related research in medical genetics and population genetics were mobilized in many corners of the world, the "less developed" countries on either side of the Berlin Wall, and also the new post-colonial nations. They were carried out from one place to another, thanks to scientific collaborative networks who had been created with the help of the newly created system of the United Nations technical agencies, but also with investment in former colonies and national agencies. These exchanges encouraged the education of large cadres of experts in the universities of Europe and the United States, but also at regional centers built in Asia, Africa and Latin America. The networks of transnational collaboration, this time, required not only a large number of agencies, but of intermediaries and functionaries, whose main goals included the creation and reproduction of even more inclusive networks of collaboration, despite huge differences in economic income, bureaucratic and administrative traditions (or lack thereof), and infrastructures between research teams situated around the world.

The challenge of historians of science interested in postwar South East Asia, Africa and Latin America, remains to account for these movements, as part of local projects embedded in the polarization of international relations characteristic of the postwar era. Not surprisingly, a large part of studies concerning Cold War science in places outside the western countries and the Soviet Union have emphasized the role played by scientific and technological development *in the construction of national identity*—and not the other way around. This is very clear in big-science projects, such as the atomic programs of India, China, Pakistan and even Brazil, whose national "identity" was inextricably linked to scientific and technological prestige (Phalkey, 2013; Wang, 2015). However, it was also relevant for "countries worthy of attention", a group including Mexico, Turkey, Egypt, Yugoslavia, Argentina among others, which according to international agencies had basic scientific and technological infrastructure in the mid-twentieth century, and could benefit from programs of technical assistance, education and training programs. Moreover, after the international system of health and agriculture was put in place after the war, these agencies required local experts around the world, not just as an instrument of legitimation, but as practitioners of the new fields of research and application.

More important for local histories of science, however, is to focus on how some research groups and a few institutions became part of the international collaboration networks referred to by Gaudillière. As Peruvian historian of science Marcos Cueto has argued, there was excellence at the periphery. Most common, however, was the fact that a few highly localized research programs, taking place in laboratories on "less developed" or *Third World* countries, became incorporated into the international system of science, either by providing huge amounts of data and specimens for new collections—for instance, serological and molecular samples (Suárez-Diaz, 2014), or measurements on atomic fallout, meteorology and other atmospheric and environmental sciences (Mateos & Suárez-Diaz, 2015)—or by their contributions in the understanding of natural phenomena.

8.4 Concluding Remarks: From Diffusion and Spread-to Itineraries

Assimilation, diffusion, and penetration were some of the terms used by Burian et al. (1988) to refer to the movement of concrete fields of research, tools and problems into a new context. However, as I have argued, their account of the arrival of genetics, and the creation of a local relevant tradition in molecular biology, was far distant from the model of the "spread of knowledge" defended by Basalla twenty years before. There is nothing about exploration, colonization and appropriation in Gayon's reconstruction of how Mendelism arrived in France. Instead, he and his colleagues argued that genetics had to carve its way amidst relevant and fruitful local research traditions to which the new approach had little to offer. It could be argued that this is proof of the reciprocal nature of the relations between French scientists and their counterparts in Britain and the United States, and that Basalla's model only applies to exchanges between western and non-western countries. But this observation misses a crucial point related to the movement and displacement of science: the idea that knowledge can be "spread" over a blank slate, or "diffused" through a different research culture assumes that agency and drive lies only at the donor end of the exchange. This, however, is not the case, as shown by numerous accounts of the movement of science—including the arrival of disciplines and practices to France, and places far away across the Atlantic, like Mexico.

The importance of transnational collaborative networks has taken the attention of historians of science to countless other agents and intermediaries, without whom scientific knowledge and techniques could not be mobilized: local representatives of international agencies, officials coming from colonial administrations, and transnational experts taking technical and scientific knowledge to every continent. The number of travels taking place in the decades after the Second World War was enormous, and it has drawn the attention of historians to the myriad contingent ways in which such movements took place, or were halted. In the last few years historians of science—and to a minor extent yet, philosophers—have come to view the *itineraries* of science in their displacement from one place to another, problematizing the simple metaphors of spread, circulation and diffusion put in place in the late 1980s and early

1990s (Secord, 1994). The work of Gayon and others was an important contribution to the historian's tool-kit to ask new questions concerning the geography of science, and its movements.

References

Anderson, W. (2009). From subjugated knowledge to conjugated subjects: Science and globalisation, or postcolonial studies of science? *Postcolonial Studies, 12*(4), 389–400.

Anderson, W. (2014). Making global health history: The postcolonial worldliness of biomedicine. *Social History of Medicine, 27*(2), 372–384.

Basalla, G. (1967). The spread of Western science. *Science, 156*, 611–622.

Birn, A.-E. (2012). *Marriage of convenience*. Rochester Studies in Medical History.

Bonneuil, C. (2006). Mendelism, plant breeding and experimental cultures: Agriculture and the development of genetics in France. *Journal of the History of Biology, 39*(2), 281–308.

Browne, J. (2002). *Charles Darwin: A biography, Vol. 2–The Power of Place.*

Burian, R. M., Gayon, J., & Zallen, D. (1988). The singular fate of genetics in the history of French biology, 1900–1940. *Journal of the History of Biology, 21*(3), 357–402.

Burian, R. M., Gayon, J., & Zallen, D. T. (1991). Boris Ephrussi and the synthesis of genetics and embryology. In S. F. Gilbert (Ed.), *A conceptual history of modern embryology* (pp. 207–227). Springer.

Cohen, D. & O´Connor, M. (2004). Comparative history, cross-national history, trans-national history definitions. In D. Cohen & M. O´Connor (Eds.), *Comparison and history. Europe in cross-national perspective.* Routledge.

Creager, A. N. (2002). Tracing the politics of changing postwar research practices: The export of 'American' radioisotopes to European biologists. *Studies in History and Philosophy of Science Part C: Studies in History and Philosophy of Biological and Biomedical Sciences, 33*(3), 367–388.

Creager, A. N. (2009). Radioisotopes as political instruments, 1946–1953. *Dynamis, 29*, 219–239.

Cueto, M. (2007). *Cold war, deadly fevers; Malaria eradication in Mexico, 1955–1975.* Woodrow Wilson Center Press, The Johns Hopkins University Press.

Cueto, M., & Palmer, S. (2014). *Medicine and public health in Latin America: A history.* Cambridge University Press.

Fisher-Fishkin, S. (2005). Crossroads of cultures: The transnational turn in American Studies. Presidential Adress to the American Studies Association, November 12, 2004. *American Quaterly, 57*(1), 17.

Gaudillière, J. P. (1993). Molecular biology in the French tradition? Redefining local traditions and disciplinary patterns. *Journal of the History of Biology, 26*(3), 473–498.

Iriye, A. (2004). Transnational history. *Contemporary European History, 13*(2), 211–222.

Krige, J. (2006). Atoms for peace, scientific internationalism, and scientific intelligence. *Osiris, 21*, 161–181.

Krige, J. (2012). Hybrid knowledge. The transnational coproduction of the gas centrifuge for uranium enrichment in the 1960s. *British Journal for the History of Science, 45*(3), 337–357.

Livingstone, D. N. (2003). *Putting science in its place: Geographies of scientific knowledge.* University of Chicago Press.

Mateos, G., & Suárez-Díaz, E. (2015). Clouds, airplanes, trucks and people: Carrying radioisotopes to and across Mexico. *Dynamis, 35*(2), 279–305.

Mateos, G., & Suárez-Díaz, E. (2019). Technical assistance in movement: Nuclear knowledge crosses Latin American borders. In J. Krige (Ed.)., *How knowledge moves. Writing the Transnational History of Science and Technology* (pp. 345–367). The University of Chicago Press.

Miller, C. A. (2006). An effective instrument of peace: Scientific cooperation as an instrument of U.S. Foreign Policy, 1938–*1950*. *Osiris, 21*, 133–160.

Packard, R. M. (2007). *The making of a tropical disease: A short history of malaria.* Johns Hopkins University Press.

Phalkey, J. (2013) *Atomic state: big science in twentieth-century India.* Permanent Black.

Pohl-Valero, S. (2014). "La raza entra por la boca": Energy, diet, and eugenics in Colombia, 1890–1940. *Hispanic American Historical Review, 94*(3), 455–486.

Secord, J. A. (2004). Knowledge in transit. *ISIS, 95*(4), 654–672.

Seth, S. (2009). Putting knowledge in its place: Science, colonialism, and the postcolonial. *Postcolonial Studies, 12*(4), 373–388.

Suárez-Díaz, E. (2014). Indigenous populations in Mexico: Medical genetics and cultural anthropology in the work of Ruben Lisker in the 1960s. *Studies in the History and Philosophy of the Biological and Biomedical Sciences, 47*, 108–117.

Suárez-Diaz, E. (2015). La perspectiva transnacional de la ciencia. *Ludus Vitalis, 23*(43), 58–81.

Suárez-Díaz, E. (2017). Blood diseases in the backyard: Mexican 'indígenas' as population of cognition in the mid-1960s. *Perspectives on Science, 25*(5), 606–630.

Suárez-Díaz, E. (2018). Local traditions and the death of diffusion: Jean Gayon's contributions to contemporary narratives of 'the molecular vision of life.' In F. Merlin & P. Huneman (Eds.), *Philosophie, histoire, biologie: mélanges offerts à Jean Gayon* (pp. 121–134). Éditions Matériologiques.

Suárez-Díaz, E. (2019). The molecular basis of disease and evolution: A Cold War alliance. *Journal of the History of Biology, 52*(2), 325–346.

Turchetti, S., Herran, N., & Boudia, S. (2012). Introduction: Have we ever been 'transnational'? Towards a history of science across and beyond borders. *British Journal for the History of Science, 45*, 319–336.

Vargas-Domínguez, J. (2017). The "Problematic" Otomi: Metabolism, nutrition, and the classification of indigenous populations in Mexico in the 1930's. *Perspectives on Science, 25*(5), 564–584.

Wang, Z. (2015). The Chinese developmental state during the Cold War: The making of the 1956 twelve-year science and technology plan. *History and Technology, 31*(3), 180–205.

Chapter 9
Jean Gayon and the French School of Population Genetics

Michel Veuille

Abstract The history of evolutionary theory in France after 1859 and of genetics after 1900 is unique to this country, and most of it remained poorly understood until the historical studies conducted by Jean Gayon, Richard Burian, Doris Zallen and collaborators. Academic studies and teaching in evolutionary genetics in France developed only after geneticists Philippe L'Héritier (1906–1994) and Georges Teissier (1900–1972) conducted a series of experiments as of 1932. Having a dual background in descriptive zoology and mathematical biology, Teissier and L'Héritier were pioneers in experimental population genetics and were among the first geneticists to draw attention to the maintenance of polymorphism in populations as a major evolutionary issue in the 1930s. L'Héritier and Teissier used population cages ("démomètres") as the main experimental paradigm to tackle this question. This provided the basis for a scientific school that flourished in Paris after WWII. This school is characterized by its interest in a variety of selective mechanisms potentially leading to a polymorphic equilibrium, even though most of its members remained outside the debates on polymorphism that raged in USA and UK in 1950–1970. L'Héritier and Teissier were generally at odds with neutralism, with the famous exception of the collaboration of one of its members, Maxime Lamotte (1920–2007), with French theoretician Gustave Malécot (1911–1998). This school never abandoned its deep roots in zoology, and while its image in France rested on its interest in Darwinism, it was established institutionally on a network of laboratories and periodicals in zoology, which can be viewed as a source of academic power. This led to some kind of isolation of this school, which eventually dissolved in the 1970s in a rising new generation of geneticists restoring international ties. The contribution of this school to French science is considerable since it introduced three fields: genetics, Darwinism and biological statistics.

Translated and adapted from Veuille (2018a, 2018b).

M. Veuille (✉)
UMR, École Pratique des Hautes Études, Université Paris-Sciences-Lettres, CNRS, EPHE, MNHN, SU, UA Paris 7205, France
e-mail: michel.veuille@ephe.psl.eu

P.-O. Méthot (ed.), *Philosophy, History and Biology: Essays in Honour of Jean Gayon*, History, Philosophy and Theory of the Life Sciences 30,
https://doi.org/10.1007/978-3-031-28157-0_9

Keywords Gayon · Population genetics · Modern synthesis · Neo-Lamarckianism · Teissier · L'Héritier

9.1 Introduction

From the end of nineteenth century to the middle of twentieth century, French research in evolution and heredity entered a dead-end from which it eventually emerged following a singular path. After 1932 the school of population genetics established in Paris by Teissier and L'Héritier played an essential role to reinstate French research into these fields and helped it to reconnect with global science (Burian et al., 1988, Burian & Gayon, 1999, Gayon & Burian, 2000, Loison et al., 2016). These researchers established genetics as a discipline of its own at university, both in terms of teaching and research. L'Héritier and Teissier also contributed to the rise of the Modern Synthesis through their studies in experimental populations of *Drosophila.* This part of the history of science, however, long remained poorly understood, including by French geneticists and historians of biology. Jean Gayon's detailed case-studies were essential for re-constructing the successive stages of this history. Gayon's career may be said to have involved four stages: (i) the struggle for establishing Darwinism on a valid theory of heredity; (ii) the history of heredity in France in nineteenth century, in the positivist tradition of Claude Bernard and Louis Pasteur; (iii) the introduction of genetics in French universities by L'Héritier, Teissier, Ephrussi and others; (iv) the establishment of a French school of population genetics by L'Héritier and Teissier and their students. Many of these studies were conducted by Jean Gayon in collaboration, especially with Richard Burian and Doris Zallen.

Jean Gayon trained as a philosopher, but his interests were in the history of science. To this end he also trained as a biologist over a period of nine years and even completed a masters in evolutionary genetics under the supervision of Claudine Petit, a former student of Teissier. I met Gayon at the end of this period at a seminar held at the Institut d'histoire et de philosophie des sciences et des techniques in Paris. I was preparing a PhD in Teissier's former laboratory in Gif-sur-Yvette, under the supervision of another of Teissier's former students, Charles Bocquet. I considered myself a follower of Teissier, whose course at the university had convinced me (as an undergraduate interested in evolution) that population genetics was the right path for me to follow, even though it remained extremely marginal in French biology. I was dissatisfied with the diminishing role played by this school in the 1980s, which was one reason why I was interested in the history of science. Jean and I held many discussions about evolutionary genetics during his PhD studies. He published his thesis as a book in France (Gayon, 1992) and in the UK (Gayon, 1998). I also met him and Dick Burian at the beginning of their study of the foundation of French genetics laboratories in Gif-sur-Yvette in the 1950s.

Our relationship became a true collaboration in the 1990s due to a serendipitous event. My group had moved to a Paris university where I was working in the same building as the last zoology laboratory once directed by Teissier. There I interviewed his then retired successor as director of the laboratory, Guy Echallier (1925–2017), who spent most of his retirement in the laboratory as an honorary professor. He told me that since I was interested in Teissier's life, I should look at a pile of decaying cardboard boxes containing stuff abandoned under the stove pipe of the boiler room, and that the laboratory would be happy to leave it to me. Thankfully, there had been no leak of fuel! The boxes contained Teissier's notebooks as a student, along with his experimental notebooks on *Drosophila* population genetics, hundreds of pages of hand-written biometrical results, publication drafts, reprint boxes on Mendelian and non-Mendelian heredity, elements of the correspondence of Teissier with L'Héritier and Ernst Mayr, etc. It also contained his Légion d'honneur medal in its mailing envelope (the most distinguished French medal[1]) and a whole storage box of documents related to his resignation as the chairman of the CNRS during the cold war. Some of this material was eighty years old. This was a treasure trove: first-hand information on the beginnings of population genetics in France. Jean and I conducted an extensive study of Teissier's research using this material (Gayon & Veuille, 2001). It was followed by specific studies by Gayon (2015, 2016). Below I present an overview of these studies that I weave into a comprehensive narrative, a synthesis for which I take full responsibility.

9.2 Genetics in France in the Early Twentieth Century

Contrary to what the words suggest, there is no historical continuity between Jean-Baptiste Lamarck (1744–1829) and the neo-Lamarckism that prospered in France in late nineteenth century. Continuity was intentionally suggested by the statue of Lamarck erected in 1909 at the main entrance of the Muséum National d'Histoire Naturelle de Paris, to jointly celebrate the centenary of Lamarck's *Philosophie zoologique* (1809), the centenary of Darwin's birth, and the fifty years of the *Origin of species* (1859). At that time the Director of the Muséum, Edmond Perrier (1844–1921), was a leading neo-Lamarckian in French zoology. But for decades after its publication, Lamarck's *Philosophie zoologique* had been considered a speculative construct. Georges Cuvier had won against Lamarck's followers, and French biologists had endorsed the spirit of pure reliance on facts. In her *L'introduction du darwinisme en France au dix-neuvième siècle* (1974), Conry shows the resistance of the French positivist establishment to the *Origin*, a book too much reminiscent of the speculations of Lamarck's doctrine. French biologists were following the principles put forward by Claude Bernard, Louis Pasteur, and Marcellin Berthelot, based on the predominance of observed facts and of experiments over theories.

Understanding how France eventually became "Lamarckian" requires examining separately neo-Lamarckism as a view of evolution, and the inheritance of acquired

---

[1] He had been de facto awarded this medal in 1946 as the director of a governmental institution (CNRS), but, being a staunch leftist, had not agreed to receive it; the de rigueur ceremony had apparently been skipped.

characters as a view of heredity, which was often called "Lamarckian inheritance", and was inspired by Lamarck's second law after dropping the reference to the transmutation of species (Gayon, 2006).[2] The inheritance of acquired characters was widely accepted even though it was strengthened by the works of French physicians and its development in French psychiatry became influential (López Beltrán, 1992; Gayon, 2006). Moreover, it was compatible with Bernard and Pasteur's conception of the organism (Gayon, 1995, 2013). Bernard's (1865) view of biology rested on homeostasis; heredity belonged to physiological mechanisms, and physiological modifications were thought to be passed on across generations: "Organic tradition, or heredity, is the mere continuation, or memory, of the anterior states through which the organism has passed" (Bernard, 1867, p. 216).[3]

Bernard and most French biologists were non-evolutionists. They did not see evolution as a relevant scientific issue since it could not be subjected to experimental confirmation. They opposed pure mechanism to concepts like adaptation, which, they thought, implied purpose. The success of French positivism in physiology thus correlates with French lack of interest in heredity and evolution. In the long term, however, it became difficult to ignore the phenomenon of descent with modifications of species in well-established disciplines such as comparative anatomy and comparative embryology. Evolutionary studies were introduced in the 1880s by biologists under the heading of "*transformisme*", a French word initially coined by anthropologist Paul Broca (1867, p. 401), which suggested that evolution could be studied as pure fact, irrespective of any hypothesis about the mechanisms of biological change.

Two different sources of French Lamarckism have been put forward by historians. Blanckaert et al. (1979) underlined its socio-political aspects, especially among anthropologists (Blanckaert, 2009). In the 1870s several members of the Anthropological Society of Paris rediscovered Lamarck. They were most influenced by Haeckel's reading of descent theory (Blanckaert, 2009; Conry, 1974). This group called itself "scientific materialism". Among them, Martins re-edited Lamarck's *Zoological Philosophy* in 1873, and Letourneau translated Haeckel's *History of Creation* in 1874, a translation that contained the first occurrence of a new French word that was destined for great success: "*Lamarckisme*". They saw strong social consequences in evolutionary theory. Historian of science Jacques Roger thus wrote: "Undoubtedly, neo-Lamarckian ideology took specific forms in France: it contributed to the struggle for republic, against clerical and monarchist forces" (1979, p. 280).

Scientific materialists were interested in Lamarck because of the connection it established between social evolution and biological evolution, an idea that departed from Comte's positivist orthodoxy. There was also some chauvinism in their Lamarckian stance: a witness of the time and a rare French advocate of Darwin ironically

[2] On Lamarckism and neo-Lamarckism as well as Gayons' study of the inheritance of acquired characters, see the chapter by Tirard, this volume.

[3] All translations from French are mine.

commented on the patriotic fad for Lamarck: "It fell to a German, Ernst Haeckel, to tell France that *transformisme* had for father a genuine French".[4]

French zoologists and botanists became evolutionists (and Lamarckian) during the same decade as anthropologists. They were not anti-Darwinian, but considered natural selection as a secondary factor in evolution, the primary factor being the direct effect of environment and nutrition upon physiology. This effect could persist across generations through inheritance. Neo-Lamarckism could combine with the emphasis researchers in the Pasteurian and Bernardian traditions placed on physiological mechanisms (Loison, 2010).

A French (Swiss-born) population geneticist, Boesiger (1980), has suggested that the late rise of French genetics was linked to Bergson's vitalism. But vitalism was not influential in biology. Most French biologists accepted mechanism as the only valid framework. Camille Limoges was probably closer to reality when stressing their organicism: "Central to the [presuppositions of the neo-Lamarckian approach] was the conception of the organism as a whole of tightly correlated parts such that any particulate theory seemed wrong and naïve" (1980, p. 324).

Teissier, who was to introduce the Modern Synthesis to France, witnessed the inadequacy of French teaching in evolution in the early years of twentieth century. "When I was a university student", he said, "Darwinism was generally considered an outdated doctrine and no biologist in France really believed in the existence of natural selection" (Teissier, 1958, p. 62). As an experimental approach to heredity, Mendelism was most welcome in the French context. It was introduced to France by experimental botanists who regarded it as a useful method of plant engineering (Blaringhem, 1928; Boeuf, 1936). Gaston Bonnier, an experimental botanist and a neo-Lamarckian, was the editor of de Vries's first paper on the rediscovery of Mendel's laws in the *Comptes-rendus de l'Académie des Sciences de Paris* in March 1900. A horticultural company owned by the Vilmorin family played an important role in the development of breeding methods in France and organized the fourth International Congress of Genetics in Paris in 1911 (Gayon & Zallen, 1998). Therefore, genetics was taught in agricultural schools, even though this teaching did not rest on any strong research practice.

Since they were gradualist, these botanists saw no contradiction in claiming that evolution depended on Lamarckian mechanisms. Several zoologists combined their practice of genetics and cytogenetics with a sceptical view about Darwinism, or with diverging views on adaptation (Lucien Cuénot, Albert Vandel, Jean Rostand). Delage and Goldsmith (1909) wrote a clear review of the state of evolutionary theory, thus indicating that the French were fully aware of theoretical issues about heredity and evolution and knew very well the state of the art in matters of genetics and evolutionary thinking in scientific literature. Gayon and Burian (2004) found a strong contrast between their knowledge of the literature and their passivity regarding research and teaching in genetics. Cuénot was the only Frenchman whose work in genetics was internationally important before 1930 (Gayon & Burian, 2004). A

[4] "Il a fallu qu'un Allemand, Haeckel, apprît à la France que le transformisme avait pour père un véritable français"; a sentence that Duval (1886, p. 437) attributes to Lanessan.

pioneer of animal genetics, he believed he should not encourage graduate students to pursue a career in genetics in France, and, indeed, Guyénot, a *Drosophila* geneticist, moved to Switzerland to pursue his career. However, some French biologists, including André Mayer,[5] were concerned about the condition of French research on heredity, and advised two young researchers, L'Héritier and Ephrussi, to apply for Rockefeller fellowships and study genetics in the United States, which they did in 1931–1932 and 1934–1935 respectively. This was the start of French genetics. Blaringhem, a neo-Lamarckian botanist interested in Mendelism, also encouraged Gustave Malécot in his theoretical population genetics studies.

Research in population genetics thus started when Philippe L'Héritier (1906–1994), a postgraduate student at the École normale supérieure (ENS) of Paris, went to Iowa for a one-year training in experimental genetics (1931–1932). The ENS is a small and elitist institution of the French university system. At the time, the curricula to study biology included two "preparatory" courses during which students received advanced mathematical teaching. Therefore, L'Héritier was well prepared to tackle population genetics, a field which had just been placed upon a mathematical foundation, and awaited confirmation in genetics. In the United States he attended the 6th International Congress of Genetics at Cornell, where a ground-breaking session was held in evolutionary genetics with contributions by Ronald Fisher, J.B.S. Haldane, Sewall Wright, Theodosius Dobzhansky and Hermann Muller. L'Héritier returned to Paris with Drosophila lines and joined the ENS where he wrote his PhD under the supervision of Georges Teissier (1900–1972), then an assistant professor. His mind was full of ideas for testing the genetical theory of evolution. Immediately after his return to France he published a booklet presenting a short version of Fisher's (1930) *Genetical Theory of Natural Selection*,[6] leaving out Fisher's digressions on eugenics, and including some ideas from Wright. This was the first book of population genetics ever published in France, and probably the third in the world (L'Héritier, 1934) after Fisher's and Haldane's.

Therefore, not only was genetics primarily introduced to France as population genetics, but it also attracted interest from people involved in mathematical biology.

---

[5] André Mayer, like Emmanuel Fauré-Frémiet (Ephrussi's superviser) and Louis Blaringhem directed a laboratory at the EPHE (École Pratique des Hautes Études). EPHE was the main French research agency; it provided financial support, researchers and assistants to experimental laboratories, on the condition of annual reports; it had been founded by Louis Pasteur, Claude Bernard and other prominent scientists in 1868 to promote scientific research, and was based according to Marcellin Berthelot on "the absolute freedom guaranteed to each of the persons who contribute to the theoretical and practical teaching" (Fournier 1980, p. 17); its members were recruited for life on the basis of scientific excellence, and for this reason were free to encourage whatever they thought was an emergent field (Henriet & Veuille 2018; Veuille 2018a, 2018b). After WWII, EPHE was replaced in this function by several national institutes, including the CNRS (Centre National de la Recherche Scientifique), which provided support to laboratories through a renewable term of five years with evaluation; their function was also to define a national strategy of research (Veuille 2018a, 2018b).

[6] This short book was published in a collection directed by Teissier that covered all recent advances in mathematical approaches to population biology, with volumes by Lotka, Volterra, Wright, Gause and himself.

Teissier was precisely interested in quantitative biology, both at the organismic and the population levels.

9.3 Teissier's Allometry: Quantitative Growth and Populations

Teissier's first research interests were in developmental biology, as viewed from two very different angles. On the one hand, as a zoologist, he was interested in Hydrozoans, which he studied every summer at the Roscoff biological station in Britany. He went there probably every year of his scientific life. On the other hand, as a mathematician of biology, Teissier was interested in the biometrical study of growth.

Since students trained at the ENS had a strong mathematical background, Teissier was well prepared to connect the biological and the mathematical dimensions of development. Teissier's interest in allometry was initially introduced as the theory of "biological similarity", which he set forth jointly with mathematician René Lambert (Lambert & Teisssier, 1927). The two men demonstrated how space, time, and energy scales are modified in organisms of similar shape but of different size. These parameters included heartbeat, gestation duration, life duration, oxygen consumption, etc. The paper was mathematical, but the background was physiological. Teissier was neither a pioneer on this subject, nor the one who designed its mathematical formula. Gayon (2016) examined the priority issues and showed that the early works of Teissier in 1926 closely followed Huxley's (1924) interests on the subject, and so did their first books (Huxley, 1932; Teissier, 1934). But Gayon also found out that Teissier was influenced by French researchers in the early 1920s, including Lapicque, Champy, following a paper by Dubois in 1897. Teissier's work was significant in the diffusion of biometry in France and he wrote two books on the subject in the biomathematics collection which he directed (Teissier, 1934, 1937). Huxley and Teissier wrote a famous paper in three different journals and in three languages in which they coined the word "allometry" (Huxley & Teissier, 1936a, 1936b, 1936c) in reference to the change in the shape of an organ because of the differential enlargement of every part in relation to its own law of growth. One of these papers was published in *Nature*. These were foundational papers since "allometry" gave its identity to a new field.

Contrary to Galton's and Pearson's biometrical study of regression, the formula used in relative growth makes no use of correlation coefficient *r*. In a first step the observed values for two variables, x and y, are log-transformed. In a second step the slope of the relation is defined as the ratio of the standard deviations of x and y (σ_x and σ_y). In other words, both the regression and the allometry lines correspond to equation:

$$\left(x - \overline{\overline{y}}\right) = b(x - \overline{x})$$

But in the case of regression, the value of b is given by:

$$by/x = r\frac{\sigma_y}{\sigma_x}$$

whereas in the case of allometry it is given by:

$$b = \frac{\sigma_y}{\sigma_x}$$

It can be shown that the allometry line is the diagonal of the rectangle (with sides parallel to the axes) that includes the ellipse of equal probability of data. It is not coincident with the major axis of this ellipse, unless the variances of x and y are equal, in which case the allometry line becomes an "isometry line", meaning that the shape is not changed with an increasing size. Since the allometry line (known at that time in France as "*la droite de Teissier*") does not use correlation, it can be applied to any pair of variables, including those belonging to different objects or organisms. As emphasized by Gayon (2016), Huxley had shown that what mattered in this relation of two log-transformed variables was not the ratio between two sizes, but the ratio between two growth-rates. Whereas Teissier's zoological studies were carried out on Hydrozoans, which are soft-bodied invertebrates, he conducted his biometrical studies on Crustacea, in which growth is continued during adulthood and on the carapace of which fixed points can easily be defined. The growth of the whole body was initially described as a vector of bivariate coefficients which Teissier and his student Bocquet called "equilibrium profiles". Later, Teissier used a single multivariate analysis, and defined the first eigenvector as "size". Teissier applied a relative growth approach to several questions including cell cultures and plant growth (Chouard & Teissier, 1932; Ephrussi & Teissier, 1932). His conception of critical points led Hélène Charniaux-Cotton (1957) to the androgen gland, a previously unknown organ in Crustacea.

An important step in the development of Teissier's interests in biometrics was a shift from organic growth to population growth that occurred at about the time he began studying *Drosophila* populations with L'Héritier in 1932. Several of the booklets published in his editorial collection in "biometry and biological statistics" were devoted to issues in population growth and species interactions (Gause, 1935; Kostitzin, 1934; Lotka, 1934, 1939; Volterra & D'Ancona, 1934). Pearl's (1928) studies in population dynamics influenced his work with L'Héritier on the dynamics of larval competition in *Drosophila* experimental population (see below). But Teissier also worked on population growth in collaboration with his brother-in-law, and another pioneer of genetics in France, Jacques Monod.

At the time, Monod was writing his doctoral thesis on the population growth of bacterial cultures on varying substrates. Fitting mathematical theory to experimental data led them to modify the Verhulst-Pearl logistic curve (Monod & Teissier, 1936). Morin and Monod (1942) made additional changes to this model. In his thesis, Monod wrote a chapter on the theoretical work in the context of his collaboration with Teissier and noted his influence upon his own work (Monod, 1942, pp. 116–124). The dynamics of bacterial culture growth led Monod to discover "diauxy", a new phenomenon that results from genetic regulation in enzyme synthesis after a change in culture medium. This work was important on the shift in Teissier's interests from an organismic to a population approach. It is significant that while his 1934 booklet (Teissier, 1934) dealt only with the exponential stage of growth, for which Δx is a positive function of size x, his 1937 book also included a development on the approach to the plateau for which Δx decreases as x increases and finally cancels, giving the logistic curve that characterizes the approach to equilibrium in populations. Both Monod and Teissier referred to "physiological" limiting factors, whereas in population studies such models involve the carrying capacity of the environment, a critical parameter in experiments carried out at the same time using *Drosophila* population boxes.

9.4 L'Héritier and Teissier's Experimental Population Genetics

L'Héritier and Teissier constructed a new experimental device which they called a "*démomètre*", also known as a "population box" (Fig. 9.1). The idea had been suggested in a book from Pearl (1928). This device allowed both researchers to let a population of several thousand *Drosophila* stand for several generations in a stable state through a daily renewal of rearing medium. Dozens of population boxes were simultaneously cared for in the laboratory. The biometrical background of Teissier allowed him to compare the relative growth of two populations, and there is evidence that the two associates first understood their system as a mean to compare different boxes, until a fortuitous contamination induced them to let free competition operate between two alleles within the same gene pool (Gayon & Veuille, 2001).

Initially, it could have been thought that a mutant *Drosophila* was at a selective disadvantage relative to the wild type and so would be eliminated, according to the classical model of a biallelic locus with fixed adaptive values for the three genotypic classes, 1, 1-hs and 1-s. This was not always the case. In one case, L'Héritier and Teissier observed the maintenance of a polymorphism, which they interpreted as a case of heterosis. In another case, they found no certain explanation, as results were more complex. This led them to pay more attention to frequency-dependent selection and to mathematical models of the maintenance of deleterious alleles in populations (Gayon & Veuille, 2001). It once happened that a new mutant, *sepia*, appeared in their population boxes and rose to a substantial equilibrium frequency. They also carried out experiments on interactions between loci, and remarked that in a fixed *sepia* background, mutant *ebony* did better against its wild type allele. Teissier knew that these mutants were poorly adaptive. He compared French and Japanese wild type strains of *D. melanogaster* and found that they differed in many biometrical characters, notably in the number of ovarioles. He maintained them in the same environment in the laboratory, to which they adapted, and found that they tended to converge towards the same biometrical values. This induced a long-term study that lasted for decades in Gif-sur-Yvette after WWII.

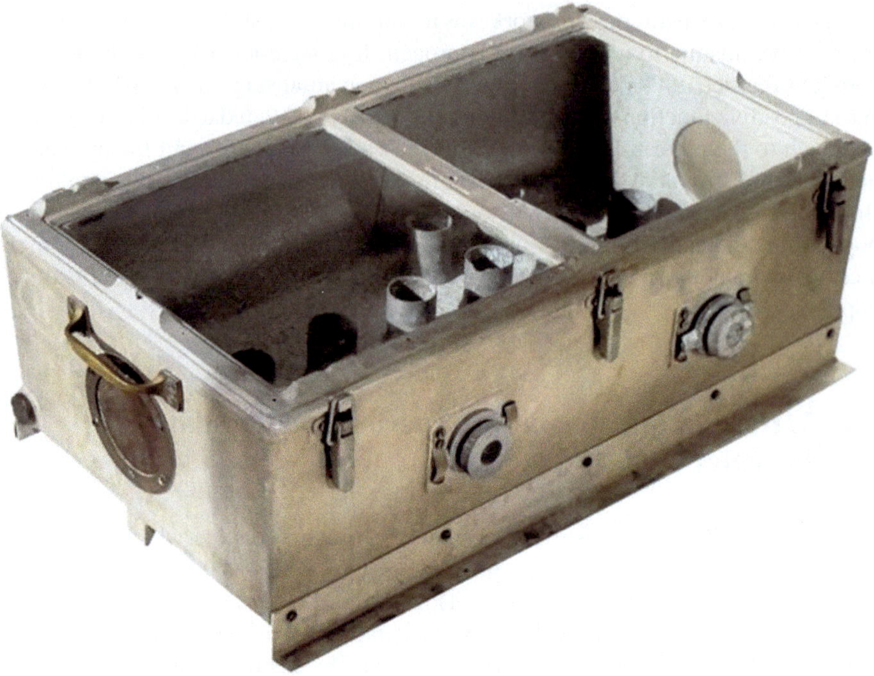

Fig. 9.1 L'Héritier and Teissier's population box ("*demomètre*"). The original model was made of wood; this model was made of aluminum after 1948; there are $7 \times 3 = 21$ pods in the bottom. Every day, a pod of old medium was replaced by a pod of fresh medium; the regular supply of food stabilized the population at a level of several thousand flies that could theoretically be maintained indefinitely; lateral extra pods were used for flash sampling the eggs laid at any moment; the collected eggs were reared in non-competitive conditions to determine allele frequencies; they provided an estimate of a fundamental parameter of theoretical population genetics: genotype frequencies at the zygote stage. (Credit: M. Veuille)

9.5 Investigating the Maintenance of Polymorphism

After the Teissier School had applied biometrical models to several population phenomena, the results of population box experiments made the maintenance of polymorphism its long-term research project. L'Héritier, who then worked in a different laboratory, decided to pursue his research on another of their common subjects: the genetics of interactions between fruit flies and viruses. On this ground Teissier established a school that helped establish the career of two generations of French geneticists.

Several famous historical debates on polymorphism in the USA and UK may lead one to see a "school" as putting forward some strong principle that its members favor for explaining genetic variation. Such schools drew attention, for instance, to the coexistence of alternative supergenes in some chromosomal regions (Ford), or to the coaptation of gene frequencies in interacting polymorphic systems (Wright), or to the benefit for the organism of being heterotic at many loci simultaneously (Dobzhansky, Wallace, Lerner). The Teissier school's approach to the "mechanisms of maintenance of genetic polymorphism", on the other hand, was eclectic. It observed that polymorphism was widespread in nature and that theory predicted several ways to establish stable polymorphic equilibria through natural or sexual selection. This school put forward several models to interpret its experiments in fruit flies on a gene-by-gene basis. In experimental devices, *ebony* was an instance of heterosis (L'Héritier & Teissier, 1937), *White* was an instance of rare-male effect (Petit, 1951), and *sepia* was an instance of frequency-dependence (a system that was further examined by a student of Claudine Petit, Anxolabéhère (1976). Teissier (1944) considered the role of lethal genes in the maintenance of balanced polymorphism. Neutrality was never considered a major factor, even though Gustave Malécot, a mathematician from ENS working in a nearby institution, was developing models in which chance, to be known later as random genetic drift, was shown to play an important part in the evolution of allele frequencies, along with selection (Veuille, 2019). Several genetic systems were examined by Teissier's students, using either laboratory experiments on fruit flies or field observations being carried out on two organisms, *Cepaea* land snails and *Sphaeroma* marine crustacea.

9.6 Teissier's Institutional Figure

World War II was an important period in Teissier's personal life,[7] but it was also important for his scientific career. First, L'Héritier and Teissier were separated, and their collaboration never resumed. Second, the war was detrimental for population box experiments, which were planned to run for years, and required daily care and had to be kept in constant environmental conditions. The experiments were interrupted and lost after the rout of the French and British armies in June 1940, and

[7] Many academics became Resistant when Pétain's collaboration regime took power in France in 1940; the communists joined them after the German invasion of Soviet Union in 1941 and set up the most active group of armed resistance, the FTPs ("*Franc-tireurs et Partisans*"). Marcel Prenant, a colleague of Teissier in the zoology department of ENS, and a notorious communist, became the head of the headquarters of FTPs. As an intellectual figure, he oversaw coordination between the communist and Gaullist resistances but was not in charge of armed groups. He recruited Teissier as his deputy, and, when caught by the nazis and deported to a concentration camp, was replaced by him. Teissier and the Gaullists succeeded in coordinating their action. Teissier became an official chief of united French resistance and co-signed the order of insurgency of Paris in August 1944, leading to the liberation of the city by the Resistance while the allied army was heading east after the Normandy landing. This action protected the city from destruction, counteracting Hitler's orders to destroy it. After WWII Teissier became a prominent leader of the French research system as the chairman of the CNRS (National Center of Scientific Research). But due to his political affiliation, pressures led to his resignation after the onset of the Cold War. He was also on a tightrope on the other side. While the Soviet Union was celebrating Lysenko, this overt French leader of scorned "Mendelism-Morganism" raised suspicion from the CP. Prenant, a zoologist, sided with geneticists and was expelled from the central committee of the party.

during the liberation of Paris in August 1944 (Teissier, 1958). Third, communication between L'Héritier and Teissier and their British and American colleagues in the early 1940s became difficult, while the Modern Synthesis was emerging. They viewed themselves as contributors to this synthesis, and Teissier resented the lack of recognition of their achievements. Their only visible contribution to the synthesis was in Timofeeff-Ressovsky's (1940) contribution to Huxley's "New Systematics", a foundational book of the Modern Synthesis. The paper presented a review of works in natural population genetics of animals by several research groups together with measures of genotypic viability values. It summed up several contributions to the Edinburgh International Congress of Genetics in 1939 and included a figure from L'Héritier and Teissier's work on the dynamics of genes *Bar* and *ebony* in population boxes. While the correspondence with Huxley involved only allometry, a connection had been established with Sewall Wright when Teissier had invited him to France in 1938 (Gayon, 2016). This was for mutual benefit. On the one hand, Wright brought the principle of population boxes to the United States, which allowed Wright and Dobzhansky (1946; Dobzhansky, 1947) to carry out such experiments on inversion polymorphisms. On the other hand, Teissier invited Wright to write a book summarizing his contribution to population genetics theory. This first book appeared about twenty years before Wright's magnum opus, published between 1969 and 1974. It was written in English and appeared in Paris just before the war (Wright, 1939). Even though it remained unknown in the UK and US due to the German occupation of France (Provine, 1971), it greatly influenced Teissier's school and Gustave Malécot's work.

After WWII, Teissier's health declined for the rest of his life but he remained a prominent member of the French scientific establishment in zoology.[8] He was an original figure due to his interests in biomathematics and his strong support for natural selection theory, then a minority stance, which made him appear as the leader of Darwinism in France. His continued involvement in systematics and in marine biology helped keeping him in contact with the French scientific community. In his official CVs, he always presented his research interests in the same order: (1) marine biology, (2) biometry, and (3) population genetics. These were also the three successive steps in his career. As the director of the CNRS (1946–1950), he had institutional power, which he used to transform this funding agency into a powerful research institution. He was simultaneously the head of three laboratories: the *Laboratoire de biométrie et génétique évolutive,* which he had built in Gif-sur-Yvette[9] (south of Paris) in 1951; the *Laboratoire de zoologie*, in Paris, which was initially located in the Sorbonne and later moved to a new scientific campus on the banks of the Seine river, where he spent most of his time; and the Roscoff Biological Station

[8] As of 1937, Ephrussi and Teissier became directors of laboratories at the EPHE, which was unusual, if not unprecedented, for simple assistant professors, meaning that their work was fully acknowledged. During WWII Teissier replaced Robert Levy both as professor and director of the ENS zoology department after his dismissal due to Pétain's so-called "laws on Jews" that barred Jews from teaching positions.

[9] This laboratory belonged to a group of three laboratories built while he was chairman of the CNRS. They were the first three genetic institutes ever opened in France (Burian, Gayon and Zallen 1988).

in Britany, the largest field laboratory in France, which he had enlarged by important works, and where he went in summer. Teissier also founded a journal, the *Cahiers de biologie marine*, edited in Roscoff, while Charles Bocquet, one of his students, was the editor of the *Archives de zoologie expérimentale et générale*, a generalist journal. Bocquet, like Teissier, was both a geneticist and an excellent zoologist, and had trained like him at the ENS. These two periodicals strengthened the position of the school in the scientific establishment. They also published the contributions of the population genetics school, with the negative consequence that these works remained rather unknown to other population geneticists worldwide. Teissier was one of the two most important members of French zoology, together with Pierre-Paul Grassé, a staunch neo-Lamarckian and his lifelong rival. Teissier was influential in marine biology, whereas Grassé's strongholds were in protistology, animal behaviour, and general zoology, in which he coordinated a prestigious encyclopedia of 17 volumes, "*Le traité de zoologie*". A festschrift attracted many contributors to celebrate the achievements of Teissier's career. It led to the publication of a big book by the Société Zoologique de France (Lamotte, 1974). The title of this book shows that Teissier's name was associated with Darwinism and variation, although its association with "animals", as though genetic variation was irrelevant in plants and bacteria, shows how much remained to be done for a full understanding of evolutionary issues in French zoology. The book mostly consisted of a survey of cases of polymorphism in a series of biological models in a variety of zoological orders taken one after another.

This school played a prominent role in teaching. It published the first textbooks of genetics (L'Héritier, 1949a) and population genetics (L'Héritier, 1949b) in France, together with the first manuals of biological statistics (L'Héritier, 1949c; Lamotte, 1957) (Fig. 9.2).

Gayon (1999) characterizes a "school" after Morrell (1979) from a series of properties. A school is financially autonomous; it develops its own style for materials and methods; it develops networks for recruiting students and for reaching out to a readership of its own through publications. This suggests that publishing is filtered by shared views on relevant issues and methods at a given time, which can only reinforce the influence of schools. A "school" thus involves communication and cooperation. It is clear that what Gayon called "Teissier's school" meets all the conditions he assigns to a school. Nothing is said of the duration of a school, however, and I will assume below that a historical account has to consider the development of a school from its origin to its end. It should also be noticed that within France, Teissier's school did not so much confront other schools from the same discipline but struggled for academic power against a background of neo-Lamarckian inertia.

9.7 Teissier's Students

Teissier had many doctoral students in zoology, four of whom were population geneticists who had joined his laboratory in the 1940–1950s. He asked the four of them to prepare a thesis on the maintenance of polymorphism. They in turn had students who

Fig. 9.2 Left to right: Helen Spurway, Georges Teissier and J. B. S. Haldane in the Zoology laboratory of la Sorbonne, Paris, in the 1950s (Credit: courtesy Françoise Benhamou)

prepared a PhD degree on topics from this school before switching to other questions, so on this account Teissier's school lasted two and a half academic generations. His doctoral students were only loosely supervised, and thus were free to conduct their research as they wanted.

The first was Maxime Lamotte (1920–2007), who worked in Teissier's first laboratory at the ENS. His study of frequency variation in the shell banding pattern of land snail *Cepaea nemoralis* (Lamotte, 1951) has remained famous for having been the focus of the first neutralist-selectionist debate, culminating at the 1959 Cold Spring Harbor meeting convened for the celebration of the hundredth anniversary of Darwin's *On the Origin of Species*. Lamotte's results (Lamotte, 1959) were presented before the main figures of the Modern Synthesis (Wright, Dobzhansky, Mayr), and were criticized by Wright as trivial, since involving no interactions between loci (Demerec, 1959, p. 86). This is also the only occasion that results from the French school were criticized by a member of Fisher's school, since its conclusions were opposed to those arrived at by Sheppard, from Ford's laboratory. Lamotte wanted to assess Wright's (1932) hypotheses on the interplay of selection and errors of sampling.[10] He chose an organism living in very small patchy colonies, whose polymorphism was considered a challenge for selective explanations (e.g. Haldane, 1932).

[10] "Errors of sampling" was Wright's initial expression for random genetic drift (Veuille 2019); the French school called it the "Wright effect" ("*l'effet Wright*"), in line with Fisher school's expression (Fisher & Ford 1950).

The neutralist stance of his interpretation was emphasized by Lamotte's collaboration with Gustave Malécot in the handling of quantitative data. Working independently from L'Héritier and Teissier, Malécot had initiated as of 1936 a purely mathematical approach to population genetics. He had a strong neutralist stance, much in advance of what would be called the "neutral theory" of the 1960–1970s. There was a deep lack of mutual understanding between the two groups, despite their being established within a stone's throw of each other in Paris's "Quartier Latin", and their being strongly influenced initially by the same British theoretician, first Ronald Fisher, and later Sewall Wright. This lack of communication resulted from their disagreement on the respective role of chance sampling and Darwinian selection in the evolution of gene frequencies. Malécot worked mostly alone; he now appears to have been ahead of his time, his influence in population genetics having been re-evaluated many times (Crow, 1995; Ishida, 2009; Ishida & Rosales, 2019; Nagylaki, 1989, Slatkin & Veuille, 2002, Veuille, 2019). Attention was drawn to the historical importance of the collaboration between Lamotte and Malécot many years later by Provine (1971) and Millstein (2008, 2009). After his PhD, Lamotte succeeded Teissier as a professor of zoology at the ENS. His initial vocation had always been that of an ecologist in African ecosystems. The International Biological Program (1964–1974) offered him an opportunity to establish a field research station in Lamto, Ivory Coast (Abbadie et al., 2005; Lachenal, 2015), where several researchers were invited to study the role of whole functional groups (termites, earthworms, fruit flies, etc.) and bio-geo-chemical cycles. He converted the laboratory of zoology of the ENS into a laboratory of ecology. Lamotte never returned to population genetics and played an important role in the development of ecosystem research in France in the 1960–1970.

The second of Teissier's population genetics students was Charles Bocquet (1918–1977) who, like Teissier and Lamotte, had trained at the ENS (Fig. 9.3). He prepared his doctoral thesis on marine crustacea in Roscoff. Teissier initially recruited him for carrying out a biometrical study in decapods using multivariate analysis, which was then a rising method. He also included him in a long-term field study with three other collaborators on the color polymorphism of isopod crustacea *Sphaeroma serratum*. This species was abundant on the Atlantic shore, and Teissier wanted to study isolation by distance based on Wright's (1939) finding that population differentiation is larger in one dimension (the coastline) than in two dimensions (the continent) as was the case for land snails. This study lasted decades and resulted in many publications providing the result of each annual collecting campaign. No synthesis of the results was ever published due to the deaths of Teissier in 1972 and Bocquet in 1977. Furthermore, the discovery of allozyme polymorphism (Harris, 1966; Hubby & Lewontin, 1966) had rendered this system obsolete. The polychromatism of these crustacea involved an epistatic series of loci, with the consequence that some genotypes remained unknown, so allele frequencies were not derived directly from phenotype counts but had to be inferred from pooled phenotypic classes assuming neutrality. Another reason that this work lasted so long is that results were puzzling. High levels of divergence were observed along the coast, even at a short distance, suggesting that local differentiation was due to genetic drift or founding events. But frequencies were locally very stable over the years, even over decades, suggesting very large population sizes (Teissier & Bocquet, 1960). Bocquet's interests switched to another polychromatic crustacean isopod which was abundant along the coasts of Britany, *jaera albifrons*. It proved to be a complex of five sympatric sibling species, three of which coexisted under pebbles of the intertidal zone in front of the Roscoff station. They perfectly illustrated the paradoxes of speciation as viewed by the Modern Synthesis, since they remained reproductively isolated on the shore outside the laboratory but yielded perfectly fertile hybrid progeny inside the laboratory (Bocquet, 1953). They shared an abundant color polymorphism, which Bocquet interpreted using two different rationales. Firstly, he said that this illustrated "Vavilov's law", according to which related species tend to show similar mutations. Second, he believed that the numerous cases of color polymorphism observed in coastal Crustacea (including *Sphaeroma, Jaera*, and other species) illustrated the fact that species living in the intertidal zone needed genetic variation to cope with a highly variable environment, although he had no causal relationship to offer for this effect. The *Jaera* model led Bocquet to redirect his research towards speciation. This initiated long-term interactions with Ernst Mayr, who visited him several times (Fig. 9.3). Bocquet directed for a time a small marine laboratory in Luc-sur-Mer, Normandy. In 1967 he succeeded the neo-Lamarckian Grassé in the chair of Évolution des êtres organisés in Paris. It was the first time since the creation of this prestigious chair in 1888 (Viré, 1979) that the holder was not a neo-Lamarckian. He also succeeded Teissier as the director of the Gif-sur-Yvette laboratory. He attracted many students at Luc-sur-Mer and Gif-sur-Yvette to continue his work on speciation in *Jaera*. In Gif he extended the topic to the search for the origin of *Drosophila melanogaster* and *D. simulans*, which were then only known to live in "unnatural" environments (Dobzhansky, 1965). With the help of systematist Leonidas Tsacas and a group of young evolutionary biologists, the laboratory discovered the African origin of these species and showed they belonged of a group of nine sibling species, all originating from Africa (Aulard, 1990; Baudry et al., 2004, 2006; Lachaise et al., 1981, 1988, 2000; Lemeunier et al., 1986; Tsacas & Lachaise, 1974). This species complex later became a major biological model in evolutionary genetics and genomics.

The third of Teissier's students in population genetics was Claudine Petit (1920–2007), whom he met as a young resistant after WWII.[11] She put forward the principle of "rare male effect" in sexual selection as a mechanism explaining the maintenance of the polymorphism of the white allele (a deleterious mutant) of *D. melanogaster* in laboratory cultures (Petit, 1951). Gayon (2015) showed the importance of this mechanism in debates on frequency-dependent selection prior to 1966. He also showed that Petit's vision of sexual selection was initially influenced by Dobzhansky's concept of "sexual isolation" in the context of speciation. She progressively developed her own interpretation in an intraspecific context that was closer to Darwin's views. Her study was eventually extended to *Drosophila pseudoobscura* by Ehrman (1967), from Dobzhansky's group. This model rested on the study of individual genes. Its

[11] She had been sentenced to 20 years in prison by a French court under Pétain's regime, then sentenced to death *in absentia* by the Germans, but she escaped and hid safely till the end of the war.

Fig. 9.3 Charles Bocquet (left) and Ernst Mayr on the beach in Normandy in the 1960s, presumably at *Les vaches noires*, a famous coastal geological site close to the Luc-sur-Mer biological station then directed by Bocquet (Credit: Courtesy laboratory EGCE, Gif-sur-Yvette)

relative importance was reassessed after the discovery of allozymes. Lewontin wrote: "A generalized intrapopulational advantage of rare males with respect to almost any genotype could not exist, if for no other reason than that every male is genotypically rare in a population with 10 percent heterozygosity and 40 percent polymorphism" (1974, p. 259). Her conclusions were also criticized by members of her own laboratory. Anxolabéhère et al. (1982) showed that her findings were flawed by the fact that her statistics did not account for the bias introduced by changes in the relative frequency of competing individuals across experiments.

The fourth of Teissier's students was Ernest Boesiger (1914–1975), who also prepared a PhD on the maintenance of genetic polymorphism by sexual selection in *Drosophila.* Instead of working, like Petit, on laboratory mutants, he worked on the natural variation of wild type flies by relating the reproductive success of males to their heterozygosity level (Boesiger, 1962). Boesiger was intellectually closer to Dobzhansky's balance school than to Teissier's maintenance of polymorphism school. Later on, he set up a quail rearing facility in order to study genetic homeostasis using means inspired by Lerner (1954). He wrote a book with Dobzhansky, *Essai sur l'évolution* (Dobzhansky and Boesiger, 1968), putting forward the main themes of the balance school for a French readership. They claimed that fitness depended on a high level of average heterozygosity, thus counteracting the effects of genetic load. This book could have raised fruitful debates among French population geneticists, yet it had little success. It was a strange association of results in *Drosophila* and reflections on mechanism and vitalism. It was also issued in 1968, the year when French universities were either empty or occupied by police since students had taken to the streets, so laboratory life was suspended. Moreover, the authors published the book in a collection of which Pierre-Paul Grassé was the editor, and they had invited him to write the foreword. Grassé was the last, long despised, and acerbic neo-Lamarckian who had struggled for decades against Darwinism in a slowly vanishing Lamarckian France, and this foreword was felt as a betrayal by the Darwinian camp. Unfortunately for the authors, Grassé's foreword denigrated the book it was intended to introduce, and denounced it as being one-sided. The book appeared after allozyme techniques had displaced the focus of attention onto a broader appraisal of polymorphism. Finally, this was the year of Kimura's (1968) neutral theory. Both Boesiger and Dobzhansky died in 1975 before answering these developments.

9.8 Successes and Weaknesses of Teissier's School

Teissier's school lasted from 1932 to 1977. Then it vanished in a rising generation that did not identify itself with its main orientations. Below I suggest a rough outline of developments that deserves a more thorough study: how did this school end, and why? It shows a bright side and a dark side. Until 1966, the results were highly topical and were followed by foreign laboratories. Wright, Dobzhansky, Haldane and Mayr visited Teissier's laboratory. His school was innovative in designing the first experimental device—the "*démomètre*"—that became a paradigm for the study of allelic replacement under natural selection. It was among the first laboratories to demonstrate empirical cases of the maintenance of genetic polymorphism, and to link them to theoretical models. One of its members elicited the first neutralist-selectionist debate. Finally, a new development in speciation studies led this school to provide the global scientific community with one of the most important research models in genetics: the melanogaster subgroup of *Drosophila*, which has ever since been extensively used in comparative genomic studies.

At a smaller scale the bright side also involves its effect upon the French scientific community, since "Teissier's school" must not be considered only with respect to other schools worldwide, but also as the Darwinian school in the oppressive Lamarckian environment of France. The Darwinians and the neo-Lamarckians united forces only in 1937, when the director of the Muséum National d'Histoire Naturelle de Paris, Paul Lemoine, wrote in an official publication of the government, *L'Encyclopédie française*, that evolution was an outdated hypothesis that was no longer used by scientists. In other periods they struggled against each other. An important conference was convened for the retirement of Grassé in 1967 (see the special issue of *L'Année biologique* Sept-Oct. 1967), at which the Darwinians and the neo-Lamarckians were given equal time and space. Teissier was famous for his rallying motto: "*Nous ne dédarwinisons pas*"![12] This context needs to be accounted for to understand the attention drawn by an experiment that at first may look like a hoax (L'Héritier et al., 1937). It consisted in lifting the lid of a population box for a given time every day and for several generations in a windy area of Roscoff, whereby winged flies were carried away by wind, whereas the wingless flies bearing allele *vestigial* were at an advantage, ending up in a flightless population. This was supposed to mimic the evolution of flightlessness in insects found in oceanic islands, which Darwin interpreted as the result of natural selection against dispersion (Gayon & Veuille, 2001; Gayon, 2014). This widely echoed experiment was in fact contradicting the neo-Lamarckian interpretation of organ degeneration in terms of use and disuse. On another occasion, L'Héritier and Teissier, although appearing as the champions of Mendelism in France, found a case of non-Mendelian transmission in a hereditary character: the sensitivity of flies to CO_2. At a time when non-Mendelian inheritance was still an oddity, they published their observations (L'Héritier & Teissier, 1937), and many other papers followed, even though they had no interpretation to put forward for a long time. This remained an enigma for years until L'Héritier finally showed that this anomaly was caused by an intracellular virus. Thus, an apparently non-Mendelian phenomenon eventually proved to be a confirmation of genetics. This showed that its authors were open to reality beyond their scientific convictions. Likewise, both L'Héritier's and Petit's groups came across transposable elements, which were an unknown form of transmissible DNA, during their genetic investigations (Picard & L'Héritier, 1971; Périquet, 1979). Today we know that not all chromosomal material follows Mendel's laws. Sex is a specific outcome of DNA evolution in part of the eukaryotic genome, and we know that most evolutionary time was spent in bacteria, which have different recombination systems. But Mendelian inheritance was once considered the only basis of heredity. Contradicting it was felt as paving the way for Lamarckism.[13]

As to the dark side of this school, it was lacking in terms of communication strategy. The German occupation of France by the Nazis, while the Modern Synthesis was emerging in the USA and UK, was certainly a handicap. But this school had no reason to publish its results exclusively in French and in the zoological journals controlled by its members. The only papers published in English by Teissier were his joint paper with Huxley and his contribution to the 1956 International Congress of Genetics in Tokyo. Bocquet had an abridged version of his thesis translated into English by Mayr for Systematic Zoology.

In 1948, Teissier was invited to speak before the International Congress of Biometrics, which he chaired. He spoke in French and published his contribution in *Biometrics* in French. Being invited to write a paper for the first issue of *Evolution*, he

[12] "We shall not dedarwinize"!

[13] Loison (2013) mentions that in his non-scientific writings, including in his academic CVs, Teissier said he believed that some form of cytoplasmic inheritance could reconcile microevolution and macroevolution. Loison adds that Teissier had no time to develop these views in academic journals or in a book. It would be worthwhile to examine this project in the specific context of debates about evolutionary mechanisms in France.

declined the honor, since he was not allowed to publish it in French (Gayon & Veuille, 2001).

All of Teissier's first generation students published their results in French. Lamotte spoke in French before the Cold Spring Harbor meeting in 1959. He had his contribution translated into English by the organizer before publication. For the third generation of the school, no paper was submitted to an English journal until Lemeunier's joint paper with Ashburner, a Cambridge professor, in the *Proceedings of the Royal Society* (Ashburner & Lemeunier, 1976). The first research paper written by a French member of this school appeared in the 1976.[14] Tenured researchers from Bocquet's laboratory published in English only after his death in 1977. It should be remarked that most French biologists published their results in French, including Nobel Prize laureates Jacob and Monod. Therefore, this is probably not a sufficient explanation to account for the progressive isolation of Teissier's school.

While the beginnings of Teissier's school owed so much to L'Héritier's one-year stay in USA in 1932, they almost never visited foreign countries again. Two population geneticists made trips to the UK while preparing their thesis (Malécot to visit Fisher, Bocquet to see Haldane). Teissier, who had confessed to Huxley that his English was poor, never went abroad, except for the International Congress of Genetics in Tokyo. There he paid a visit to the Emperor of Japan, Hirohito, who was passionate about the same group of marine invertebrates as him, hydrozoans, and the emperor returned the visit privately some years later on the occasion of an official trip to France. The first students of this school who went to the United States as postdocs were Anxolabéhère, Périquet and Veuille in the early to mid-1980s. The self-isolation of this school is well illustrated by the fact that when Dobzhansky visited Teissier's laboratory in the 1960s, Teissier did not show up. Dobzhansky, however, talked with Claudine Petit and Boesiger and got a sense of what was going on in Paris in polymorphism studies (Gayon & Veuille, 2001). Both maintained relationships with American colleagues.

Another weakness was the lack of generality of the issues raised by Teissier's school after WWII. This is clearly illustrated by the contents of the Festschrift book edited by Lamotte (1974), which consisted of a catalog of specific studies across zoological groups with absolutely no theoretical common thread, as though his favorite students were not connected together by a common heritage. Only two contributions in the book addressed the general question of the cause of polymorphism. This book illustrates the time when genes were known only from crosses, and therefore were highly linked to model species. Each research material needed a specific expertise to be developed, and the questions raised were those offered by contingent gene polymorphisms. After 1966, the members of this school did not realize the wealth of information provided by widespread Mendelian loci like allozymes to answer general questions.

---

[14] As a graduate student, I was rebuked by Bocquet for submitting a manuscript in English to a Canadian journal. I only then found out that I had crossed a line that no researcher in the laboratory had ever crossed.

Finally, another weakness of this school was a lack of cohesion. Except for Petit and Bocquet, who had some consideration for each other, Teissier's pupils showed little mutual support. Boesiger's approach, based on Dobzhansky's balance theory, remained obscure to all. They were ready to admit that existing polymorphic systems were the result of contingent properties that allowed each allelic system to survive, but they could not even consider that heterozygosity itself was a primary parameter of genomic evolution. They were unaware of the debates on genetic load going on in the United States. As a result, Boesiger, whose group was hosted by the Gif laboratory, was marginalized. Lamotte had an influent institutional position in ecology with the opening of an ecology section in the CNRS. And as the chairman of the audit committee of Bocquet's laboratory, he dismissed his research projects, and obtained the termination of the laboratory after Bocquet's death, the following year. The laboratory was kept afloat thanks to a year of union action led by the junior member of the laboratory. At that time the school hosted several young researchers, but none of them was sufficiently advanced in their career to take over the direction of the laboratory. The direction of the Gif laboratory was handed over to a newcomer in evolutionary genetics who had prepared a PhD in vain search of Lamarckian heredity in *Drosophila* (David, 1961) and had eventually rallied Mendelism. He symbolized the new course of French evolutionary research. In the 1980s, the Gif and Paris laboratories progressively dropped the themes of Teissier's school. The researchers from Teissier's school who studied the interplay of selection and polymorphism at the molecular level did it on renewed themes (Montchamp-Moreau et al., 2001; Solignac et al., 1986; Veuille & King, 1995; Veuille et al., 1998).

9.9 The Beginnings of the French Community in Population Genetics

Although Teissier's school was the most structured element of French population genetics, it was not the only component of French research in genetics. Mendelism had been taught in horticultural research as of the organization of the fourth International Congress of Genetics in Paris in 1911, and the main agricultural school (Institut National Agronomique, INA) had since taught genetics. A professor at the INA and former attendant of L'Héritier's teaching, Georges Valdeyron, taught population genetics to botanists (Valdeyron, 1961), and a group of students in agricultural research (Gachet et al., 1961) circulated a French translation of those chapters from Falconer's (1961) Introduction to Quantitative Genetics that "did not duplicate the courses of PRs Teissier and L'Héritier". Later, these researchers followed Malécot's teaching (Gillois, 2002) together with French researchers in human genetics, of whom Jacquard (1974, 1977) was the most prominent figure. Valdeyron was interested in the evolution of sex and reproductive strategies in plants and worked in an ecology institute in Montpellier. There he supervised several evolutionary botanists who established collaborations with John Maynard-Smith, including Isabelle Olivieri,

Pierre-Henri Gouyon, Finn Kjellberg, Myriam Valéro and others. This was another source for French evolutionary biology. In Montpellier another group initially set up by paleontologist René Lavocat led to a new research group involving Nicole Pasteur, a mosquito geneticist who had worked in Dobzhansky's laboratory, and Louis Thaler, a paleontologist who had worked in Simpson's laboratory (Michaux, 2018a, 2018b). Thus the new generation of Teissier's students could join their counterparts in other laboratories. The development of allozymes as genetic markers that was occurring at that time contributed to unite the field of evolutionary genetics and other population studies (Pasteur et al., 1987). Following the succession crisis in Gif, two meetings were organized in turn by laboratories from Teissier's school (Paris in 1979 and Gif in 1980). They were the first two of a series of annual meetings of French population geneticists, the so-called PPD,[15] which was directly inspired by the model of the Population Genetics Group (PGG) in the UK, for about forty years. This forum circulated scientific information across French universities. This was the end of Teissier's school and the beginning of a new community which renewed the themes of population genetics in France.

Acknowledgements I thank Claude Lévi for discussions about the Teissier years. I owe much to the late Lily Joly for help and discussion. I thank Jean-Baptiste Grodwohl for comments on an earlier version of this paper. I thank Françoise Benhamou, Jean Luc Da Lage, Catherine Montchamp-Moreau and Laure Kaiser-Arnauld for iconography. I thank the documentation department of EPHE for keeping the Teissier archives.

References

Abbadie, L., Gignoux, J., Le Roux, X., & Lepage, M. (Eds.). (2005). *Lamto: structure, functioning and dynamics of a savanna ecosystem.* Springer.

Anxolabéhère, D. (1976). Heterosis overdominance and frequency-dependent selection in *Drosophila melanogaster* at the *sepia* locus. *Evolution, 30*, 523–534.

Anxolabéhère, D., Goux, J.-M., & Périquet, G. (1982). A bias in estimation of variabilities from competition experiments. *Heredity, 48*, 271–282.

Ashburner, M., & Lemeunier, F. (1976). Relationships within the melanogaster species subgroup of the genus Drosophila (Sophophora). I. Inversion polymorphisms in *Drosophila melanogaster* and *Drosophila simulans*. *Proceedings of the Royal Society B, 193*, 137–157.

Aulard, S. (1990). Polymorphisme chromosomique de *Drosophila melanogaster* en Afrique et dans les iles de l'Océan Indien. Paris: Doctoral Thesis, P. and M. Curie University.

Baudry, E., Viginier, B., & Veuille, M. (2004). Non-African populations of *Drosophila melanogaster* have a unique origin. *Molecular Biology and Evolution, 21*, 1482–1891.

Baudry, E., Derome, N., Huet, M., & Veuille, M. (2006). Contrasted polymorphism patterns in a large sample of populations from the evolutionary genetics model *Drosophila simulans*. *Genetics, 173*, 759–767.

Bernard, C. (1865). *Introduction à l'étude de la médecine expérimentale.* Baillière & fils.

Bernard, C. (1867). *Rapport sur les progrès et la marche de la physiologie générale en France.* Imprimerie impériale: 216.

[15] PPD: "*petit-pois déridé*" ("unwrinkled pea") is tribute to Mendel, but also means "laughing pea", a deliberate joke typical of the relaxed and humorous mood of this new period.

Blanckaert, Cl. (2009). *De la race à l'évolution. Paul Broca et l'anthropologie française (1850–1900).* L'Harmattan.

Blanckaert, C., Diara, A., Dougherty, F., Fisher J.-L. et al. (Eds.). (1979). *Les néo-lamarckiens français. Revue de synthèse*, 95–96.

Blaringhem, L. (1928). *Principes et formules de l'hérédité mendélienne.* Gauthier-Villard et Cie.

Bocquet, C. (1953). Recherches sur le polymorphisme naturel des *Jaera marina* (Fabr.). (Isopodes Asellotes). Essai de systématique évolutive. *Archives de zoologie expérimentale et générale, 90*, 187–450.

Boesiger, E. (1962). Sur le degré d'hérétozygosité des populations naturelles de Drosophila melanogaster et son maintien par la sélection sexuelle. *Bulletin biologique de France et de Belgique, 96*, 3–122.

Boesiger, E. (1980). Evolutionary biology in France at the time of the evolutionary synthesis. In E. Mayr & W. B. Provine (Eds.), *The evolutionary synthesis. Perspectives on the unification of biology* (pp. 309–321). Harvard University Press.

Bœuf, F. (1936). *Bases scientifiques de l'amélioration des plantes.* Lechevalier.

Broca, P. (1867). *Congrès international d'anthropologie et d'archéologie préhistorique*, 2e session.

Burian, R., & Gayon, J. (1999). The French school of genetics: From physiological and population genetics to regulatory molecular genetics. *Annual Review of Genetics, 33*, 313–349.

Burian, R., Gayon, J., & Zallen, D. (1988). The singular fate of Genetics in the History of French Biology, 1900–1940. *Journal of the History of Biology, 21*, 357–402.

Charniaux-Cotton H., (1957). *Croissance, régénération et déterminisme endocrinien des caractères sexuels secondaires d'Orchestia gammarella (Pallas). Crustacé amphipode.* Thèse.

Chouard, P., & Teissier, G. (1932). Variation de l'intensité de croissance chez les plantules de melon au cours du développement et en fonction de la quantité de réserves disponibles. *Comptes rendus de l'Académie des sciences, série des sciences de la vie, 194*, 1976–1978.

Conry, Y. (1974). *L'introduction du darwinisme en France au XIXe siècle.* Vrin.

Crow, J. F. (1995). Motoo Kimura (1924–1994). *Genetics, 140*, 1–5.

Darwin, C. (1859). *The origin of species by means of natural selection.* John Murray.

David, J. (1961). Influence de l'état physiologique des parents sur les caractères des descendants; étude chez *Drosophila melanogaster* Meigen. Thèse.

Delage, Y., & Goldsmith, M. (1909). *Les Théories de l'évolution.* Flammarion.

Demerec, M. (Ed.) (1959). Genetics and the twentieth century Darwinism. *Cold Spring Harbor Symposia on Quantitative Biology, 24.*

Dobzhansky, T. (1947). Adaptive changes induced by natural selection in wild populations of Drosophila. *Evolution, 1*, 1–16.

Dobzhansky, T. (1965). "Wild" and "domestic" species of Drosophila. In H. G. Baker & G. L. Stebbins (Eds.), *The genetics of colonizing species* (pp. 533–546). Academic Press.

Dobzhansky, T., & Boesiger, E. (1968). *Essai sur l'évolution.* Masson.

Duval, M. (1886). *Le darwinisme.* Delahaye & Lecrosnier.

Ehrman, L. (1967). Further studies on genotype frequency and mating success in Drosophila. *American Naturalist, 101*, 415–424.

Ephrussi, B., & Teissier, G. (1932). Étude quantitative de la croissance des tissus. I. *La Croissance Résiduelle. Arch. Exp. Zellforschung, 13*, 1–29.

Falconer, D. S. (1961). *Introduction to quantitative genetics.* Oliver & Boyd.

Fisher, R. A. (1930). *The genetical theory of natural selection.* Clarendon Press.

Fisher, R. A., & Ford, E. B. (1950). The "Sewall Wright Effect." *Heredity, 4*, 117–119.

Fournier, P. (1980). *La terre et la vie : un siècle de recherches.* Paris: Publications de la 3e Section de l'École Pratique des Hautes Études.

Gachet, G., Genier, G., Huguet, L. (1961). Traduction partielle et simplifiée de 'Introduction à la génétique quantitative' par D. S. Falconer. Hand-out, date unknown posterior to 1961.

Gause, G. F. (1935). *Vérifications expérimentales de la théorie mathématique de la lutte pour la vie.* Hermann & Cie.

Gayon, J. (1992). *Darwin et l'après-Darwin. Une histoire de l'hypothèse de sélection dans la théorie de l'évolution*. Kimé.
Gayon, J. (1995). Les premiers pastoriens et l'hérédité. *Bulletin de la société d'histoire et d'épistémologie des sciences de la vie, 2*, 193–204.
Gayon, J. (1998). *Darwinism's struggle for survival*. Cambridge University Press.
Gayon, J. (1999). On the uses of the category of style in the history of science. *Philosophy and Rhetoric, 32*, 233–246.
Gayon, J. (2006). L'hérédité des caractères acquis : origines d'un terme. In P. Corsi, J. Gayon, G. Gohau, & S. Tirard (Eds.), *Lamarck, philosophe de la nature* (pp. 105–163). Presses universitaires de France.
Gayon, J. (2013). Claude Bernard et l'hérédité. In F. Duchesneau, J.-J. Kupiec, & M. Morange (Eds.), *Claude Bernard. La méthode de la physiologie* (pp. 115–132). Éditions rue d'Ulm.
Gayon, J. (2014). Une expérience de sélection naturelle en conditions naturelles : L'Héritier, Neefs et Teissier, 1937. In L. Loison (Ed.), *Le laboratoire CNRS de génétique évolutive de Gif. De part et d'autre de l'œuvre de Georges Tessier* (pp. 65–80). Hermann.
Gayon, J. (2015). Sexual selection in the French school of population genetics: Claudine Petit (1920–2007). In T. Hoquet (Ed.), *Current perspectives on sexual selection. What's left after Darwin*? (pp. 65–81). Springer.
Gayon, J. (2016). Biométrie et génétique des populations. Échanges franco-britanniques dans l'entre-deux guerres. *Bulletin d'histoire et d'épistémologie des sciences de la vie, 23*, 7–26.
Gayon, J., & Zallen, D. (1998). The role of the Vilmorin company in the promotion and diffusion of the experimental science of heredity in France, 1840–1920. *Journal of the History of Biology, 31*, 241–262.
Gayon, J., & Burian, R. (2000). France in the era of Mendelism. *Comptes-rendus de l'Académie des sciences, série des sciences de la vie, 323*, 1097–1106.
Gayon, J., & Veuille, M. (2001). The genetics of experimental populations: L'Héritier and Teissier's population cages (1932–1954). In R. Singh et al. (Eds.), *Thinking about evolution. Historical, philosophical & political perspectives* (pp. 77–102). Cambridge University Press.
Gayon, J., & Burian, R. (2004). National traditions and the emergence of genetics: The French example. *Nature Reviews Genetics, 5*, 150–156.
Gillois, M. (2002). The scientific work of Gustave Malécot (1911–1998); our common heritage. In M. Slatkin & M. Veuille (Eds.), *Modern developments in theoretical population genetics. The legacy of Gustave Malécot* (pp. 7–19). Oxford University Press.
Haeckel, E. (1874). *Histoire de la création des êtres organisés d'après les lois naturelles*. Traduction par le Dr. Ch. Létourneau, introduction par Charles Martins. Reinwald.
Haldane, J. B. S. (1932). *The causes of evolution*. Longmans Green.
Harris, H. (1966). Enzyme polymorphism in man. *Proceedings of the Royal Society of London, B: Biological Sciences, 164*, 298–310.
Henriet, P., & Veuille, M. (2018). Les origines. In P. Henriet (Ed.), *École Pratique des Hautes Études: invention, érudition, innovation, de 1868 à nos jours* (pp. 25–45). Somogy.
Hubby, J. L., & Lewontin, R. C. (1966). A molecular approach to the study of genic heterozygosity in natural populations. I. The number of alleles at different loci in *Drosophila pseudoobscura*. *Genetics, 54*, 546–595.
Huxley, J. S. (1924). Constant differential growth ratios ad their significance. *Nature, 114*, 895–896.
Huxley, J.-S. (1932). *Problems in relative growth*. Lincoln Mac Veagh, The Dial Press.
Huxley, J. S., & Teissier, G. (1936a). Terminology of relative growth. *Nature, 137*, 780–781.
Huxley, J.-S., & Teissier, G. (1936b). Terminologie et notation dans la croissance relative. *Comptes-Rendus Hebdomadaires De La Société De Biologie, 121*, 934–937.
Huxley, J.-S., & Teissier, G. (1936c). Zur Terminologie des relativen Grössenwachtums. *Biologisches Zentralblatt, 56*, 381–383.
Ishida, Y. (2009). Sewall Wright and Gustave Malécot on isolation by distance. *Philosophy of Science, 76*, 784–796.

Ishida, Y., & Rosales, A. (2019). The origins of the stochastic theory of population genetics: the Wright-Fisher model. *Studies in History and Philosophy of Biological and Biomedical Sciences, 79.*

Jacquard, A. (1974). *The genetic structure of populations.* Springer.

Jacquard, A. (1977). *Concepts en génétique des populations.* Masson.

Kimura, M. (1968). Evolutionary rate at the molecular level. *Nature, 217*, 624–626.

Kostitzin, V. A. (1934). *Symbiose, parasitisme et évolution (étude mathématique).* Hermann.

Lachaise, D., Lemeunier, F., & Veuille, M. (1981). Clinal variation in male genitalia in *Drosophila teissieri* Tsacas. *The American Naturalist, 117*, 600–608.

Lachaise, D., Cariou, M.-L., David, J. R., Lemeunier, F., & Tsacas, L. (1988). Historical biogeography of the *Drosophila melanogaster* species subgroup. *Evolutionary Biology, 22*, 159–225.

Lachaise, D., Harry, M., Solignac, M., & Lemeunier, F. (2000). Evolutionary novelties in islands: Drosophila santomea, a new melanogaster sister species from São Tomé. *Proceedings of the Royal Society of London b: Biological Sciences, 267*, 1487–1495.

Lachenal, G. (2015). At home in the postcolony: Ecology, empire and domesticity at the Lamto field station, Ivory Coast. *Social Studies of Science, 46*, 877–893.

Lamarck, J.-B. (1809). *Philosophie zoologique.* Dentu.

Lambert, R., & Teisssier, G. (1927). Théorie de la similitude biologique. *Comptes rendus de l'Académie des sciences de Paris, 184*, 121–123.

Lamotte, M. (1951). Recherches sur la structure génétique des populations naturelles de *Cepaea nemoralis* (L). *Éditions Du Bulletin Biologique De La France Et De La Belgique, Suppl., 35*, 1–238.

Lamotte, M. (1957). *Initiation aux méthodes statistiques en biologie.* Masson.

Lamotte, M. (1959). Polymorphism of natural populations of *Cepaea nemoralis. Cold Spring Harbor Symposia on Quantitative Biology, 24*, 65–86.

Lamotte, M. (Ed.). (1974). *Le polymorphisme dans le règne animal, volume publié par la Société zoologique de France à la mémoire de Georges Teissier*. Mémoires de la Société zoologique de France n° 37.

Lemeunier, F., David, J. R., Tsacas, L., & Ashburner, M. (1986). The melanogaster species group. In M. Ashburner, H. L. Carson., & J. N. Thompson Jr. (Eds.), *The genetics and biology of drosophila V. 3* (pp. 147–256). Academic Press.

Lerner, M. (1954). *Genetic homeostasis.* Oliver & Boyd.

Lewontin, R. C. (1974). *The genetic basis of evolutionary change.* Columbia University Press.

L'Héritier, P. (1934). *Génétique et évolution, analyse de quelques études mathématiques sur la sélection naturelle.* Hermann.

L'Héritier, P. (1949a). *Traité de génétique, T. 1: Le mécanisme de l'hérédité; hénétique formelle.* Presses Universitaires de France.

L'Héritier, P. (1949b). *Traité de génétique, T. 2: la génétique des populations.* Presses Universitaires de France.

L'Héritier, P. (1949c). *Les méthodes statistiques dans l'expérimentation biologique.* Éditions du CNRS.

L'Héritier, P., Neefs, Y., & Teissier, G. (1937). Aptérisme des insectes et sélection naturelle. *Comptes rendus de l'Académie des sciences, 204*, 907–909.

L'Héritier, P., & Teissier, G. (1937). Une anomalie physiologique héréditaire chez la drosophile. *Comptes-Rendus De L'académie Des Sciences, 205*, 1099–1101.

Limoges, C. (1980). A second glance at evolutionary biology in France. In E. Mayr & W. Provine (Eds.), *The evolutionary synthesis: Perspectives on the unification of biology* (pp. 322–328). Harvard University Press.

Loison, L. (2010). *Qu'est-ce que le néolamarckisme? Les biologistes français et la question de l'évolution des espèces.* Vuibert.

Loison, L. (2013). Georges Teissier (1900–1972) and the modern synthesis in France. *Genetics, 195*, 295–302.

Loison, L., Gayon, J., & Burian, R. (2016). The contributions—and collapse—of Lamarckian heredity in Pasteurian molecular biology: 1. Lysogeny, 1900–1960. *Journal of History of Biology, 50*(1), 5–52.
López Beltrán, C. (1992). *Human heredity 1750–1870; The construction of a domain.* PhD Dissertation, University of London.
Lotka, A. J. (1934). *Théorie analytique des associations biologiques. Première Partie: Principes.* Hermann.
Lotka, A. J. (1939). *Théorie analytique des associations biologiques. Deuxième Partie: Analyse démographique et application particulière à l'espèce humaine.* Hermann.
Michaux, J. (2018a). Louis Thaler (1930–2002). In P. Henriet (Ed.), *École Pratique des Hautes Études : invention, érudition, innovation, de 1868 à nos jours* (pp. 172–173). Somogy.
Michaux, J. (2018b). René Lavocat (1910–2007). In P. Henriet (Ed.), *École Pratique des Hautes Etudes : invention, érudition, innovation, de 1868 à nos jours* (p. 189). Somogy.
Millstein, R. (2008). Distinguishing drift and selection empirically: The great snail debate of the 1950s. *Journal of the History of Biology, 41*, 339–367.
Millstein, R. (2009). Concepts of drift and selection in the great snail debate of the 1950s and early 1960s. *Transactions of the American Philosophical Society, 99*, 271–298.
Monod, J. (1942). *Recherches sur la croissance des bactéries.* Hermann & Cie.
Monod, J., & Teissier, G. (1936). *Comptes-Rendus De L'académie Des Sciences, 202*, 162–164.
Montchamp-Moreau, C., Ginhoux, V., & Atlan, A. (2001). The Y chromosomes of *Drosophila simulans* are highly polymorphic for their ability to suppress sex-ratio drive. *Evolution, 55*, 728–737.
Morin, F., & Monod, J. (1942). Sur l'expression analytique de la croissance des populations bactériennes. *Revue scientifique, 5*, 227–229.
Morrell, J. B. (1979). The chemists breeders: The research school of Liebig and Thomson. *Ambix, 19*, 1–46.
Nagylaki, T. (1989). Gustave Malécot and the transition from classical to modern population genetics. *Genetics, 121*, 103–118.
Pasteur, N., Pasteur, G., Bonhomme, F., Catalan, J., & Britton-Davidian, J. (1987). *Manuel technique de génétique par électrophorèse des protéines.* Lavoisier.
Pearl, R. (1928). *The rate of living.*
Periquet, G. (1979). Study of some factors maintaining the 'atrophie gonadique' character in experimental populations of *Drosophila melanogaster*. *Genetica, 51*, 27–32.
Petit, C. (1951). Le rôle de l'isolement sexuel dans l'évolution des populations de *Drosophila melanogaster*. *Bulletin biologique de France et de Belgique, 85*, 392–418.
Picard, G., & L'Heritier, P. (1971). A maternally inherited factor inducing sterility in *D. melanogaster*. *Drosophila Information Service, 46*, 54–54.
Provine, W. B. (1971). *The origins of theoretical population genetics.* The University of Chicago Press.
Roger, J. (1979). Les néo-lamarckiens français [présentation]. *Revue de synthèse, 95–96*, 279–282.
Slatkin, M., & Veuille, M. (Eds.). (2002). *Modern developments in theoretical population genetics. The legacy of Gustave Malécot.* Oxford University Press.
Solignac, M., Monnerot, M., & Mounolou, J. C. (1986). Mitochondrial DNA evolution in the melanogaster species subgroup of Drosophila. *Journal of Molecular Evolution, 23*(1), 31–40.
Teissier, G. (1934). *Dysharmonies et discontinuités dans la croissance.* Hermann.
Teissier, G. (1937). *Les lois quantitatives de la croissance.* Hermann.
Teissier, G. (1944). Équilibre des gènes létaux dans les populations stationnaires panmictiques. *Revue scientifique, 82*, 145–154.
Teissier, G. (1958). *Titres et travaux scientifiques.* Prieur et Robin.
Teissier, G., & Bocquet, C. (1960). Génétique des populations de *Sphaeroma serratum*; stabilité du polychromatisme local. *Cahiers De Biologie Marine, 1*, 221–230.
Timofeeff-Ressovsky, N. W. (1940). Mutations and geographicla variation. In J. Huxley (Ed.), *The new systematics* (pp. 73–136). Oxford University Press.

Tsacas, L., & Lachaise, D. (1974). Quatre nouvelles espèces de la Côte d'lvoire du genre *Drosophila*, groupe melanogaster et discussion de l'origine du sous-groupe melanogaster (Diptera:Drosophilidae). *Annales de l'Université d'Abidjan: Écologie, 7*, 193–211.

Valdeyron, G. (1961). *Cours de génétique de l'I.N.A.* Paris: Centre de Documentation Universitaire et S.E.D.E.S réunis.

Veuille, M. (2018a). La section des sciences naturelles et de physiologie. In P. Henriet (Ed.), *École Pratique des Hautes Études : invention, érudition, innovation, de 1868 à nos jours* (pp. 59–71). Somogy.

Veuille, M. (2018b). Jean Gayon et l'école française de génétique des populations. In F. Merlin & P. Huneman (Eds.), *Philosophie, histoire, biologie: mélanges offerts à Jean Gayon* (pp. 135–150). Éditions Matériologiques.

Veuille, M. (2019). Chance, variation and shared ancestry: Population genetics after the synthesis. *Journal of the History of Biology, 52*, 537–567.

Veuille, M., & King, L. M. (1995). Molecular bases of polymorphism at the *Esterase-5B* locus of *Drosophila pseudoobscura. Genetics, 141*, 255–262.

Veuille, M., Bénassi, V., Aulard, S., & Depaulis, F. (1998). Allele-specific population structure of *Drosophila melanogaster Alcohol dehydrogenase* at the molecular level. *Genetics, 149*, 971–981.

Viré, M. (1979). La création de la chaire d'étude de 'L'évolution des êtres organisés' à la Sorbonne en 1888. *Revue De Synthèse, 95–96*, 377–392.

Volterra, V., & D'Ancona, V. (1934). *Les associations biologiques étudiées au point de vue mathématique.* Hermann.

Wright, S. (1932). The roles of mutation, inbreeding, crossbreeding, and selection. In D. F. Jones (Ed.), *Proceedings of the sixth international congress of genetics* (pp. 356–366). Genetics Society of America.

Wright, S. (1939). *Statistical genetics in relation to evolution.* Hermann & Cie.

Wright, S., & Dobzhansky, T. (1946). Genetics of natural populations. XII. Experimental reproduction of some of the changes caused by natural selection in certain populations of *Drosophila pseudoobscura. Genetics, 31*, 125–156.

Chapter 10
Jean Gayon on the "Modes" of Cultural Evolution

Jorge Martínez-Contreras

Abstract Cultural evolution is the theory of evolution by means of natural selection in the realm of culture. The development of cultural evolution over the last few decades has gained great importance. In this chapter, I analyze Jean Gayon's critical analysis of what he calls the "modes" of cultural evolution (Gayon, 23:139–150, 2005). These "modes", for Gayon, are sets of research programs inspired by the same paradigm; they take for granted the existence of a close relationship between biological and human social phenomena. In addition, each of these modes seeks to connect the biological, the social, and even the physical sciences. One of the main differences between these modes is the importance they give to biology in anthropology. Since cultural evolution has grown into a field of great importance today, reflecting on Gayon's critical appraisal regarding the strengths and weaknesses of these modes takes on a particular relevance. In conclusion I put forward some of my own criticisms of the work of Gayon regarding his assessment of the modes of cultural evolution; I suggest that a new study carried out in the spirit of the one undertaken by Gayon in 2005 could make a major contribution to the development of CE theory.

Keywords Cultural evolution · Darwinism · Epidemiology of representation · Sociobiology · Naturalism · Ethology · Anthropology

10.1 Introduction

Cultural evolution (CE)—or sociocultural evolution for some authors—is the theory of evolution by means of natural selection, not only in biology, but also in the realm of culture. The latter term is defined as the set of information changing (or not) over

---

Translated from Martínez-Contreras (2018).

J. Martínez-Contreras (✉)
Universidad Autónoma Metropolitana, Mexico City, Mexico
e-mail: pascalo696@me.com

P.-O. Méthot (ed.), *Philosophy, History and Biology: Essays in Honour of Jean Gayon*, History, Philosophy and Theory of the Life Sciences 30,
https://doi.org/10.1007/978-3-031-28157-0_10

time, which is transmitted from generation to generation through learning, imitation, and so on, and which impacts both individuals and societies.

The development of cultural evolution over the last few decades—i.e., the unified study of human and non-human cultural phenomena as part of evolution—has gained great importance, bringing into its fold natural, social, and humanistic interpretations of what culture is. For instance, towards the end of 2016, a "Sackler Colloquium" devoted to cultural evolution took place at the University of Irvine, in collaboration with the National Academy of Sciences. It was called: "The Extension of Biology Through Culture". This scientific event was co-organized by Francisco J. Ayala—perhaps the last living representative of the Modern Synthesis–, Marcus W. Feldman, and Andy Whitman, and brought together experts in the study of various animal species, but also prehistorians studying ancient human societies. One year later, the Cultural Evolution Society was founded. This international scientific society was promoted notably by biologists, philosophers, and anthropologists, alongside other specialists (Peter J. Richerson and Dan Sperber are its President and President-elect, respectively). We know the importance of ad hoc scientific societies that arise when a discipline or set of disciplines seems to have reached a certain maturity. The existence of cultural evolution as an object of collective study by a whole group of old and new disciplines no longer seems to be debated, except among traditional thinkers opposed to evolutionism, in particular religious theorists. But this was not the case previously: naturalism had fallen out of favour among many social thinkers who feared, in part through ignorance of the disciplines that make up this field, that the biological study of humans could justify racism. In reality, quite the opposite occurs because there is no better protection against racist positions than the study of evolution; human physical variations are merely ecological adaptations. If these beliefs, which are not necessarily religious ones, are still around, they are no longer seen as scientific arguments against CE. The ever-growing set of disciplines that come together to form the core of CE—in something like the way that the Modern Synthesis constitutes the core of Darwinism—are now concerned with analyzing various cultural phenomena, but no longer in the exclusive context of human societies, because culture also exists in many animal societies, and very possibly also in plants, as shown by recent research on communication between trees (Simard, 2018; Wohllebe, 2015). No doubt a new integration will come later. But the scientific importance of CE is now well established in contemporary science.

In this paper, I want to analyze some of Jean Gayon's works on the study of cultural evolution. In 2005, he published a critical analysis of what he called the "modes" of cultural evolution in *Ludus Vitalis* (Gayon, 2005). These "modes" or "trends" for Gayon are sets of research programs inspired by the same paradigm; they take for granted the existence of a close relationship between biological and human social phenomena. Each of these modes seeks to connect the biological, the social, and even the physical sciences. One of the main differences between them is the importance they accord to the role of biology in anthropology. Since CE has grown into a field of great importance today, reflecting on Gayon's critical predictions regarding the strengths and weaknesses of these modes takes on a particular relevance. In what follows, I will venture some of my own criticisms of the work of Gayon regarding his assessment of the modes of cultural evolution.

10.2 Four Modes of Cultural Evolution

The Darwinian study of cultural evolution in human and non-human animals has been developing steadily for over half a century, in particular since the publication of Gayon's text on the subject.[1] The idea of CE is based on the idea that certain animal species, particularly, but not exclusively, *Homo sapiens*, have the ability to receive, develop, integrate, and pass on cultural data from one generation to the next. Initially, the problem was to know if the mechanisms of natural "biological" selection could be applied to the realm of culture. Although Darwin did not use this terminology, the early ideas of CE are nevertheless considered to derive directly from the intuitions of the English naturalist. The current idea of cultural evolution differs, however, from that of the "evolutionary anthropologists" of the late nineteenth century, as I will argue below with reference to Gayon's critical inquiry. Indeed, at that time, evolutionism was bound up with the idea of progress, as shown in the writings of philosophers and sociologists such as Hegel, Marx, and Comte, the most important representatives of this belief. Contemporary evolutionary theory of culture, by contrast, no longer assumes or adopts the idea of progress. This is an important difference because it is now widely accepted that evolution by means of natural selection is incompatible with the idea of progress or teleology.

In his article, Gayon critically analyzed what he calls the "major modes" of CE. His paper does not appear at all dated, more than a decade after its publication; the problems present in the "trends" he discerns remain the focus of current debates between "traditional" cultural anthropologists and evolutionary anthropologists and biologists, even if some of these trends seem to be gaining the upper hand, scientifically speaking, as I will show at the end of this chapter. To begin with, Gayon notices the large number of disciplines that the epistemological concept of EC brings together. EC refers to a set of disciplines, to which I will add a few more, without claiming to be exhaustive: sociology, economics, ethnology, biological and social or cultural anthropology, linguistics, ecology, ethology, paleoanthropology, prehistory, cognitive science, neuropsychology, and evolutionary psychology, among others. Gayon thinks as a philosopher when he brings out the naturalistic assumptions at the foundation of CE. As he reminds us, contemporary naturalism has nothing to do with earlier forms of naturalism, which were connected to the philosophical doctrines of atheism and materialism. Today, philosophical naturalism draws on science and empirical data to address philosophical problems: it is no longer an ontology but a

[1] For instance, the field developed from the discovery of culture in Japanese macaques (*Macaca fuscata*), and the famous young female Imo (1957), who discovered how to clean potatoes with water, a behaviour later transmitted to almost the whole of her troop. The field was then extended to the study of different chimpanzee cultures (see Martínez-Contreras, 2011).

methodology, as he points out. I share this way of doing philosophy, especially in the context of the philosophy of biology.

After making these points, Gayon turns to the four modes of CE. I will first summarize his criticisms and then ask whether, after more than fifteen years, one (or a few) of these modes of CE dominate current debates between experts in the field. This will lead me, in turn, to raise a criticism of Gayon's own critiques. Regarding the first mode, he writes:

> The first type of cultural evolution refers to studies on the origins of the cultural capacity of the human species: when and in what circumstances did it appear, and what are its biological bases? This question comes before any speculation about the effective diversity of cultures. The postulate here is that the members of the human species share a genetic heritage that makes them able to assimilate, modify and transmit all that is commonly counted as essential to "cultural behaviour" in humans: symbols, language, cognitive categories, a theory of mind, techniques, ethical rules, etc. The problem is to provide a precise list of universal cultural aptitudes, to identify their genetic and psychological bases, and to reconstitute the environmental circumstances within [which] these aptitudes appeared and were maintained in the course of evolution. (2005, p. 140)

Discussing this mode of CE, Gayon explains that animal psychology, ethology and anthropology are among the most important disciplines for determining the specificity of human culture; he also reminds us that "culture is not specific to humans" (2005, p. 140). The most important author for this mode of cultural evolution is Dan Sperber, according to whom, in Jean Gayon's words, "what is specific to the human species is that culture structures all aspects of human behaviour in all circumstances. In other animals, culture intervenes occasionally and only for certain kinds of behaviour" (Sperber, 2003, in Gayon, 2005, p. 141). I believe that Sperber (1996) did not sufficiently study what is known today about non-human animal societies (e.g., chimpanzee societies); otherwise, he would not claim that culture only occurs "occasionally" in other animals. In any case, Gayon notices that this program focusses on events that predate the present diversity of human cultures, because it seeks above all to identify universals that were established when Hominids became cultural animals. The program is also "neutral" concerning the coupling or decoupling between the present evolution of the human species and its current cultural history. Finally, this evolution towards culture would have occurred on a timescale measured in 100,000 s of years (we now know that *Homo sapiens* is at least 300,000 thousand years old, and we would not be surprised to find, in the future, even older traces of its presence). For Gayon, this "mode" is the most consensual one among the four he analyzes in this article. Regarding the second mode, he states:

> The second theoretical mode of cultural evolution is the opposite of the previous one in terms of time scale. It consists of the various attempts that have been made over the last three decades to describe and explain cultural change with the help of concepts or models inspired by the schemes of population genetics. Here the methodological postulate is that of a deep analogy between genetic and cultural transmission and, as a consequence, between genetic evolution and cultural evolution. The analogy is constructed in the following way. Population genetics relies upon the existence of discrete replicating units which sometimes mutate and diffuse within and between populations due to processes such as natural or sexual selection, random drift, and migration. (2005, pp. 141-142)

For Gayon, the most important representatives of this mode are Luigi Luca Cavalli-Sforza and Marcus Feldman (1981) in the fields of "epidemiology of representations", the diffusion of technical innovations, scientific change, linguistics, and economics, together with the philosopher David Hull on the evolution of cultural items (Hull, 1988). In philosophy, the study of the origin of ethics followed a similar path and gained ever greater importance. Gayon calls this program *analogical* because it doesn't conflict with the development of quantitative social sciences, and it disputes the claim that cultural phenomena can be reduced to or explained by evolutionary biology. The rate of change of this mode of evolution is very rapid compared to the one studied in population biology. Indeed, if gene replication depends on the reproduction of organisms, this is not the case in this analogical mode, where there is vertical transmission, from generation to generation, through teaching/learning and imitation in a social context. However, this same social context also allows for horizontal cultural transmission, as occurs in the use of language—regarding its rules, but also its colloquial expression–, the display of emotions, moral codes, administrative standards, etc., and horizontal transmission is in turn much faster than vertical transmission. Analogously speaking, cultural phenomena can arise in the same way that genes express themselves: by selection, drift, or migration. I would add that we can also speak of mutations, as in the case of the "Eureka" phenomenon in science and technology. But in this regard, Hull (1988, cited by Gayon) points out that the intentionality of an individual scientist is not sufficient to produce new social objects: the best idea does not survive if other individuals do not appropriate and disseminate it. Finally, for Gayon, proponents of this mode of CE are interested in phenomena occurring over a very short timeframe, sometimes a generation, but sometimes also much less. Moreover, he reminds us that this mode is only said to be "evolutionary" in an analogical way, insofar as it borrows certain concepts from the biological sciences for use in the social sciences.

The third mode analyzed by Gayon is the one that seems to favor a certain *parallelism*, and not an analogy in the strict sense, between biological phenomena and cultural phenomena: "It consists of showing that in certain cases the genetic evolution of human populations and their cultural transformation are parallel. [...] Parallel evolution of gene pools and culture means that in a given population these two features are correlated" (2005, p. 143). Gayon, again, takes the works of Cavalli-Sforza et al. (1998) as a model, but he focuses here on texts he himself co-wrote with Piazza from 1988 onward. These authors have compared and related the notable resemblance between the distribution of linguistic families and the dispersal of human populations originating from 100,000 and 150,000 years ago to the present day (though we now know that we must speak of at least 300,000 years): languages and peoples seem to tell the same story, at least over long historical times (millennia). As we also know, one of Ernst Mayr's major arguments—which Gayon does not cite—is that reproductive isolation is one of the most important mechanisms of speciation. The same thing happens with languages: languages tend to diversify as soon as there are insurmountable obstacles, be they geographic or political. What matters is the level of separation the obstacle produces, whether partial or total (and sometimes completely separating populations, as in the case of the insular isolation highlighted by Wallace and Darwin): reproductive fissions are similar for organisms as well as for cultures. Cultural separation, in turn, can promote biological separation. As far as its temporality is concerned, this mode would be intermediate between modes 1 and 2: much faster than mode 2, which acts at the level of a generation at most, but slower than the evolution of the prehistoric phenomena studied in mode 1. Gayon grants this mode a temporality of a thousand years.

Finally, Gayon examines the mode that seems to be the most problematic, both in theoretical and ethical terms. This mode concerns the coupling of biological evolution and CE. The main representative here is Edward O. Wilson (1975, 1978, with Lumsden, 1981) and his concept of sociobiology. For Gayon: "the central hypothesis of Sociobiology is that, in all species, social behaviour is designed by natural selection. [...] Wilson applied this hypothesis to the species where social behaviour attains an exceptional degree of complexity, the human species" (2005, p. 144).

It is especially in his above-mentioned book co-authored with the physicist/mathematician, Charles J. Lumsden, that Wilson discusses problem of CE in depth. These authors propose a general theory of gene-culture coevolution, and therewith also the concept of "culturgens", which refers to the transmission of behaviours that generate culture. Although they admit that cultural items are transmitted by imitation and learning, they leave the door open to certain epigenetic phenomena controlled by genes. The difference with other modes is the possibility for cultural phenomena to be transmitted exclusively through imitation and learning. Wilson's critique targets the works of those that postulate that the biological emergence of culture allowed humans to evolve culturally. Wilson says this would be a "gene-Promethean" hypothesis: "a group of Promethean genes has freed the human mind from the other genes" (cited in Gayon, 2005, p. 145). And Gayon reminds us of Wilson's slogan in *On Human nature* that "Genes hold culture on a leash" (Ibid.). For Wilson, indeed, any culture must consider its environment: "the hypothesis of a purely cultural transmission is not likely because it would imply that human populations have been and are unable to distinguish in a given environment cultural items which are adapted and cultural items which are not" (2005, p. 145). The general conclusion is obvious: the emergence of culture did not end the biological evolution of humans; on the contrary, culture is a major element of a still ongoing biological evolution: "gene-culture coevolution is an inevitable property of the human species" (Ibid.). Wilson estimates that the fixation of culturgens in a given population takes fifty generations or a thousand years: cultural history thus falls into the realm of biological evolution, although Wilson does not provide evidence for this claim. The possible existence of gene-culture coevolution constitutes the most radical thesis among the four modes of CE studied by Gayon.

Following the presentation of this typology, Gayon reflects on all of these four modes. Let me summarize the main observations made. The first mode or set of research programs attempts to explain the emergence of culture as a new global environment for humans. This program is not about historical cultural change as such; it is about past events that would have led humans to acquire culture: it is the idea of evolution *towards* culture. It thus differs from the other three programs, which all admit a certain relationship between cultural and genetic change.

The analog program (by Cavalli-Sforza & Feldman, 1996) is for Gayon the weakest of the four because it is based on a simple analogy between CE and biological evolution. As to the program that observes some parallelism (by Cavalli-Sforza et al., 1996) between CE and biological evolution, it implies that cultural and biological changes are sometimes caused by the same phenomena. Concerning sociobiology, Gayon considers this program to be the strongest: here, EC and biological evolution are not only correlated, for they result from the same causes. Also concerning this fourth mode, Gayon analyses some of the debates that are still going on within and around it.

A first debate occurred among researchers in the social sciences opposed to quantitative methods, and in particular among philosophers or sociologists working in the tradition of hermeneutics, the aim of which is to "understand" rather than to "explain" human behaviour (Heidegger, Sartre, Gadamer, and Ricœur, among others, are good examples of this way of thinking about humans). But this was not the most significant controversy. A second one came from those who conceptualized culture as an autonomous phenomenon. This idea had detrimental consequences in the development of anthropology in the twentieth century because it created a staunch opposition between social anthropology and physical or biological anthropology, as if humans were not as much animals as they were cultural beings. Contrary to what one might think—especially in Mexico, where these two disciplines are still separate—Lévi-Strauss opposed this dichotomy, which he called metaphysical in the sense given to this term by Comte (that is to say, an idea that has become obsolete due to the advent of the sciences). For Lévi-Strauss, the mapping of cultural phenomena should be put in direct relation with the mapping of biological phenomena (Lévi-Strauss, 1983, cited in Gayon). A third opposition to the culturgens thesis came from those who opposed the ecological paradigm to the genetic paradigm. Jared Diamond (1997, cited in Gayon)—who considers himself a sociobiologist–, in particular, suggested that it is ecological phenomena and geographic obstacles to the migration of humans and their domesticated species, rather than genetic predispositions, that explain the major trends observed in the human species.

The fourth debate that raged in the wake of the culturgens hypothesis came more from misunderstandings and malicious intentions than from scientific debates as such. While Gayon doesn't characterize it in this way, this last debate concerns accusations of racism made against Wilson—and sometimes even Cavalli-Sforza–, of which they weren't guilty at all. For some, however, to study physical adaptations of ecological origin—such as skin color—implies racism, while it is quite the opposite: such melanistic adaptations show that humans were not capriciously "created" from different people, with diverse colors, but that, as in the case of the melanin of the skin, it is a question of adaptations to the intensity of the ultraviolet radiation occurring in a given location on the planet. The ideas of "superior" and "inferior" races were nothing but arguments in favor of exploiting people and the development of slavery. However, Gayon limited his study of sociobiology to its theoretical core: in anthropology, he provided a view of sociobiology as one of the most modest forms of the biology-culture relationship; for example, forms of culture more likely to interact with biological aspects of humans: disease, nutrition, basic technical skills, and affective dispositions. I do not know if Gayon ever changed his mind on this issue. But what must be emphasized is that there are no human cultural domains which cannot—in a scientific context, free from any religious or spiritual compromise—escape the possibility of being studied in relation to biological phenomena, including religious and spiritual beliefs, which, like all beliefs, can only exist in a brain whose neurons interact with each other through electrical and chemical messages.

10.3 Conclusion

To conclude, Gayon underlines in his paper his commitment to naturalism and empiricism (see also Gayon, 1999). This is the commitment of philosophers and scientists to rely much more than they do now on the methods and theories of the natural sciences. If a criticism can be addressed to this important set of critical analyzes regarding the four modes of cultural evolution proposed by Gayon, it concerns precisely that point I raised at the beginning of this chapter, namely the presence of cultural evolution in animals, but also Gayon's effort to limit the sphere of application of sociobiology to humans. The Sackler Colloquium, for instance, clearly showed that culture exists and is transmitted vertically, by learning and by imitation—from generation to generation–, and horizontally, in many species of mammals and birds. An example of the latter way of transmitting these cultural objects is the case of humpback whales, for only a few populations among these cetaceans sing in a "dialect" originating in a given population, though this dialect may be observed thousands of kilometers away in other populations of the same species. On the other hand, studies of prehistory also point out an important link in humans between culture and biology (Foley & Lahr, 2011).

A new study carried out in the spirit of that undertaken by Gayon in 2005 could undoubtedly make a major contribution to the development of CE theory.

References

Cavalli-Sforza, L. L., & Feldman, M. W. (1981). *Cultural transmission and evolution: A quantitative approach*. Princeton University Press.

Cavalli-Sforza, L. L., & Feldman, M. W. (1996). *Gènes, peuples et langues*. Odile Jacob.

Cavalli-Sforza, L. L., Menozzi, P., & Piazza, A. (1996). *The history and geography of human genes*. Princeton University Press.

Cavalli-Sforza, L. L., Piazza, A., Menozzi, P., & Mountain, J. L. (1998). Reconstruction of human evolution: Bringing together genetic, archaeological, and linguistic data. *Proceedings of the National Academy of Sciences, 85*(16), 6002–6006.

Diamond, J. (1997). *Guns, germs and steel: The fate of human societies*. Norton.

Foley, R. A., & Mirazón, L. M. (2011). The evolution of the diversity of cultures. *Philosophical Transactions of the Royal Society of London b, Biological Sciences, 366*(1567), 1080–1089.

Gayon, J. (1999). Sélection naturelle biologique et sélection naturelle économique: examen d'une analogie. *Économies et Sociétés, 35*, Hors-Série, 107–126.
Gayon, J. (2005). Cultural evolution: A general appraisal. *Ludus Vitalis, 23*, 139–150.
Hull, D. (1988). *Science as a process. An evolutionary account of the social and conceptual development of science*. The University of Chicago Press.
Lévi-Strauss, C. (1983). *Race et culture*. In *Le regard éloigné (1971)*. Plon.
Martínez-Contreras, J. (2011). O modelo primatológico de cultura. In P. C. C. Abrantes (Ed.), *Filosofia da* biologia (pp. 224–240). ARTMED Editora S.A.
Martínez-Contreras, J. (2018). Critique des critiques de Jean Gayon sur l'idée de différents 'modes' d'évolution culturelle. In F. Merlin & P. Huneman (Eds.), *Philosophie, histoire, biologie: mélanges offerts à Jean Gayon* (pp. 305–316). Éditions Matériologiques.
Simard, S. W. (2018). Mycorrhizal networks facilitate tree communication, learning and memory. In F. Baluska, M. Gagliano, & G. Witzany (Eds.), *Memory and learning in plants* (pp. 191–213). Springer.
Sperber, D. (1996). *La contagion des idées*. Odile Jacob.
Sperber, D. (2003). Culture et modularité. In J.-P. Changeux (Ed.), *Gènes et culture* (pp. 277–299). Odile Jacob.
Wilson, E. O. (1975). *Sociobiology: The new synthesis*. Belknap Press of Harvard University.
Wilson, E. O. (1978). *On human nature*. Harvard University Press.
Wilson, E. O., & Lumsden, C. J. (1981). *Genes, mind and culture*. Harvard University Press.
Wohlleben, P. (2015). *Das geheime Leben der Bäume: was sie fülen, wie sie kommunizieren*. Ludwig.

Chapter 11
On Revisiting Wallace's 1858 Theory of Natural Selection

Jonathan Hodge

Abstract Alfred Russel Wallace's 1858 essay on natural selection is often related by historians to his earlier views in the 1840s, and to his much later reminiscences around 1900. It is interpreted here as an amplification and transformation of a conjectural sketch composed two or three years before 1858 in his *Notebook* 4. Like the sketch, the essay was more brains-on bookwork than hands-on fieldwork. Wallace's theory was not a group-selectionist theory. As an individual-selectionist theory about self-helping wild animal life, it supported, by contrast, his socialist view of human society. He never rejected Darwin's analogy between artificial and natural selection, accepting it most probably on first encountering it.

Keywords Wallace · Darwin · Gayon · Natural selection · Struggle for existence · Domestic and wild animals · Socialism · Canine varieties conjecture

11.1 Introduction

In both his wonderful books—I know them in their English versions, *Darwinism's Struggle for Survival* (1998) and *Knowledge of Life Today: Conversations on Biology* (Gayon and Petit, 2019)—Jean Gayon compares and contrasts very instructively the two theories of natural selection arrived at independently by the first Darwinians: Charles Darwin and Alfred Russel Wallace. My paper honours Jean's life and work by revisiting Wallace's theory.

The subtitle of Jean's first book is *Heredity and the Hypothesis of Natural Selection*; and its main theme is that the hypothesis of natural selection was not originally consistent with any credible account of heredity. Only in the twentieth century did natural selection receive such support.

In both books Jean suggests that from the beginning there were two divergent Darwinian positions regarding heredity. Darwin tried unsuccessfully to find convincing support for natural selection from generalisations about heredity. But

---

J. Hodge (✉)
University of Leeds, Leeds, UK
e-mail: m.j.s.hodge@leeds.ac.uk

P.-O. Méthot (ed.), *Philosophy, History and Biology: Essays in Honour of Jean Gayon*, History, Philosophy and Theory of the Life Sciences 30,
https://doi.org/10.1007/978-3-031-28157-0_11

Wallace, Jean held, did not engage this issue, and for a good reason. Theorising about heredity was about how individual parents produce individual offspring resembling and differing from their parents and siblings. Darwin's theory was about selection among such individuals, and so needed integrating with heredity theory. But, Jean says, Wallace's theorising was, in today's terms, group-selectionist, about selection of varieties not individuals, and did not require this integration. Jean sees Darwin and Wallace relating in quite divergent ways to his historiographical theme. Darwin is initially the theme's central figure, while Wallace is not concerned with its principal issues.

In revisiting Wallace's 1858 theory I will be questioning this contrasting of Darwin and Wallace, and suggesting that Jean's historiography can be enhanced by giving Wallace a more positive role. Great teachers live on in the work of their protégés. I would be delighted if some of Jean's students were to join me in honouring Jean by taking up this topic. Knowing Jean and his family, as I and my family have done over many years, has been a cherished pleasure and valued privilege. My warm thanks to our editor for this opportunity to commemorate an inspiringly brilliant scholar and an endearingly admirable person.

11.2 The Contexts of a Canine Varieties Conjecture (CVC)

Here I prepare to introduce Wallace's February 1858 essay (henceforth 58E) as a textual descendant of an earlier notebook sketch. 58E was Wallace's first statement of his theory of natural selection. This label is disputable as the essay has no analogy between artificial and natural selection. But Wallace's theory ascribes adaptive change to naturally-caused selection rather than instruction, and is in this general sense a theory of natural selection. Wallace sent the essay to Darwin hoping that Lyell would also see it. He did. Thanks to Lyell and Hooker, it appeared in the famous joint—Darwin and Wallace—publication of summer 1858 (Darwin & Wallace, 1858).

There are few clues from the 1850s directly indicating why Wallace's essay was written as it was. So historians have often related the essay to his preoccupations in the 1840s—he was born in 1823—and to his much later recollections around the end of the century (Jones, 2002; McKinney, 1972; Moore, 1997). But I will suggest here that the insights gained from these remote sources can be supplemented, once we see how 58E was written as a much amplified and transformed version of an earlier canine varieties conjecture (CVC).

Wallace's Malayan years had begun in April 1854. The CVC was composed in his *Notebook* 4 (MS pp. 39–40) most probably in the second half of 1855 (Costa, 2013a). The CVC does not include any theorising about naturally-selective causation. How then did Wallace arrive at his 1858 theory of natural selection? He did so by having new thoughts in the years after the conjecture and before the essay.

Giving this answer to this question requires us to look briefly at the early 1850s context of the CVC. For this preliminary purpose two texts are indispensable: the first is Wallace's "Sarawak law" paper written early in 1855 (Wallace, 1855). It was titled "On the law which has regulated the introduction of new species" and was published in London later that year. Second is a sequence of some twenty pages (34–54) in *Notebook* 4 where Wallace is agreeing and disagreeing with the fourth (1835) edition of Charles Lyell's *Principles of Geology*, the book he was most influenced by throughout these early Malayan years. His preoccupation with this dialectical task probably went back to the mid-1840s. For it was then that Wallace—most decisively influenced, it seems, by Robert Chambers' anonymously authored *Vestiges of Creation* (1844)—first dissented from Lyell's rejection of the transmutation of species (or evolution, as we now say).

The 1855 paper is often read as mainly disagreeing with Edward Forbes. But this reading can mislead. Certainly, Forbes is disagreed with at length and by name, as Lyell is not. For Lyell is only named as having successfully solved the problem of how species become extinct. Certainly, too, the paper takes its title verbatim from Lyell, and opens by endorsing comprehensive views on biogeography in relation to geology, views that readers would have recognised as Lyell's controversial teachings.

However, Wallace argues that the law regulating the introduction of new species is this: "*Every species has come into existence coincident both in space and time with a pre-existing closely allied species*" (Wallace, 1855). In this paper, Wallace does not contrast this law with any thesis of Lyell's. But he does in 1857 in writing on the fauna and flora of the Aru Islands, and in quoting and explicitly disagreeing with Lyell (Wallace, 1857). The matter in dispute had been evident to Lyell himself when he studied the 1855 paper. Wallace was read correctly, by Lyell, as a covert transmutationist countering his, Lyell's, view that the law determining where and when any species originated was adaptation: each species being created at a place where there is the soil and climate and other conditions to which it is best adapted (Wilson, 1970). Quite generally, then, adaptation is, for Lyell, what determines the representation of species, and hence of supra-specific kinds of life, at different times and places on the earth's changing surface now, in the past and into the future.

Wallace's Aru Islands paper explains why his 1855 law should replace Lyell's law. Lyell's adaptationism predicts wrongly that regions with similar conditions will have had similar species originate in them, while regions with differing conditions will have had differing species doing so. Wallace's law makes no such prediction, because it assumes that a species originates when and where it does because of the presence there of an earlier very similar ancestral species. Lyell read Wallace correctly as implying that the lines of affinities between successive species are irregularly branching because of common ancestries and divergent descents.

For Wallace, his law was a specification of an even more comprehensive generalisation. The "great law" regulating the peopling of the earth with animal and vegetable life is that "every change shall be gradual" (Wallace, 1855, p. 195).

Beyond this certainty lay unfinished business. Geology, Wallace says, gives us positive proof of the extinction and production of species although not disclosing how either has taken place. The *modus operandi* of extinction has however been well illustrated by Lyell. Geological changes, although gradual, must occasionally change external conditions enough to make the continued existence of some species impossible. To discover how extinct species have been replaced by new ones is, Wallace declares, the most difficult and most interesting problem in natural history; and his formulating of the law determining when and where any species can and did originate may, he hopes, be considered as a step towards a complete solution of this problem (Wallace, 1855, p. 190).

Pages 34–54 in *Notebook* 4 tell a tale of two books: Wallace's and Lyell's. Wallace's was planned in these pages and referred to by its provisional title: *On The Organic law of Change*; but it would never be completed. Page 34 is headed: "Notes *from Lyell's 'Principles'*". The following page is headed: "Note *for Organic law of* change". Wallace's principal theme was what this title announced and what the Sarawak law paper had declared: gradual change, Lyell-style. More precisely, it would be gradual change for the organic realm, wrought by causes as active at present as throughout the past, causes which Lyell himself had not searched for as consistently and successfully as he had done for the inorganic world. Lyell's book was then a twofold precedent for Wallace's. Lyell's teaching is to be vindicated but also corrected and improved in accord with Lyell's own principles and ideals (Costa 213b).

Two clusters of issues concerned Wallace when contemplating long times and wide spaces: whether gradual change over time was progressive, as Lyell had denied; and whether life was distributed geographically according to adaptive providence, as Lyell held, or determined by ancestral precedents as Lyell denied. At much smaller scales Wallace was countering Lyell's insistence that species could only diverge into varieties, and so not enough to give rise to new descendant species. On page 44, soon after writing the CVC, Wallace reflected that many of Lamarck's views are quite untenable and easily controverted; but not so "the simple question of a species being produced in time from a closely allied distinct species," which may continue existing "as long or longer than its offshoot."

11.3 An Unlimited Varietal Succession Conjecture

Wallace's canine varieties conjecture, CVC, on MS pages 39–40, is explicitly opposing Lyell from the start: "Lyell, says that varieties of some species may differ more than other species do from each other without shaking our confidence in the reality of species." But why, Wallace asks, should we have any such confidence in the limited mutability and so stability and distinct reality of species? Isn't this confidence just a prejudice, he urges, like the prejudice in favour of the earth's stability, a prejudice Lyell has argued against so ably? The evidential question is addressed directly: "In fact what positive evidence have we that species only vary within certain limits?" Wallace continues by arguing that the burden of proof is on Lyell in defending this limited variation thesis.

> Let us suppose that every variety of the Dog but one was to become extinct & that one say the spaniel, to be gradual spread over the whole world, subjected to every variety of climate & food, and domesticated by every variety of the human race. Have we any reason for supposing

> that in the course of ages a new series of varieties quite distinct from any now existing would not be developed,—& then should the same process be repeated & one of these varieties farthest removed from the original, again be spread over the earth & be subjected to the same variety of conditions, does it not seem probable that again new varieties would be produced, & have we any evidence to show that at length a ["stop" deleted] check[inserted] would be placed on any further change & ever after the species remain perfectly invariable <under any circumstances whatever>. Those who advocate variation within definite limits must suppose so, though the only ground for their opinion is that the varieties which have been produced under the influence of man have certain limits–<But we have no proof how the varieties of dogs were produced. All varieties we know of are produced at birth the offspring differing from the parent. This offspring propagates its kind. Who can declare that it shall not produce a variety which process continued at intervals will account for all the facts>. (Costa, 2013a, MS pp. 39–40)

The bracketed words in this text seem to have been inserted some time after the main entry but most likely in 1855 or 1856.We may take them as integral with the conjecture as a whole as we see how it proceeds from opening to ending.

From the start varieties are at the centre of Wallace's attention because Lyell has put them there. Wallace supposes that every variety of dog except one becomes extinct. And the one variety that has not gone extinct, the spaniel variety say, is then living like a species; for it is gradually spreading far and wide and being influenced by every variation in climate and food and by domestication by every variety of humans. As with such a species, we would suppose that in long ages a new array of varieties would arise from the spaniel variety, the new varieties being different from any dog varieties now existing. Suppose next that all of these new varieties, except one, were to become extinct; and suppose that this one variety is among those that are least like the original parent stock from which the spaniel and other varieties diverged; and suppose again that this one variety spreads as widely and is influenced by equally varied conditions as the spaniel variety was; then it surely seems probable that, as before, new varieties would be produced, with all but one supposed to become extinct. And so on, again and again.

Do we have any evidence that eventually there would be a check on any further change, and that a species would remain forever perfectly invariable under any circumstance? Lyell and other advocates of variation within definite limits must suppose this, although the only ground for their view is that varieties produced under human influences have such limits. But there is no proof as to how dog varieties were produced. All known animal varieties, wild or domestic, are produced at birth. An offspring differs from its parents and eventually propagates its distinctive characteristics; for these characteristics are hereditary, and the variety breeds true. We should not assume, as Lyell does, that such a variety will never go on to produce another new variety. If reiterated at intervals such a process of varietal succession would, Wallace concludes, account for all the facts.

11.4 What the Essay Has That This Canine Varieties Conjecture Lacks

In its very title—"On the Tendency of Varieties to depart indefinitely from the Original Type"—58E fits with the CVC in explicitly arguing for an endless succession of varieties departing unlimitedly from the original type, and so for a reiterated process of varietal succession which can explain all the facts.

But 58E addresses causal-explanatory tasks that the CVC does not address. In the CVC, no causes are invoked for the extinguishing of those varieties that become extinct; nor for ensuring that the surviving varieties will be among those furthest in character from the original specific type; nor for ensuring that varieties of wild species tend to vary unlimitedly in the wild, while domestic varieties do not if they leave domesticated living on the farm for life in the wild.

We'll see that these causal-explanatory tasks are precisely those that Wallace sees himself as successfully addressing in 1858, and doing so with a single cluster of reasonings about a single "principle". This is the "struggle for existence" in the wild, a principle, a lawful cause, quite absent in the life of domestic animals on the farm.

His most conspicuously triumphant moment in the essay comes when, in his italics, "*progression and continued divergence*" are explicitly "deduced" from (a) the general laws that are corollaries of this struggle- for- existence- in- the- wild principle and (b) the frequent occurrence of varieties in wild animals. This causal-inferential conjunction is decisively significant because (b) is in the CVC and is, he says in 1858, accepted by all naturalists; while (a) is not in the CVC and is what gets most of the early pages in 58E. We may say, then, that Wallace reached his 1858 theory in two stages: first, by arriving at the CVC reasoning and, second, by arriving at his 1858 theory by integrating (b) which was consensual, and already in his CVC, with (a) which was neither consensual nor in that conjecture, and was then of his own conceiving after writing the CVC and before writing 58E (Wallace, 1858b, p. 59).

These textual details discredit any construction narrative for the 1858 theory that has Wallace arriving at the theory in a single simple insight, in one swift shift, from complete ignorance to comprehensive knowledge about the origins of species. These same details also clarify how 58E agrees with the CVC while going far beyond it. To see next what this clarification can do for us, we do best to focus first on the invocation in (b) of the frequent occurrence of varieties in animals living in the wild.

The notes added at the end of the CVC are indispensable here. For they include three claims about all known varieties: (i) that they are produced at birth, when (ii) an offspring differs from its parent, and when (iii) this offspring propagates its type, its distinctive character. The first two claims concern how the first individual of a new variety relates to its parents, the third how it relates to its offspring. Putting these three claims together, Wallace is saying that the difference between such an individual and its parents is present at birth, and so not acquired after birth, and is a hereditary difference to be passed on, in true breeding, to the following generations.

To confirm that 58E is conformed throughout to this threefold thesis, we need to look at its first three paragraphs, especially at what is said about "permanent or true varieties." For here Wallace endorses the widespread belief in such varieties; and he tacitly invokes this belief as decisive for his treatment of indefinite varietal succession in the central section of his paper, and so in his deduction there of unlimited progression and divergence in wild animal life.

His short "Note on the Theory of Permanent and Geographical Varieties," published very early in 1858 in *The Zoologist,* prepares us for this endorsement and invocation (Wallace, 1858a). For, he argues there—with no reference to geographical varieties—that the doctrine of permanent varieties, and the doctrine of specially created unvarying species, are inconsistent with each other. The special creation doctrine accepts, he says, that varieties can have their origins in natural divergences within species, whereas species can only originate supernaturally. But, he continues, if some varieties are taken to be distinguished, as species are, by permanent character differences, then those varieties are distinguished from species only in the magnitude or degree—and not in the permanence or true-breeding—of their character differences. So, varieties and species are not two different kinds of entities, and should not be ascribed to quite different kinds of causation. Unsurprisingly, in 58E he assumes throughout that, contrary to the doctrine of specially-created unvarying species, both varieties and species are permanent, true-breeding races, and so should be referred to equally natural causes.

This assumption allows us to understand why Wallace explains at his essay's very opening what arguments are targeted for rebuttal. He focuses right away on "one of the strongest arguments" in favour of the original and enduring distinctness of species. This argument holds that varieties of domesticated animals are unstable and, left to themselves in the wild, return to the normal form of the parent species. This instability is assumed moreover to be a tendency of all varieties, including those varieties occurring among wild animals living in the wild; and it is seen as a provision for preserving the unchanging stability of the originally created distinct species.

This argument has led, Wallace says, to a very general belief in the stability of species. However, he reflects that there is also a common belief in "permanent or true varieties." These are "races of animals which continually propagate their like, but which differ so slightly (although constantly) from some other race, that the one is considered to be a *variety* of the other." There is usually no way to tell which is the variety and which the original species, except in the rare cases when one race shows itself to be the original species by producing an offspring unlike its own members and resembling those of the other race. Now, Wallace continues, to admit the existence of such varieties may seem incompatible with belief in the permanent invariability of species. But this difficulty has been resolved, he reports, by assuming that such wild animal varieties have strict limits, and can never again vary further from the original type, although they may revert to it. From the analogy of domesticated animal varieties reverting in the wild, reversion of wild animal varieties is thought very probable.

The whole argument assumes, Wallace emphasises, that varieties of wild species, varieties occurring and living in the wild, are governed by the same laws—regarding their permanence or further variation—that govern feral domestic animal varieties which have originated on farms but escaped since to life in the woods. But the very object of his essay is, he declares, to show that this same-laws assumption is false. For, in the state of untamed, undomesticated nature, there is a general principle which causes many wild varieties to survive the parent species, and to give rise to successive variations departing further and further from the original type; and, moreover, this very same principle also produces in those feral, domesticated, escaped-to- the- wild, animal varieties the very opposite tendency: to revert, that is, to the parent form. His next paragraph starts by identifying this principle, this lawful cause: "The life of wild animals is a struggle for existence."

11.5 The Struggle for Existence in the Wild

In preparing to read 58E as aimed at this object and as a descendant of the CVC, we must focus next on some textual details which can lead to broader contextual issues. The CVC was aimed directly only at the wild-variety-limits thesis itself. 58E is also aimed at this thesis, but at this thesis as supported by the domestic-variety-reverting-in-the-wild thesis. And here the new invocation of the struggle for existence in the wild, absent in the CVC, has a double role. It has one role in driving indefinite varietal succession in wild animals; and it has another role in confirming the domestic-variety-reversion-in-the-wild thesis while discrediting this thesis as support for the wild-variety-limits thesis. These two roles are announced in advance in the essay's first three paragraphs, so that readers will know what is coming in the rest of the essay.

Wallace's readers are made aware right from the beginning that a main issue is the contrast between wild and domesticated life. There is no such preoccupation in the CVC. All that is said there is that the only ground for opposition to unlimited variation is "that the varieties which have been produced under the influence of man have certain limits." In a notebook entry very soon after the CVC, Wallace is more expansive on these issues: changes brought about "artificially in short periods may have a tendency to revert to the parent stock," although this, he adds in an inserted note "in animals is not proved". Reversion, he writes, is considered "a grand test of a variety" meaning presumably a reliable indication that a variety is not a species. But, he continues, when the change "has been produced by nature during a long series of generations, as gradual as the changes of Geology," it could be "permanent & thus true species be produced" (Costa, 2013a, MS p. 45). By October 10th, 1856, we know that Wallace was advising Darwin to keep "domestic varieties & those appearing in a state of nature, distinct," in Darwin's words thanking Wallace in May the following year (Wyhe & Rookmaker, 2013, p. 129).

So, contrasting wild and domestic animals—crucial to the very object of the essay—was no new concern of Wallace's in February 1858.We do not know when he first came to the essay's comprehensive views of these contrasts; but, once again, on this topic there was probably no single swift shift from ignorance to knowledge a few days before mailing 58E to Darwin.

After announcing its very object—namely, refuting and replacing the assumption that varieties, in wild animals and in domestic animals living in the wild, are governed by the same laws regarding their permanence or further variation—the essay proceeds to unfold its argumentation over what we may distinguish as three main divisions: (I) a first division preparing for the deduction of endless progress and divergent change; (II) a central second division climaxing in that triumphant deduction, and a final third division (III) dealing with (i) domestic animal life (ii) a replacement for Lamarck's account of adaptive progress, and (iii) a survey of all the general facts that the essay's theory can explain.

Going back now to the essay's three opening paragraphs, it is striking that what follows, in the first division, is a sequence of four paragraphs occupying four pages with no mention of varieties. These four pages are explicitly about generalisations concerning not varieties but individuals and species and the struggle for existence in the wild. Only when these fundamental generalisations are in place can they be applied to the life of wild varieties in the state of nature, as those applications are developed in the second and third divisions of the essay.

These distinctions between these divisions of the essay are essential to the essay's composition and the structure and strategy of its entire argument. But Wallace's terminology is notably consistent throughout his paper. And so, before tracking those distinct movements, we do well to see how this consistency is achieved. We may start here with what Wallace has to say about variations in his central division (II). For that division begins by considering what a variation is. A variation in an offspring is a modification of and so a deviation from parental characteristics, as when a brown offspring is born to grey insect parents. A variety is produced only when a variation is hereditary and so is the beginning of a new race, a new permanent and true-breeding generational succession. Such a generationally persistent variety is a race. And a species is a race. A variety just is one or more individuals which share, and so are distinguished by, one or more hereditary peculiarities. And so too is a species. Races that are species are more different from each other than races that are varieties. The members of a race, whether a variety or a species, are distinguished from members of another race only by their hereditary peculiarities; so races, whether varieties or species, are not also distinguished as geographical or reproductive communities. Such community distinction possibilities are not mentioned.

What Wallace says about the entities signified by these terms is also strikingly consistent. One variation can be adaptively advantageous, another disadvantageous, in individual struggles for existence; and so one variety can be better adapted than another. And likewise, one species can be better adapted than another. A better adapted race of birds, whether a variety or species, is so because its individual members are better adapted—better able especially to find food, avoid predators and survive harsh seasons—than the members of another race, another variety or species. For two races, whether species or varieties, to be competing simply is for their individual members to be competing with members of the other race as well as with members of their own. The members of an adaptively, competitively superior race will inevitably become more numerous than the members of another adaptively inferior race. Existence, duration, lasting, surviving are what is prolonged and preserved when the finding of food and other necessary conditions of existence are met. One individual can outlast or outlive another but one individual cannot outnumber another. One race, a species or variety, can outlast and consequently outnumber another only if its individuals do so. The outlasting of better over worse adapted individuals is the cause not the effect of one race lasting longer and being more numerous than another. For an individual, for a variety and for a species, the struggle for existence is a struggle not to get into existence but to stay in existence. Varieties and species succeed in doing so only as long their members do. Individuals are members of varieties or species but do not have members. If an individual succeeds in its struggling for existence this is because it has done what it has to do—with its organisation and habits, with its wings, say, and its migrations—if it is to thrive in its circumstances, another favourite Wallace term. Failure in an individual's struggle is death; for a variety or species it is extinction, terminal loss of population.

A leading theme throughout 58E is that individual wild animals have to struggle to exist unaided by individuals of their own variety or species or by those of any other race; while domestic animals do not have to struggle at all because helped—most obviously fed, protected and sheltered—by their human carers. There is, Wallace emphasises, only self-help in the wild. Success in surviving depends entirely on the individual's unaided efforts and energies.

11.6 Humans, Society and Nature

Before going to the three divisions in Wallace's paper we should first relate this leading theme to two striking absences in the essay. Plants are absent and so are humans.

Plants appear as food for animals and humans, but the essay's theory of varietal succession is never explicitly applied to plants, not even in the paper's last sentence where there is the one mention of "organized beings". The most likely reason for plants being excluded is that Wallace sees no contrast between struggling self-helped-only wild plants in the state of nature and domestic plants cared for by humans; no contrast for plants equivalent to the contrast so fundamental for his theorising about animal life.

Humans appear as growers of crops, carers of some animals and threats to others; but the essay's theory is never extended to humankind. This deliberate omission is most probably because of another decisive contrast, one that will loom large in Wallace's 1864 paper on natural selection and man. The contrast traces very likely to his reading, around 1853, of Herbert Spencer's *Social Statics* (1851), the socialist-inclining book cited as uniquely instructive in that 1864 paper. Consider civilised human life, now and even more when the socialist ideals shared by Spencer and Wallace are realised in the future. Humans, unlike animals, are naturally social and sympathetic. So, by contrast with wild animal life, there is help for struggling humans. But unlike domesticated animals, humans are aided by con-specific help, from other social and sympathetic humans, who are all collectively self-domesticating.

These themes are manifestly of decisive contextual significance for any interpretation of 58E. To see why, forget 58E for a moment and consider the assimilative projection deployed when any account of nature is aligned with a normative view of society; a view that sees competition, say, as a driver for progressive change in both human and animal life. Here the social ideal is supported by a likening, a comparison, of human and animal life.

Wallace in 58E is not doing any such aligning, likening and comparing; indeed he is doing exactly the opposite. His 58E theory is implicitly but unmistakably contributing to his socialist view of human progress, not by comparing but by contrasting civilised self-domesticating human life and entirely self-aiding wild animal life.

I must then reject Michael Ruse's view of Wallace. Silently discounting Kottler (1985) and Bulmer (2005), Ruse insists that Wallace is a group selectionist. As in society so in nature, Ruse argues: Wallace, the lifelong follower of socialist Robert Owen, values collective, cooperative, group actions over competitive, individual endeavours. For Ruse, Wallace is to be contrasted with Darwin, ever a liberal aligned with Adam Smith and therefore an individual selectionist. The latest version of Ruse's take on Wallace is in the book *Debating Darwin* which he has co-authored with his friend Robert Richards. This is a friendly but dialectical co-authoring because these two friends strongly disagree over how to interpret Darwin and Wallace. I am friends with both of them and I am glad to acknowledge that they are both right, although only right in thinking the other is wrong. Especially relevant here is Richards's refutation of Ruse's contrast, between a Smithian individual selectionist Darwin and an Owenite group selectionist Wallace; a refutation matched by Ruse's countering of Richards's mistaken take on Darwin as a German romantic (Richards & Ruse, 2016).

Jean Gayon, although inclining to read Wallace as a group selectionist, does not join Ruse in tracing this to any of Wallace's political commitments. My own view differs from Jean's and from Michael's. I think Wallace's selectionism about animals in the wild is individualist in its self-help-only conception of the struggle for existence, and in its hereditarian conception of variations, varieties and species. And I think that Wallace's selectionism is individualist because he was a socialist aligned with Owen and Spencer. His views, on self-domesticating, social, sympathetic, cooperative, altruistic humans, were contrasted so thoroughly with his views of wild animals, precisely in order that his views about humans could be supported by these very contrasts.

11.7 The Decisive Population-Ecology Generalizations

We can now return to 58E and to the first two of its three main movements or divisions. Both end with commanding generalisations. The opening movement, filling those four pages where varieties are not mentioned, ends with two italicised generalisations which are going to be applied to varieties from the very opening of the next movement.

The first generalisation says that the total animal population of any region or country is generally stationary because kept down by periodic food shortages and other checks. The second says that the relative abundance or scarcity of the individuals of the region's various species is entirely due to their organisation and consequent habits. For some organisations and habits make it harder to procure food and secure safety, while others make it easier. Any such difference, in difficulty or facility, is therefore balanced by a proportionate difference in the population existing in a given area.

These two population-ecology generalisations—as we may call them with permissible anachronism—are pivotal for the whole essay. Only if they are understood can Wallace proceed, he says, to consider varieties which these generalisations apply to directly and importantly. We will focus on this crucial transition to varieties in due course; but we need first to see how Wallace prepares for it over the previous pages.

After the paper's third paragraph announcing the object of the whole paper, a fourth paragraph succinctly summarises how the four pages will proceed in building toward the two population-ecology generalisations. They will move from the struggles of individuals to the populational abundance of some species and rarity of other closely-allied and therefore very similar species.

The full exertion of all their faculties and energies is required of wild animals to preserve their own existence and provide for their infant offspring, procuring food in the harshest seasons and avoiding predation are the primary conditions determining the existence of individuals and entire species. These conditions determine too the population of any species; and, by carefully considering their circumstances, we may be able to explain what is otherwise so inexplicable: why closely allied species often differ so much in their abundance or rarity.

In following Wallace through what he previews succinctly here, notice what he will concentrate on most often: birds, their food in severe seasons and their predators. Notice too what, he will insist, is irrelevant in explaining any species' populousness: clutch or brood or litter numbers. His insistence is probably due to his wish to oppose Lyell's thinking that prolificness (a Lyell word) was directly relevant to populousness (Costa, 2014, p. 197).

A familiar Malthusian contrast shows the irrelevance. The contrast is between the unchanging actual numbers in a population subject to such checks as food limitation and predation, and the potential geometrical increase in numbers in the absence of all such checks. Even one pair of a bird species with a small average clutch size could have vast numbers of descendants in only a few generations if those checks were absent; showing that it is the checking and not low prolificness that keeps actual population numbers constant and limited. The mathematics of that checking confirms the conclusion. Even if the progeny in a species is only twice as numerous as the parents, then, whatever the average number of individuals living in any country, twice that number must die each year; and large broods will be superfluous in keeping up that average number of survivors. On this irrelevance of fertility, Wallace features a species of birds that he has never seen in a country he has never visited: the North American passenger pigeons who get more 58E words than any other species. These birds, he says, raise only one or two offspring a year but, despite predation, their vast flocks fill the skies thanks to the widespread and plentiful availability of their food, and to their powerful flying in migrating to distant supplies when faced with local shortages.

On the annual dying in the wild, most important is what the mathematics does not consider. Because the individual survival and existence of each animal depends upon itself, those dying will most often be the weakest: the very young, the old and the diseased, while those surviving longest will be the most healthy and vigorous. Wallace is explicit: it is because life in the wild is an unaided individual struggle for existence that the dying is not random. These illustrative individual differences are not of course hereditary. And that is entirely appropriate. Wallace is not yet applying to true-breeding varieties what he has established about individuals and species.

What is more, before getting to varieties, he has to show how what goes on among the individuals of a species must occur among the several allied species of a group: namely, that those species best adapted in relation to their circumstances must reach and maintain high numbers, while the less adapted must diminish in numbers and in extreme cases become extinct. Between these extremes there will be species presenting various degrees of capacity to preserve life; and this is how the abundance or rarity of species can be explained. Our ignorance will generally prevent us tracing all the effects to their causes; but if we knew enough about the various species' organisation and habits, and could measure the degrees of their capacity to sustain life in all the surrounding circumstances, then the proportionate abundance of individuals that inevitably results could be calculated.

Wallace's reasoning here is revealing for three reasons: first, this passage immediately precedes the two italicised population—ecological theses that will be applied to varieties. Second, we can anticipate how that application will proceed: namely, by treating the various varieties of a single species as he has here treated the various closely allied species of a genus or family. Third, we can learn from comparing this paragraph in Wallace's paper with a remarkably similar passage in a book of Darwin's that Wallace had access to in the Malayan years. The book is the second, 1845, edition of Darwin's *Beagle* voyage *Journal of Researches*; and the passage—at the end of the eighth chapter and newly written for this edition—discusses Lyell's account of species extinctions (Darwin, 1845, pp. 175–176). In the 1839 edition Darwin had argued for a non-Lyellian, neo-Brocchian intrinsic species mortality theory of species extinctions. But here he returns to an acceptance of Lyell's theory. In doing so Darwin emphasises two themes central to Wallace's division (I): one, although Darwin does not mention him by name, is Malthus's contrast between actual checked stable population numbers and the potential geometrically-increasing numbers without such checks. The other is the causal ignorance theme: namely, that we rightly assume that population differences between similar species must be due to differences in their adaptedness and in their circumstances, although we cannot expect to learn at all easily just how the decisive causes are having their effects.

Wallace surely had memories of reading Malthus about a decade and a half before writing his essay. But he may have drawn also on a much more recent reading of Darwin's Malthusian support for Lyell's account of species extinctions, and been following Darwin in not naming Malthus in his essay.

11.8 Unchanging and Changing Conditions

The opening paragraph of the central division (II) of the essay is long. It considers first what happens in constant conditions, and then what happens when changes occur. Variations from the typical form of a species must often affect individual habits or capacities. Even a change of colour making for more or less visibility may affect safety. More consequentially a decrease or increase in limb strength would affect food procurement. Indeed, most changes would increase or decrease the powers of prolonging existence. What is more, weaker legs or wings will cause a loss of population; while any species producing a variety with a slightly increased ability to preserve existence must gain in numbers. There may be individual exceptions but on an average the rule will always hold good. So, all varieties fall into two classes: "those which under the same conditions would never reach the population of the parent species; and those which would in time obtain and keep a numerical superiority".

Now let the physical conditions become worse, thanks to a drought or invading locusts or any change making existence more difficult for a species to maintain and extinction harder to avoid. Obviously, of all the individuals composing this species, those forming the least numerous and most feebly organized variety would suffer first; and with continued severe pressure this variety would soon be extinct. If the same causes persist, the parent species would suffer next, gradually losing numbers and, if the adverse conditions recur, becoming extinct. Eventually, the most superior variety would alone remain, and when favourable conditions returned would rapidly increase in numbers and occupy the places of the extinct species and its extinct varieties.

Throughout this reasoning, starting with varying individuals and going on to varieties and species, nothing is happening to varieties and species except the aggregated consequences of what is happening to their members, in their individual and more or less successful unaided struggles for existence. It is likewise in the reasoning taking Wallace to his triumphant deduction. He continues with an echo of the CVC. The variety he says, would now have replaced the species. But here he adds: "of which it would be a more perfectly developed and more highly organised form." The echo is in the assertion of replacement only; the claim for more perfection and higher organisation is provided by the post-CVC thinking about progress through struggle. Likewise with what follows: the superior, better adapted variety would be more prolonged in its individuals' existence and in the existence of its race. It could not revert to the original form, for that form is inferior and could not compete for existence with the new superior one. Even if the original type continues to reproduce itself the new variety must continue to be preponderant in numbers and, if adverse conditions recur, alone survive.

The CVC claimed that a variety that becomes the only one not to go extinct will be living like a species in increasing in range and numbers and facing varied conditions and circumstances. In living like a species, it may be presumed to vary like one. So, as Wallace says now, the new, improved and populous race might eventually give rise to new varieties, exhibiting, as in the CVC, several diverging modifications of form. The CVC is however transcended in the closing sentence and a half of the deduction: any of the modifications that tend to increase the facilities for preserving existence must by the same general law become in their turn predominant numerically. Here progress and continued divergence are deduced from the general laws which regulate the existence of animals in a state of nature, and from the undisputed fact that varieties do frequently occur.

The result will not be invariable. Wallace is only arguing, he says, for certain varieties having a tendency to maintain their existence longer than the original species. And the consequences of this tendency must ensue; for although the doctrine of chances can never be trusted on a limited scale, the high numbers of wild animal individuals ensure that the outcome over long periods of time will converge on what theory demands. With this statistical defence of his theory in place, Wallace can now move to the third and final division of his expository task.

11.9 The Closing Movement

Turning to domesticated animals—and how varieties produced among them are affected by the principles just dwelled upon—Wallace insists on an essential difference in the condition of wild and domestic animals. Among the wild ones their well-being and very existence depend on the full exercise and good health of all their senses and powers, while among domestic animals these are only partly exercised and sometimes entirely unused. The consequences of this difference are that, in a wild species, if any variety has increased power or capacity in any organ or sense this leads directly to increased numbers and the outlasting of inferior varieties, while in domestication any such increased powers and capacities are useless and unused. He continues by drawing further corollaries from this contrast. All variations in a domestic animal have an equal chance of survival; and none that would be disadvantageous in the wild are so under domestication. Short-legged sheep and pouter pigeons are varieties that could never arise in nature because the first steps toward those forms would lead to the extinction of the race. Any such races can avoid extinction in the wild only by reverting to the type of the original wild stock.

From what Wallace says here, together with what he has said earlier in deriving his theory, it follows that for all domestic varieties in the wild there are possible only the same two fates that face the least adapted variety in a wild species when conditions become more difficult: reversion or extinction. In a wild species, however, there will usually be at least one variety that is better able to survive in the worsened conditions than is the original parent species form. On the return of favourable conditions, such a superior variety will thrive and increase so as to become an improved replacement for the original species and, as in the CVC, unlimited varietal succession will proceed. The reversion of domestic varieties in the wild does not refute the possibility of this succession, because these consequences of the struggle for existence are quite different for wild and domestic varieties. What applies to the one is inapplicable to the other. Domestic animals are "abnormal, irregular, artificial," depending on human care for their very existence and subject to varieties that never occur, and indeed never can occur, in the wild. In sum, no inferences about wild varieties can be drawn from domestic varieties. Having here achieved the very object of his paper regarding such inferences, Wallace closes out his exposition by displaying his theory's explanatory virtues.

In contrasting his explanation of adaptive progress with the customary British misrepresentation of Lamarck he can now show his theory at work concerning real life examples. He says Lamarck's hypothesis—that progressive changes in species have resulted from animals' willed attempts to modify their structure and habits—has been easily and often refuted by writers on varieties and species, and the whole question thereby taken to be settled. But Wallace's view shows, he says, that similar results must be produced by principles constantly at work in nature. He dwells on three examples: the retractile claws of avian raptors and feline carnivores; giraffes' necks; and the concealing colouration of various animals.

Those necks were not lengthened by wilful stretchings, but because any varieties occurring with a longer neck than usual secured food over the same ground as their short-necked companions, and on the first scarcity of food were able to outlive them. Among many animals many colourings have occurred, and those races best adapted to concealment from their enemies have survived the longest. In these examples some individuals are outliving other individuals, and some varieties and so some races are outliving other varieties and races. But there is the usual causal priority: the winning and losing among individuals is causing racial winning or losing, not the other way round. The pseudo-Lamarckian theory refuted here gave causal priorities to individual desires and actions, and Wallace's alternative matches them with its own individual causal priorities.

Looking ahead Wallace sees three benefits from acceptance of his theory. It explains the balanced compensations seen in wild animals: strong wings for fast fleeing compensating for weak weapons for fending off predators. Harking back to the Sarawak law paper, there is an explanation for the many lines of divergence from a central type; and for more recent types being more specialised in structure.

His final paragraph implicitly distinguishes the theory itself as such from its explanatory deployments and virtues. The theory is about what happens to animals including domestic animals in the wild: namely, a tendency in nature for some kinds of varieties to depart unlimitedly from the original type; it is a tendency due to the same principle, the struggle for existence, that causes domestic varieties to revert in the wild. That is the theory. Here is what it can explain. "This progression, by minute steps," and hence this theory, agrees with and can therefore, he implies, explain all the phenomena presented by organized beings, their extinctions and successions in past ages and all the modifications of form, instinct and habits that they exhibit.

11.10 Concluding Remarks

I began this paper with reflections on Jean Gayon's historiographical theme of heredity and natural selection. I return to it here. Wallace's theorising makes claims about differences, most prominently the differences between what happens to wild and to domestic animal varieties in the state of nature. Some of the wild ones progress, all the domestic ones revert or go extinct. Again, in the woods there is struggling with only self-help; while, thanks to human caring, no farm animal is struggling unaided. In society people struggle, but with help from other social and sympathetic humans. As I argued in Sect. 6 of this paper, Wallace's 58E supports his socialism not by likening civilised human and wild animal life, but by contrasting them. Just as contexts can elucidate texts so can texts illumine contexts. Of all his contrasts this one reveals most directly where he was coming from ideologically. Wallace's theory of wild animal life, as grounded in his populational ecology and his hereditarian take on varieties, was individualist; and through this contrast his theory could support his socialist philosophy of civilised human life.

His contrasts are not about hereditary variation. All the variation relevant to the theory is hereditary, whether in humans or in animals. All the contrasts are about circumstances and struggling or not struggling to survive with or without help in those circumstances. So what differ are not the processes producing hereditary variation. What differ are the contrasting fates of hereditarily varying individuals, varieties and species in contrasting circumstances in the woods, on the farm or in the city. All circumstances are relational and include for wild animals not merely the food, weather and predators, but also their unaided individual interactions with these indispensable resources and ubiquitous threats.

Wallace wrote very little about heredity before doing so at length in *Darwinism*. We do know that in the mid-1840s he was reading in the books of the two ethnologists, James Cowles Prichard and William Lawrence, books published earlier in the century (Costa, 2014, p. 45; McKinney, 1969). And it seems very likely that this reading convinced Wallace once and for all that acquired characters are not inherited, that, in later labelling, "Lamarckian" inheritance does not happen. If so, his enthusiastic response in *Darwinism* to August Weismann's alternative to all "Lamarckian" theories of heredity is unsurprising (Wallace, 1889, p. 440). There was in the 1850s no theory like Weismann's available. So, on the question of the causes of hereditary variation it seems that Wallace relied on Prichard and Lawrence's teaching that, with a greater variety of environmental conditions, more hereditary variations are produced; but without the variations being acquired as inheritable adaptive responses to those environmental conditions.

In two respects heredity was for Wallace a highly relevant topic bearing directly on his theorising over the three years, from the early 1855 Sarawak law paper through to the February 1858 essay on natural selection. First, it bore on the CVC question about how all varieties, wild or domestic, are produced. Second, it bore on his accepting that races which are permanent and true varieties differ only in degree not in kind from races that are species.

This acceptance simplified the heredity topic for Wallace. In his *Notebook* 4 he responds in various ways to various proposals in the literature concerning variations, varieties and whether or not their peculiarities are inherited. But in the essay he was content to say—at the crucial moment when making the deduction central to his theory—that all he was assuming was the undisputed fact that varieties are always occurring; and he could leave unsaid that these were permanent and true races as discussed at his paper's opening.

My aim here has been to complicate narrative issues while not making conceptual questions murkier. By inserting the CVC between the Sarawak law paper and the 1858 essay, we can, I suggest, get welcome complications that move us away from any swift shift story about knowledge replacing ignorance in a single moment of Malthusian insight. And we gain in conceptual clarity by seeing more precisely what causal-explanatory challenges the CVC raised but left unresolved, and how their resolutions in the essay were grounded in new contrasts between wild and domestic animal life. The dual role for the struggle for existence in 1858 was not just filling gaps in the earlier CVC; because the need to counter the domestic-variety- reversion-in- the- wild objection to unlimited wild-varietal succession was not confronted in that conjecture. Equally, that conjecture had no direct bearing on the inclusion or exclusion of mankind in relation to any general varietal succession theorising. The self-help-only-in-the-woods versus other-help-on- the-farm thesis complicated the contextual ideological issues in fresh ways beyond where the CVC left them.

Do these narrative complications and conceptual clarifications make implausible any consummating Malthusian moment story, any story that has Wallace remembering what he had learned from reading Malthus over a dozen years before; and that has him quickly seeing how to move all the way, from what is left unexplained in the CVC to what is comprehensively explicable with the essay's roles for the struggle for existence in wild animal life? Implausible, yes perhaps, but downright incredible, no. In questioning the single, simple, swift shift tales that may tempt Wallace's biographers, as they tempted Wallace himself especially in his old age, I have been very aware that the documentary resources, published and otherwise, are far from settling these matters. My questionings are not designed for uncritical acceptance but to prompt further critical questioning.

As for two differences between Darwin and Wallace: first, Wallace saw his theory as a varieties and species theory. Much of the time, as in the CVC, he was inferring conclusions about species from generalisations about varieties. But on a crucial cluster of issues the inferring ran the other way. In unchanging conditions two closely-allied species can be in competition in the same region; and if one is better adapted than the other, then their population numbers will be different but stable and constant. Those numbers will change when conditions change for the worse; and the competitive winning of the superior and terminal losing of the inferior species will await such an adverse shift in conditions. This assumption Wallace transfers to the various varieties of a species, so that such changes in conditions are needed to drive their populational winning and losing. This transference was sanctioned for Wallace by his view that varieties and species are races that differ only in the amount of the differences between them. This view was consistent with the CVC's key argument that a single surviving variety would be living like a species and so varying like one. In Wallace's reasoning about varieties, there is a clear contrast with Darwin's view of winning and losing among con-specific varieties not requiring any driving by environmental changes (Bulmer, 2005).

And, second: looking at the developments from CVC to 58E has brought out very prominently how fundamental to the essay was the thesis of that famous sentence discrediting all inferences from domestic to wild varieties. Given that sentence, it is not surprising that one sometimes reads historians saying quite unequivocally that Wallace never accepted Darwin's analogy between natural and artificial selection. Unequivocal pronouncements are often valuable prompts for sharpening clarifications. And this one does need to be met with an unqualified denial. For what is true is that Wallace never rejected the analogy. Go to his book of 1889, *Darwinism* (Wallace, 1889). There in Chaps. 5 and 6 is the analogy, exactly as Darwin was first developing it in early 1839 and ever thereafter. But is there not a famous sentence in the book's preface (at p. vi) that is plainly negative about the selection analogy? No: that sentence is not about any analogical comparing of natural and artificial selection; for it is not about selection at all, but about hereditary variation. It is negative about Darwin's relying unduly on inferring conclusions about hereditary variation in the wild by generalising from facts about hereditary variation in domesticated animals. And so it is not a reflection about selection but about the material for selection.

But could Wallace be consistent in welcoming Darwin's analogy when he first met it in 1859–1860? Yes, for three reasons. First, he had not given any thought until then to artificial selective breeding of domestic animals. So he could simply learn what Darwin had to teach him on this. Second, Darwin on artificial selection did obviously, albeit unknowingly, contradict Wallace's thesis that the variations of domestic animals have the same chance of survival unlike those in the wild. But, Wallace, in replacing that thesis with Darwin's views, did not have to change his mind about wild animals nor about domestic animals living in the wild. Third, Darwin has artificial selection making domestic animal varieties suited to men's fancies and uses, and so not usually making them better able to struggle in the wild, indeed far from it in many cases. On this contrast between domestic and wildlife, there was again no irreconcilable clash with Wallace.

But do we have any grounds for thinking that Wallace did welcome Darwin's analogical comparisons and contrasts? Yes, we do. Only in one text, namely 58E, does Wallace develop his wild versus domestic life contrasts, and never thereafter; which suggests very strongly that he simply set them aside after reading Darwin, and replaced them in his natural selection theorising with Darwin's analogical comparisons and contrasts. In his 1870 book, *Contributions to the Theory of Natural Selection* (Wallace, 1870), the analogy is invoked in an essay first written in 1867; and is only barred in the very slightly modified version of 58E printed there.

But in welcoming the analogy was Wallace not giving up his theory? No: his 58E contrasts were not such that those contrasts constituted his theory. His reasoning about animal life in the wild was his theory, and his reasoning about domestic animals was refuting an objection to that wildlife theory. This reasoning was not a component of the theory but a pre-emptive countering of a countering of it.

Darwin's selection analogy was not his theory either, but support for his theory. The existence of selection in the wild was secured by his reasoning from the consequences of the struggle in the wild, as in Wallace. So, the analogy did not establish that selection was going on in the wild; it established not the existence but the powers of the selection due to this struggle. For this struggle, being more precisely, comprehensively, consistently, persistently and enduringly selective, could have the power to make not only varieties as artificial selection has done, but new species as artificial selection has not (White et al., 2021).

Papers on Wallace often become papers about Darwin. Before this paper does, let me end it with a promise to prepare a sequel on how the two Lyellian naturalists resembled and differed from one other. They converged independently on very similar theories; and it is still a good question to ask why it was those two naturalists and no others who did. A good question, but one for another time, when this paper's views about Wallace's thinking from 1855 to 1858 may perhaps be useful.

Acknowledgements I am much indebted to discussions with Jim Costa and John van Wyhe. Their invaluable studies of Wallace have greatly enhanced our knowledge of his Malayan years. The section on references in the superb edition of *Notebook* 4, at pp. 544–545, is very informative on the indispensable online resources. The *Alfred Russel Wallace page* website directed by Charles Smith guides one to the others. The two multi-authored volumes edited by Charles Smith et al. include important new papers. The text of this paper has benefited greatly from Professor Charlotte Sleigh's professional editorial expertise. Many thanks too to George Beccaloni, especially for decisive guidance on Wallace's debts to Malthus, and to Greg Radick for characteristically cogent suggestions for crucial improvements to the final version.

References

Bulmer, M. (2005). The theory of natural selection of Alfred Russel Wallace FRS. *Notes and Records of the Royal Society, 59*(2), 125–136.

Costa, J. T. (2013a). *On the organic law of change. A facsimile edition and annotated transcription of Alfred Russel Wallace's "Species Notebook" of 1855–1859.* Harvard University Press.

Costa, J. T. (2013b). Engaging with Lyell: Alfred Russel Wallace's Sarawak law and Ternate papers as reactions to Charles Lyell's *Principles of geology. Theory in Biosciences, 132*(4), 225–237.

Costa, J. T. (2014). *Wallace, Darwin and the origin of species.* Harvard University Press.

Darwin, C. R. (1845). *Journal of researches* (2nd ed.). John Murray.

Darwin, C. R., & Wallace, A. R. (1858). On the tendency of species to form varieties, and on the perpetuation of varieties and species by natural means of selection. *Journal of the Proceedings of the Linnean Society: Zoology, 3*(9), 45–62.

Gayon, J. (1998). *Darwinism's struggle for survival. Heredity and the hypothesis of natural selection.* Cambridge University Press.

Gayon, J., & Petit, V. (2019). *Knowledge of life today. Conversations on biology: Jean Gayon interviewed by Victor Petit.* Wiley.

Jones, G. (2002). Alfred Russel Wallace, Robert Owen and the theory of natural selection. *British Journal for the History of Science, 35*(1), 73–96.

Kottler, M. J. (1985). Charles Darwin and Alfred Russel Wallace. Two decades of debate over natural selection. In D. Kohn (Ed.), *The Darwinian heritage* (pp. 367–432). Princeton University Press.

Lyell, C. (1835). *Principles of geology* (4th ed., 4 vols). John Murray.

McKinney, H. L. (1969). Wallace's earliest observations on evolution: 28 December 1845. *Isis, 60*(3), 370–373.

McKinney, H. L. (1972). *Wallace and natural selection.* Yale University Press.

Moore, J. (1997). Wallace's Malthusian moment. The common context revisited. In B. Lightman (Ed.), *Victorian science in context* (pp. 290–311). Chicago University Press.

Richards, R. J., & Ruse, M. (2016). *Debating Darwin.* Chicago University Press.

Spencer, H. (1851). *Social statics.* John Chapman.

van Wyhe, J., & Rookmaker, K. (Eds.). (2013). *Alfred Russel Wallace: Letters from the Malay Archipelago.* Oxford University Press.

Wallace, A. R. (1855). XVIII.—On the law which has regulated the introduction of new species. *Annals and Magazine of Natural History, 16*(93), 184–196.

Wallace, A. R. (1857). XLV.—On the natural history of the Aru Islands. *Annals and Magazine of Natural History, 20*(121), 473–485.

Wallace, A. R. (1858a). Note on the theory of permanent and geographical varieties. *Zoologist, 16*(185–186), 5887–5888.

Wallace, A. R. (1858b, August 20). III. —On the tendency of varieties to depart indefinitely from the original type. *Journal of the Proceedings of the Linnean Society: Zoology, 3,* 53–62.

Wallace, A. R. (1864). The origin of human races and the antiquity of man deduced from the theory of natural selection. *Journal of the Anthropological Society of London,* 2, clviii–clxx.

Wallace, A. R. (1870). *Contributions to the theory of natural selection.* Macmillan.

Wallace, A. R. (1889). *Darwinism. An exposition of the theory of natural selection with some of its applications.* Macmillan.

White, R. M., Hodge, M. J. S., & Radick, G. (2021). *Darwin's argument by analogy: From artificial to natural selection.* Cambridge University Press.

Wilson, L. G. (Ed.). (1970). *Sir Charles Lyell's scientific journals on the species question.* Yale University Press.

Part III
Case-Studies on Biological Concepts and Theories

Chapter 12
Phenomenology and the Species Problem: The Need for Dialogue Between Traditions

Phillip R. Sloan

Abstract This chapter explores the relevance of insights drawn from the Continental tradition of Phenomenology for the solution of the long-standing "species problem" in the philosophy of biology. Returning to the roots of Continental Phenomenology in the work of Edmund Husserl, rather than to its later developments, the paper situates the discussion of the species concept in relation to the concepts of "intentionality" and the "life-world" as developed by Husserl. Current conflicts surrounding the interpretation of the meaning of "species" in biological discourse are described in terms of two different intentional "thematizings" of the life-world, both of which satisfy theoretical scientific interests. One of these deals with the interest in the synchronic classification of the world. The other concerns the diachronic historical development of life. It is argued that this Phenomenological analysis give a more satisfactory interpretation of the species problem than that which has been generated by conflicts between advocates of "individualist" and "class" conceptions of natural species. The larger goal of this chapter is to highlight the importance of Jean Gayon's work in developing a dialogue between Continental and Analytic approaches to the philosophy of biology.

Keywords Species problem · Phenomenology · Edmund Husserl · Jean Gayon · Intentionality · Reductionism

> Contemporary naturalism, positivism, and other forms of empirically oriented positions are characterized by the fact that they hold, more or less explicitly, that the categories of natural science are sufficient and proper for the interpretation of the full range of human experience […] The descendants of the phenomenological tradition view Man in a more nearly transcendental frame. His mode of existence is not discontinuous with nature but is qualitatively different from that of the lower animals and the rest of nature […] Whereas the inert objects of the world and lower animals are primarily in the world of their observers, the scientists, human beings are molders and shapers of their own reality; they possess a world

Adapted from Sloan (2018).

P. R. Sloan (✉)
University of Notre Dame, Notre Dame, IN 46556, USA
e-mail: psloan@nd.edu

P.-O. Méthot (ed.), *Philosophy, History and Biology: Essays in Honour of Jean Gayon*, History, Philosophy and Theory of the Life Sciences 30,
https://doi.org/10.1007/978-3-031-28157-0_12

defined by their ideas, their attitudes, their interpretive and evaluational schemas. (Natanson, 1969, in Straus et al., 1969, p. 88)

12.1 Introduction

It is my distinct honor to contribute to this volume dedicated to the landmark work in the history and philosophy of science of Jean Gayon. As contributions to this volume demonstrate, the considerable work of Gayon moved in many directions and had many layers, ranging from historical studies, to work in areas of rhetoric, general philosophy, and the epistemology of science, bringing the special contributions of French thought into these discussions. He was also engaged in ethical issues raised by the life sciences. My reflections that follow are not treated in the way Jean pursued these issues in his own writings, but they grow out of my many conversations with him over the years, and the insights I gained from him in pursuing connections among traditions of inquiry.

In offering my own contribution to this volume, I have been inspired by Jean Gayon's ability to work across the major intellectual divide that separates "Continental" history and philosophy of science from perspectives that have developed in the Anglophone literature.[1] In many ways the inquiries dominated by the heritage of Kant, Hegel, Husserl, Émile Meyerson, Maurice Merleau-Ponty, Gaston Bachelard, Georges Canguilhem, Michel Foucault, and Jacques Derrida have had modest connection with the internal dialogue within Anglophone history and philosophy of science, and this has been particularly true in the philosophy of the life sciences. This essay is an effort to advance a dialogue between these traditions in the philosophy of biology, a dialogue now being pursued by several other scholars.[2]

There are indeed identifiable historical reasons for this lack of connections. The heritage of Logical Empiricism, combined particularly in the United States with aspects of endemic American Pragmatism, created an environment generally hostile to the metaphysical inquiries of German philosophy, and to European philosophy more generally. Marburg neo-Kantianism formed the main exception. As the specialized field of analytic philosophy of biology took rise in the 1960s, it developed generally out of the heritage of Logical Empiricism and did this with a vengeance. The original Logical Empiricist rejection of metaphysics—whether this be the speculations of German Idealism, substance vitalism, or reductive materialism—has been replaced by a scientific materialism that has become a kind of consensus position in many discussions in the philosophy of life science.[3] With this has been a loss of contact with continental philosophy of biology that extends from Kant (Brenner &

[1] See the useful collection in Gutting (2005).

[2] See Pierre-Olivier Méthot's valuable analysis of some of these issues in his (2023 in Bianco et al., 2023).

[3] On change from the original "semantic" physicalism of Schlick, Neurath, Carnap, Hempel, and the Wittgenstein of the *Tractatus*, to the ontological reductionism manifest in identity theory and emergentism, see the overview by Ansgar Beckermann (1992).

Gayon, 2009; Grene, 1965). Although there are clear exceptions to this—the work of Hans-Jörg Rheinberger, Lenny Moss, and Phillip Honenberger as the most prominent recent voices in Anglophone literature pursuing these issues (Honenberger, 2015a; Moss, 2015 in Honenberger, 2015b, 2020; Rheinberger, 1997, 2010)—there is still a lack of significant interchange. Jean Gayon was a leader in efforts to develop a more fruitful dialogue between traditions, particularly between select strands of French philosophical thought and analytic philosophy of biology, drawing particularly on heritage of Auguste Comte, Émile Meyerson, Alexandre Koyré, Gaston Bachelard, and Georges Canguilhem.

In his 2009 review article reviewing the field of the philosophy of biology, Gayon summarized a quantitative survey of articles in the main journal for Anglophone philosophy of biology, *Biology and Philosophy*. This demonstrated, at least at that time, the narrowly "regionalist" concerns of this subdiscipline, with focus on five primary topics: evolutionary theory, taxonomy, species and group concepts, ecology, and genetics (Gayon, 2009 in Brenner & Gayon, 2009). This he contrasted with French discussions that have been concerned with a broader integration of the philosophy of biology with the history of philosophy and the history of science, an approach currently characterized in the literature as "historical epistemology," a standpoint explored by other essays in the current volume.[4]

This chapter is offered as a contribution to this inter-tradition dialogue with elaboration of some of the importance of Phenomenology, as originally developed by a philosopher of importance to both French and German philosophy of science—Edmund Husserl—for the philosophy of biology. I should note at the start of my essay that appeal to the Phenomenological tradition here is not something that Jean pursued. The portraits on the wall of his Paris office were those of Georges Canguilhem and Gaston Bachelard, not of Edmund Husserl and Maurice Merleau-Ponty. Nonetheless, Jean was able to see a draft of this essay and make encouraging comments on it.

My interest in assisting in this dialogue arises from a concern with the contemporary development of more aggressive versions of strong biological reductionism, now commonly extended to claims about the human being as an entity totally explicable by a combination of molecular genetics, sociobiology, bio-robotics, and cognitive neuroscience. Although each of these themes has been around for some time, it is the new intersection of these, and the new levels of synergy between them created by dynamic research programs in biology, that underlies my concern. This has significant implications for ethical questions, for our self-understanding of what it is to be human, for the possibilities of the technological and political manipulation of human beings, and ultimately for the status of scientific reason itself. Apart from blunt pronouncements of these views by scientists such as Francis Crick (Crick, 1994), we also see the development in the last decades of a more philosophical "ruthless" reductionism by philosophers of science following in the footsteps of recent work by experimental neuroscientists like Eric Kandel (Bickle, 2003).

---

[4] See Méthot, Brenner, and Duchesneau, this volume. To be sure, *Biology and Philosophy* now deals with an expanded range of issues.

Such claims involve what Husserl and the subsequent phenomenological tradition would consider to be extensions of the "natural attitude," which are then refined and developed in the direction of scientific realism to comprise "replacement ontologies" (Husserl, 1972 [1913], Chaps. 2–3).[5] We then have the problem of A. S. Eddington's famous "two tables." One is that of our common experience. The other is the "scientific" table which "is the only one which is really there" (Eddington, 1929, quoted in Hempel, 1966, p. 78). We also see presentation of rational arguments intended to convince us of the truth of these replacement ontologies which, if accepted, involve paradoxes concerning the status of the knowledge claims being advanced.

Questioning of such claims does not require any rejection of the impressive body of empirical research in the life sciences that might be relevant to bioreductionist claims, nor does it mean to deny the realist intent in such scientific inquiry. What such claims have ignored is the presence of the human actor in science. It is my claim that a phenomenological perspective gives us the ability to step outside these frameworks of analysis, and it provides us with the needed concepts that can recognize the validity of bioreductionist inquiry while at the same time it enables us to retain the importance and fundamental role of the human knower in the scientific process itself.

To be sure, several of the considerations to which I am directing attention have arisen in partial ways from an internal self-critique within analytic philosophy itself. Philosopher John Searle, for example, seems independently to have rediscovered the need for a richer meaning of intentionality, a topic that was developed in detail by Brentano and Husserl at the opening of the twentieth century (Jacob, 2019; Searle, 1983, 2012 in Salice, 2012a). Thomas Nagel has recently put forth arguments concerning the role of "mindedness" in nature in relation to Neo-selectionist Darwinian theory that are similar to those developed fifty years previously by Hans Jonas (Jonas, 1966, Chap. 2; Nagel, 2012). That this pushback is developing from within Analytic philosophy itself suggests the deficiency of some of its categories of analysis when these bear on issues concerning biology, consciousness, and human contributions to the knowledge of the world. They point to the need to rethink certain assumptions more adequately. My claim in this chapter is that grounds for this rethinking are available through a re-engagement with at least some of the dimensions of Continental thought. More specifically, I will argue that such discussion can be productive in developing a richer analysis of issues specifically in the philosophy of biology.

Previous philosophers—notably Michael Polanyi and Marjorie Grene—have developed arguments along the lines of those I am presenting (Grene, 1965, 1974; Polanyi, 1964).[6] These have, however, been often dismissed as "subjectivizing" science, or as introducing irrationalism into scientific discourse, and their arguments seem to have little purchase on discussions within Anglophone philosophy of science,

---

[5] For a clear, uncomplicated, and incisive introduction to phenomenology, I recommend Robert Sokolowski's overview (Sokolowski, 2000).

[6] For a current overview of Marjorie Grene's arguments, I am indebted to Phillip Honenberger (2021, 2023).

particularly in the regionalized domain of the philosophy of biology.[7] They also lack some of the attention to important distinctions in the concept of intentionality that enable a more rigorous analysis of the issues.

To re-invigorate this discussion, I return directly to Husserl and to the rigor with which he explored, over a century ago, the relation of scientific inquiry to the human subject. Such a return to Husserl's original program bypasses some of the difficulties that are inevitable when later developments in Continental philosophy by Heidegger, Levinas, Foucault, and Derrida are also introduced. I also find the rigor and non-oracular style of Husserl more likely to engage analytic philosophy of science. Husserl was originally trained as a mathematician, not as a philosopher, and his work emerges from the same time-horizon and Germanic intellectual and cultural framework as the work of Frege, Mach and early Logical Empiricism, and it exemplifies much of the same rigor and concern with logical analysis, empiricism, and the focus on the philosophy of science. But the insights developed by Husserl provide, in my view, a valuable corrective to the developments of strong reductionism, behaviourism, and genetic and biophysical determinism that were inspired by, but then have been extended beyond, the original Logical Empiricist "physicalist" program.

I limit my focus to two key issues. First, I briefly summarize some of the insights into the role of "intentionality" in scientific knowing as understood in the phenomenological tradition, and the relation of this rich sense of intentionality to Husserl's notion of the "life-world" (*Lebenswelt*). Secondly, I will focus this more precisely on the way in which intentionality can be brought to bear on the issues of contention within the philosophy of life science surrounding the so-called "species problem," a topic to which Jean Gayon has devoted several of his own writings since the beginning years of his contributions to the philosophy of biology.

[7] On the negative reception of Polanyi's *Personal Knowledge* by the philosophical community, see Nye (2014), Chap. 8. Marjorie Grene's philosophical contributions to the philosophy of science, and specifically the philosophy of biology, have been more favorably assessed, but have often generated controversy. See Auxier and Hahn (2002), Part 2; and Gayon and Burian (2007). I have been unable to locate any literature that directly engages the similarities and differences between the discussions within the Phenomenological literature and the arguments of Polanyi and Grene. The extensive effort to develop a more creative interface between phenomenology and life science, edited by Anna-Teresa Tymieniecka (Tymieniecka, 1999), has no index entry to either name. Documented readings by Polanyi of Husserl's works seem to date only from late in Polanyi's career although more research is needed. There are two pages of notes on what seems to be Husserl's "Philosophy as Rigorous Science" dated August 5, 1963 in the Polanyi Archives at the University of Chicago. In these comments he criticizes Husserl, Heidegger, Sartre, and Jaspers, for failing to recognize "non-conceptual" knowledge. He also critiques Husserl for his failure to include a concept of "effort or commitment" in his understanding of intentionality (University of Chicago Michael Polanyi archives, Box 23, folders 4, 16, fol. 1, cited with permission of the University of Chicago Library). I thank Professor Mary Jo Nye for this valuable lead (personal communication), and Dr. Xiaoxing Jin for assistance in obtaining this material. In one text composed a year after the date of this document, Polanyi speaks of his concept of "indwelling" as "being-in-the-world," and that this has been studied by "Existentialism and phenomenology...under other names." He then comments that "we must re-interpret such observations now in terms of the more concrete structure of tacit knowing" (Polanyi, 1964, "Preface," p. xi).

12.2 Intentionality and Scientific Knowing

As is well-known, concern with the intentionality of consciousness and its fundamental role in knowledge of the world, was brought into modern prominence by Franz Brentano (1838–1917) at the end of the nineteenth century in seminal works that deeply influenced the early Edmund Husserl. Husserl indeed considered intentionality to be "the main phenomenological theme" (Husserl, 1972 [1913], p. 222). Although there is a burgeoning literature surrounding the topic of intentionality, the meaning of the concept varies widely and only some of these usages map on to Husserl's understanding (Byers, 2002; Cairns et al., 2001; Jacob, 2019; Smith & McIntyre, 1982).

In much of the contemporary Anglophone literature in the philosophy of science and the philosophy of mind, intentionality has been interpreted as a matter of having "beliefs" and "desires." This interpretation underlies Daniel Dennett's discussions, for example, where intentionality is explicated in terms of "folk" psychology and reduced to issues in systems theory and sub-personal cognitive psychology (Dennett, 1987, Chap. 3). The issue is then to determine how such beliefs and desires can be related to a world defined by a totalizing biophysical materialism, which is assumed to constitute the consensus position among contemporary philosophers of mind. A more nuanced presentation is that given by Pierre Jacob, who defines intentionality as "the power of minds and mental states to be about, to represent, or to stand for, things, properties and states of affairs" (Jacob, 2019).

The primary advocate for a more robust role for intentionality in contemporary analytic philosophy, John Searle, has pushed back in some respects against some of the Dennett-style reductionism and Turing-machine behaviourism, and in doing so he has independently developed an analysis of intentionality, especially that associated with speech acts, topics that also occupied Brentano and Husserl previously. Like Husserl, Searle sees intentionality as a central problem in philosophy and develops the point that intentionality is not reducible simply to having "beliefs and desires." Human intentionality is instead a "matter of directly acting causally on the world and having the world act causally on us" (Searle, 2012, in Salice, 2012a, p. 19).[8] Perceptively, Searle also sees that intentionalities are not simply isolated units, but

[8] Searle commented earlier that he simply ignored the rich tradition of prior discussion, "partly out of ignorance of most of the traditional writings on Intentionality...in the relentless pursuit of my own investigations" (Searle, 1983, p. ix). His notion of intentionality, although much richer than that found in Dennett, for example, still does not address the more comprehensive set of issues dealt with by Husserl from the date of the composition of *Logical Investigations* and beyond. For Husserl, the intentionality assumed by Dennett, and to a large degree by Searle, would only be an intentionality of the "natural attitude," and does not bear on the primary intentionality of consciousness. On this distinction see Husserl (1970a [1913], II, V: paras. 16–20, pp. 576–86). For remarks on Husserl and Searle, see Smith and McIntyre (1982, p. 225) where it is concluded that there are broad points of agreement. Where I would see a primary difference is in Searle's assumption of the primacy of scientific materialism. Husserl's primary analysis of intentionality requires suspending such assumptions in the "bracketing" of the natural attitude.

that they take place against complexes of other intentionalities or networks that may be made simply background (Ibid.).

But in the end, the issue remains for Searle a matter of determining how to relate a stronger and more adequate sense of intentionality to a comprehensive scientific materialism which, as with Dennett, is taken as established:

> The central problem in philosophy in the present era is the problem with reconciling a certain conception we have of ourselves with what we know about the world from the hard sciences. … Granted [the] conception of the universe as consisting of mindless, meaningless, physical particles, the fundamental question in philosophy is, How do we square this conception with the conceptions that we have of ourselves as conscious, intentionalistic, free-will having, rational, moral, language-using agents? The biological phenomenon of intentionality is one of, or perhaps the crucial notion, in mediating the relationship between the human reality and the more basic reality of physics and chemistry. (Ibid., p. 10)

As we see in this quotation, the material world, as defined by physics, is deemed "more basic" than any human reality. The issue then facing Searle is to find some place within this framework for the obvious intentionality of consciousness in its many manifestations. The best that seems to be allowed by those advocating this metaphysics of scientific materialism is some kind of non-reductive physicalism that can make room for emergent properties.[9]

The radical approach developed by Husserl, however, was to position such appeals to scientific materialism in a complex relationship to a *transcendental* standpoint in which the intentional inquiries of science and the philosophy of science are themselves placed within the framework of the more general intentionality of consciousness considered in itself.

Husserl's critique of naturalism, and its commonly attendant scientific materialism in relation to intentionality, dates from his foundational *Logical Investigations* (1970a), first published in 1901–1902, with a general programmatic statement of his position first delivered in his 1911 essay "Philosophy as Rigorous Science." The full development of his arguments would then appear in his later lectures of the 1930s as the final presentation of his philosophical program in the *Crisis of European Sciences and Transcendental Phenomenology*, written in three parts between 1934 and 1937 and first published in German from his manuscript notes in 1954, with the English translation appearing in 1970 (Husserl, 1970b). Since space does not allow detailed development of Husserl's arguments in detail, I will rely on indicated literature for further elaboration and will offer here only some summary comments of relevance on the way in which I interpret Husserl and those in his tradition to have approached the relation of intentionality to scientific knowledge claims. I will then apply aspects of these arguments to a specific issue in the philosophy of biology.

---

[9] See the collection in Beckermann et al. (1992). It is indicative of the lack of dialogue between traditions that there is no mention of Husserl or Continental discussions in this entire series of essays. More comprehensive is the collection by Salice (2012a). This is framed by Searle's introductory essay (pp. 9–22), which continues to develop the theme of intentionality in relation to a consensus scientific materialism. But the volume also has the important discussions by Salice (2012b) and Besoli (2012) that develop the phenomenological themes in dialogue with analytic philosophy.

The fundamental difference between Husserl's starting point and the philosophical developments that created much of the Anglophone analytic tradition hinges on the primacy given to the phenomenon of consciousness and self-reflection *itself* as something to be considered apart from all our "natural" attitudes, including that of our scientific understanding of the world. This argument went through stages of development in Husserl's thought, with the introduction in his *Ideas* of 1913 of his concept of "bracketing" of the intentionalities of our natural attitude (Husserl, 1972 [1913], para. 32, pp. 99–100). Robert Sokolowski has put this distinction of natural and transcendental perspectives succinctly: "When we enter into the phenomenological attitude, we crawl out of the natural attitude, rise above it, theorize it, and distinguish and describe both the subjective and the objective correlates that make it up" (Sokolowski, 2000, p. 50).

In his *Cartesian Meditations* (1931), Husserl returned to radical Cartesian doubt as the means of discovering the primacy of the *ego cogito.* Given frequent criticisms that this grounded his philosophical project in solipsism, which cannot supply a sufficient grounding for scientific realism, Husserl clarified and developed his project to affirm his belief in world that could not be claimed to be a reduction of experienced reality to mental phenomena (Ameriks, 1977). The final statement of these issues is in the elaboration of his program with the introduction of the concept of the *Lebenswelt* in the *Crisis.* This work affirmed the unassailable reality of the external world, and related his phenomenological position to history, social life, and natural science.

Crucial for seeing how Husserl's interpretation of intentionality differs from that which might be articulated within physicalism, is the distinction he seems to have derived from both William James and Kant between the "empirical" and "transcendental" egos, a distinction then developed more fundamentally in his later work. The empirical ego is that dimension of human consciousness and self-awareness which might indeed be the subject of a causal cognitive neuroscience. The further dimension of this—revealed by the "phenomenological reduction"—is the intentional consciousness that stands *behind and logically prior to* any such inquiry. This is the transcendental ego. It is here where the full meaning of Husserlian intentionality is properly located. This renders Husserl's intentionality something more fundamental than having "beliefs" or "desires." It is what is designated as "categorial" intentionality.[10] Husserl's distinction of noetic and noematic experience, developed in *Ideas I*, makes beliefs and desires *modes of* a more primary meaning of intentionality.[11] This categorial intentionality is the fundamental act of consciousness, pervading all reflections on objects and the world, both those encountered within the "natural" attitude, and also those objects of pure consciousness when belief in the world itself is suspended, or bracketed in Husserl's terminology. Such intentionality

---

[10] As Sokolowski summarizes this point: "What happens in categorial intentions is that the things we perceive become elevated into the space of reasons, the domain of logic, argument, and rational thinking" (2000, p. 94).

[11] I am indebted for this clarification to Philippe Huneman. The important discussion of *Ideas I*, Chap. 9 deals with this distinction.

is *the condition of* rather than the *object of* any explorations of a causal scientific nexus.[12] As he puts this in his Fourth "Cartesian Meditation":

> There is but one possible method, the one demanded by the essence of intentionality and of its horizons; it becomes clear that the transcendental ego (in the psychological parallel, the psyche) is what it is solely in relation to intentional objectivities. Among these, however, are necessarily included for the ego existing objects and, for him as related to a world, not only objects within his (adequately verifiable) sphere of immanent time but also world Objects. It is thus an essential property of the ego, constantly to have systems of intentionality. (Husserl, 1999 [1931], p. 65)

Husserl's recognition of the existence of a more fundamental intentionality from which the intentionality of scientific inquiry is derived, standing as an act of the transcendental ego, is a central point that I wish to emphasize in this discussion. This means that the refined scientific inquiry into mechanisms, causes, and processes as the "ontological" or "realist" focus of scientific inquiry, is taking place within the natural attitude, and is always subsidiary to this primary intentionality of the human knower that can be isolated as pure categorial intentionality. This means that the claims of strong causal determinism defined and discovered in the empirical domain cannot therefore undermine the freedom of investigation, precisely because such pursuit presupposes the presence of the free and thinking self behind this inquiry. But unlike the solipsism of Descartes, this always means the self in relation to some object of knowledge. As Joseph Kockelmans elaborated on this point:

> [...] if we want to discover what the essence of intentionality, and consequently of our consciousness, really is, we must try to discover our own real presence to ourselves and our presence to things and the world, which is inseparably connected with this self-presence [...]; in every concrete, conscious act the unity of the world appears to be already constituted before this unity is explicitly posited as such in a cognitive act of identification [...]. (Kockelmans, 1966, p. 39)

12.3 Abstracting from the Life-World

Intentionality, as a fundamental act or *noesis* of an abiding ego that stands apart but always in relation to its objects, or *noemata*, defines the way in which scientific interests and the naturally given world of experience can be seen to interact. The inquiries of refined modern technoscience, with its instrumental mediations, theoretical structures, manipulative techniques, and social organizations, are typically directed, especially in the experimental life sciences, to the search for underlying causal mechanisms rather than to the pursuit of mathematical laws and ideal mathematical objects, topics which formed Husserl's focus in discussions of science (Craver & Darden, 2013; Machamer et al., 2000). These inquiries involve the *intentional thematizings* of a more primordial horizon or "world-about-me," which Husserl

[12] This does not mean that Husserl denies causality, but that empirical causal relations are part of the "bracketing" in the phenomenological reduction. I thank my colleague Karl Ameriks for this clarification.

originally described with such terms as "indeterminate reality," or "continuous ring around the actual field of perception" in the *Ideen*.[13] By the composition of the *Crisis* in the mid 1930s, Husserl had developed these themes in a more historical and social direction with his concept of the primordial *Lebenswelt*.[14] These themata selected out within the "life-world" constitute "regions" of inquiry that define a limited ontological domain.[15]

Husserl's sustained development of the concept of the surrounding, and pre-given life-world in his last philosophical writings, added a new emphasis on the historically, socially, and embodied aspects of existence that he had not previously developed in any detail in his earlier abstract reflections. For many interpreters, this introduced a new dimension into his program which escaped the potential for solipsism and subjective Idealism latent within his earlier writings. With his concept of the life-world, Husserl moved this a considerable distance from a simple acknowledgement of the obvious fact that we live our mundane existence in a world of things, associations, physical events and other persons, a world that is accepted without reflection as the locus of our normal activity. Husserl's late explicit attention to the domain of the original givenness of things in consciousness of ordinary existence as an ontological grounding of human existence, draws attention to the structure and function of this comprehensive a priori, something more fundamental than the rules of logic or Kantian categories. This life-world supplies a framework for understanding subsequent specific thematic or "regional" analyses (Husserl, 1970b [1936], para. 36). The life-world provides the starting point for all scientific inquiry, the framework within which the normal intentionalities of practical life are manifest.[16] As Husserl explains:

> It belongs to what is taken for granted, prior to all scientific thought and all philosophical questioning, that the world is – always is in advance – and that every correction of an opinion, whether experiential or other opinion, presupposes the already existing world, namely, as a horizon of what in the given case is indubitably valid as existing [...] Objective science, too, asks questions only on the ground of this world's existing in advance through prescientific life. Like all praxis, objective science presupposes the being of this world, but it sets itself the task of transposing knowledge which is imperfect and prescientific in respect of scope and constancy into perfect knowledge [...]. (Husserl, 1970b [1936], para. 28, pp. 110-111)

Philosopher Michael Friedman summarizes succinctly this Husserlian concept and the way in which it represents a development beyond the arguments of the *Ideas* and the *Cartesian Meditations*. The life-world,

---

[13] Husserl (1972 [1913], paras. 27–30).

[14] It is generally concluded that this development was in response to Heidegger's discussions set forth in *Sein und Zeit* of 1927, especially the discussion of *Dasein* and his concept of Being-in-the-World (Heidegger, 1962 [1927], Pt. I, ii. 78 ff). David Carr (Husserl, 1970b, "Introduction,") instead puts more emphasis on the importance of Wilhelm Dilthey for generating this new concern. There is no mention of Heidegger in the entire text of the *Crisis* although there are several to Dilthey. I will not attempt to deal with this issue here. On Husserl and Dilthey, see Tillman (1976).

[15] Husserl (1970b; § 9–10), and Kockelmans (1966, pp. 166–167).

[16] Although employing different vocabulary, John Searle's distinction of "Network" from "Background" makes some of these same points (Searle, 2012 in Salice, 2012a, pp. 18–20).

> […] is just the world as it is ordinarily experienced or lived by each of us from our own subjective-relative points of view. It is the world now spread out around me in space and time, for example, containing physical bodies simply as they are perceived by me. It is a world in which I myself am embodied and may move, accordingly, among the physical bodies spread out around me in space […] In so doing, moreover, I undertake various intentional activities and practical projects, each associated with various ends or values, including such projects as the arts, crafts, and sciences – all with their various "techniques," both practical and theoretical. (Friedman, 2010 in Hyder & Rheinberger, 2010, p. 104)

Husserl highlights the philosophical significance of this life-world for any subsequent theoretical reflection. In his view, there is the need to understand the life-world as a "universal problem for philosophy." It forms the framework from which the intellect dynamically extends outward to engage in various specific thematic inquiries in response to human intentionalities.

One of the most important of these regions to which Husserl gave considerable attention in his late work, is that of natural science. This involves "a particular type of purposeful activities and purposeful accomplishments" that "lie within the universal framework of the life-world into which all accomplishments flow and to which all human beings and all accomplishing activities and capacities always belong" (Husserl, 1970b [1936], p. 138n).

The detailed researches of natural science, as investigations assuming Husserl's natural attitude, certainly push beyond ordinary sense and common-sense experience. They also generally ignore the pre-given surround of the life-world in order to pursue a highly abstract and detailed inquiry into the natural world. As Lee Hardy summarizes Husserl on this, "it is wholly appropriate, and perhaps even necessary, for the sciences to assume the existence of that which they seek to know" (Hardy, 2014, p. 63).

This thematizing of the life-world can be carried out at a fairly concrete level, as in field ecology, paleontology, animal behaviour studies, or taxonomy—adopting what I have termed elsewhere the standpoint of "natural-historical realism" (Sloan, 2015 in Sloan et al., 2015). Or one can pursue these inquiries through a highly mediated inquiry, employing complex instrumentation and elaborate experimental arrangements, as we find in high-energy physics and in the most reductive and detailed inquiries of molecular biology, investigations that give only indirect access to living phenomena mediated by theories and instruments that constitute complex conceptual schemes. But both such inquiries—those at the level of natural history, and those pursuing the complexities of biophysics— involve human intentionality that reaches out from this life-world.

12.4 Thinking About Species

These general reflections provide a framework and vocabulary against which I wish to address a persistent issue in the philosophical analysis of biology, the so-called "species problem," a topic which Jean Gayon addressed in several writings from both a historical and conceptual framework (1992a in Gayon & Wunenburger, 1992,

1992b in Gayon, 1992c, 2008 in Ludwig & Pradeu, 2008, 2014). This long-standing conflict has involved a conceptual puzzle over the units of evolution, the meaning of "individuality," complex questions of meaning and reference in classification, and the determination of the relation of "species" to other entities defined in formal classificatory systems, particularly as this relates to evolutionary theory. The problem has led to the advancement of a vast number of species concepts in contemporary philosophical literature, many of which appear incompatible with one another. Although this has been a long-standing issue in biology, pre-dating Darwin (Richards, 2010; Sloan, 1987 in Roger & Fischer, 1987, 2009 in Ruse & Richards, 2009), the issue has been raised to a new level by the consequences of the 1974 proposal by biologist Michael Ghiselin of a concept of biological species as "individuals" rather than as logical classes or sets (Ghiselin, 1974, 1997). The result has been substantial, but for many, still inconclusive debate concerning the conceptualization of a natural species in biology (Dupré, 1999 in Wilson, 1999; Ereshefsky, 1992, 2017; Ghiselin, 1997; Honenberger, 2015a; Hull, 1999 in Wilson, 1999; Kitcher, 1984, 1987, 1989 in Ruse, 1989; Pigliuccci, 2003; Reydon, 2005; Richards, 2010; Stamos, 2003; Wilson, 1999). Thomas Reydon's useful summary of positions falling generally either into "class" or "entity" conceptions, captures much of this diversity (Reydon, 2005). Jean Gayon generally fell into the group supporting the class conception, concluding in a 1992 paper criticizing the individualist thesis, that "it is reasonable to reserve the appellation 'individual' to two types of particular things: bodies and persons. A species is neither one nor the other" (Gayon, 1992b in Gayon, 1992c, p. 487).

Much of the conceptual snarl surrounding the species concept is to be traced to the heritage of a long and complex historical tangle that is beyond the scope of this paper to discuss (Richards, 2010; Stamos, 2003; Wilkins, 2009; see review in Sloan, 2013). But ignoring these historical issues, I suggest that these on-going debates can be disentangled from their historical complexities if we grant the importance of intentionality as discussed in the first section of this paper, and apply to this some of the distinctions of the analysis in the phenomenological tradition. To give this concreteness, I will return to some issues I have developed in greater depth elsewhere (Sloan, 2002 in Auxier & Hahn, 2002), but for which I was not at that time able to find a satisfactory solution. I approach this now with a more specific attention to the issue of intentionality and its relation to the life-world concept as Husserl has developed these themes.

There are two levels at which this topic needs to be discussed to see the importance of the points I am making. One is the distinction between the intentionality as involved in the natural attitude where our typical scientific discourse is taking place, and the "categorial" intentionality that involves the "transcendental" perspective of Phenomenology.

To attend to the first of these intentionalities, we are concerned with the intentionality that is a thematizing of the life-world for certain limited purposes. The intentionality involved in classification can be used as an example. This expresses a fundamental drive of human interest in the world, one that underlies language and conceptualization as well as scientific inquiry, and involves the distinction, description and naming of natural objects and the effort to distinguish natural kinds. Husserl developed some general points surrounding the theory of classification in his discussion of the eidetic reduction to essences known by consciousness, and the connection of this concern of consciousness with the *eide* revealed by this phenomenological reduction to knowledge of the world (Husserl, 1972 [1913], § 2–12). But this is left at a very general level and the explication of how such analysis applies to the specific issue under examination here needs to be given greater specification.

Classification, as applied specifically in biological science, reflects a range of different intentionalities—medical, ecological, theoretical, convenience of reference—dependent on the human interests involved. Since the work of Linnaeus, modern theoretical life science has also been engaged in a fundamental effort to determine a natural system of arrangements, meaning one that describes and categorizes the fundamental natural kinds in nature, as distinct from a utilitarian classification that may reflect only immediate and practical human needs and interests. It also seeks to make ontologically valid divisions among groups in such a natural system, and depict accurately their real, rather than arbitrary, relationships. Systematization in this sense has little immediate connection with evolutionary theory and has historically been grounded on such foundations as the metaphysics of creation in either religious or Platonic (*Timaeus*) meanings, or on a metaphysics of being (Aristotle). It is also generally an a-temporal activity, meaning that historical process and temporal change are typically irrelevant. There is no intrinsic relation with some concept of historical evolution of forms.

A second primary intentionality of theoretical biology involves the interest in explicating the historical relation of forms, the causes of changes of species over time, and the selective processes that have been behind evolutionary change. This interest is profoundly historical in focus. To employ Kant's important distinctions that have been lost in the historical discourse, the first inquiry concerns issues involved in the *description* of nature (*Naturbeschreibung*), and the other those concerned with the developmental *history* of nature *(Naturgeschichte*) (Kant, 2013 [1777] in Mikkelsen, 2013, pp. 70–71). Darwin, to be sure, drew on both of these inquiries in claiming in Chap. 13 of the *Origin* that "if our collections were nearly perfect, the only possible arrangement, would be genealogical. Descent being on my view the hidden bond of connexion which naturalists have been seeking under the term of the natural system" (Darwin, 1964 [1859], p. 449; Sloan, 2009 in Ruse and Richards, 2009).

To give this discussion concrete focus, I employ again a specific example I have used in more detail to illustrate this difference in intentionalities—the description of a new species of freshwater fish of the kind that can be taken from a standard taxonomic journal like *Copeia*, the major organ of ichthyological taxonomy (Sloan, 2002 in Auxier & Hahn, 2002).[17] This example involves the proposal for scientific acceptance of a previously undescribed form of the Madagascar fresh-water fish genus *Bedotia*, discovered during routine stream sampling. This was presented in

[17] As an autobiographical note, I spent several years doing this kind of taxonomic work. I should emphasize that this is an example from vertebrate taxonomy and issues in the classification of many invertebrates, bacteria, viruses and plants raise additional complicating questions that I will not attempt to address in this essay.

2001 as a new species, *masoala* (Sparks, 2001). The proposal of this new species for acceptance involved the examination of the "type" specimen of the closely similar species *B. marojery*, held in a museum collection, along with examination of its "paratypes"—preserved members of the same species which display the official range of variation in the defining characters of this type specimen.

Designation of the new species then involved demonstration that the ranges in the variations on the morphometric properties of the new form, based on measurements taken on twenty-seven different characters, could be shown statistically to cluster as a group outside the ranges defining *B. marojery*. To be sure, no one character is deemed necessary and sufficient for this distinction, although certain of these—coloration, number of vertebrae, gill raker count and number of lateral line scales—are singled out as the primary identifying characters of the new species. The new form, with designated "holotype" and "paratype" specimens, along with descriptions of the ranges of characters displayed by sixty-eight additional specimens, is thereby officially proposed as a new and separate species of the genus *Bedotia*.

The practice employed here separates one form from a closely related one by means of a limited number of defining characters, all of which are universally present in all specimens of the class, but which all individually vary statistically. To move away from a misleading critique of such taxonomic practice as involving "typology" and "essentialism," common terms of approbation in some circles,[18] this classification involves a *historically-contingent* essentialism which has no commitment to a claim that this form is perpetuated eternally, or that it conforms to some kind of transcendental archetype.

It is important to observe that very little in this museum practice has altered since the time of Cuvier and Valenciennes, with the exception of the introduction of statistical methods, stemming particularly from the work of British biometricians in the early 1900s. These methods can involve simple statistical procedures such as T-tests and measures of standard deviation, or more sophisticated computerized methods. These methods allow mathematical analysis of relations of means, ranges, and central tendencies displayed in the selected morphological characters of the two groups. By such methods, species can be separated, and their degrees of affinity to other species can be worked out. The understanding of a "species" here as the taxon, designated by a proper name, within the class category "Species," comprised of the individual specimens grouped in this collection as the class extension, seems unproblematic.

The debate over "individuality" at this level seems to have been created by the complexity that in some areas of biology, particularly invertebrate zoology and botany, the meaning of individual is highly complicated. Is a colonial bryozoan an individual entity or a complex collection of individuals with some material interconnection? Where in the life cycle of slime molds (Phylum Amoebozoa), which can show radically different phases and exist either as individual sporangia, or as

[18] For an influential statement see Hull (1965). For critical comments on this "essentialism story" see Honenberger (2015a); Winsor (2006).

combined amoeba, is one to define the individual? Defining the entities being classified here is not at all obvious. But for most of our practical life within a world of macroscopic vertebrate and invertebrate forms, these unusual cases are not typical and we can classify without major problems.

I emphasize that this act of rational classification, in spite of these complexities, is one intentionality of scientific reason, a particular thematizing of the life-world. This act of classification functionally stands in a position primordial to many other scientific intentionalities, a first stage in much of subsequent theory construction, particularly in biology. Once established, the result of this original classificatory activity can then be made subsidiary to another focal interest of more general theoretical interest, that of understanding the historical relations of forms and their evolutionary relationship in time. In this case, the prior systematization of rational taxonomy is a *precondition of* a subsequent inquiry into the way in which the currently recognized six different species of the genus *Bedotia* have emerged in evolutionary history in relation to the geological and ecological changes taking place on the island of Madagascar. It can even lead to an inquiry into an earlier splitting of members of the genus *Bedotia* from ancestral members of the teleost family *Bedotiidae* at a period when Madagascar was connected to the African mainland. In this case, a cladogram of the evolutionary and historical relations of these forms can be constructed that is profoundly historical. Our concern is with issues of common descent, and the effects of ecological changes that have been important in the speciation patterns of these forms.

This second intentional action of human agency might thereby focus on the "entity" conception of species as a material historical lineage, linked by reproduction. A "species" in this sense can be conceptualized spatio-temporally and described by its ability to respond to selective pressures and diversifying causes over time. Genetic analyses of the relations of these forms within an evolutionary framework can also supply further information on the historical relations of these forms, and the timing of lineage splitting. To be sure, controversies between cladists and biological species adherents may take place over details in evolutionary discussions, and there can be disputes arising from differences in the conclusions drawn from categorization by gross morphology and those based on underlying genetic affinities.[19]

Two points I highlight here. The distinction between these two scientific intentionalities, and the various subdivisions that might be expressed within them, is not, within the framework I am discussing, a matter of deciding which account is superior to the other or claiming that one is "essentialist" and the other not.[20] What distinguishes them at this level is the *intentional act* behind these two alternative thematizings of the world. Taxonomic classification is the original *focal* act that is a necessary condition of the evolutionary inquiry. Such classification is needed if we are even to ask the second order of questions about evolutionary relationships. The subsequent focus on evolutionary history and relationship renders the original

[19] I thank Phillip Honenberger for important comments on this point. Of course, even to do such genetic analyses implies a prior recognition of the forms to be so analyzed.

[20] For comments on the needless and related "monist" and "pluralist" dispute see Kitcher (1987).

classification subsidiary or backgrounded, and in view of this new focus, the original classification can even in turn be revised and explicated more deeply.[21]

In both cases, the distinctions being made can be claimed to have a realistic intent. It is not, in the defective way that Buffon had posed the question in the eighteenth century, a debate between those holding to "arbitrary" class groupings understood as universals, and those committed to the study of "real" historical lineages understood as spatio-temporal individuals (Sloan, 1987 in Roger & Fischer, 1987). The claims of the "new essentialists" who wish to defend traditional conceptions of definable natural types is, in my view, warranted and unproblematic (Oderberg, 2007, esp. Chaps. 5, 9). For the same reasons, it is possible to define coherent morphological types and distinguish groups at different levels of the Linnean hierarchy by specifying group-constituting properties above the species level, and even defining the units of evolution in terms of higher morphological units than the level of species (Newman, 2020 in Meincke & Dupré, 2020).

But for other, more theoretically-based, intentionalities, this static classificatory framework and its claims to epistemic realism may be replaced by one that is concerned with diachronic evolutionary relationships which are changing in time, but which also makes a claim to ontological realism. Husserl's conception of thematization allows for multiple perspectives of a single reality (Husserl, 1970b, III, p. 38). This diachronic view can feed back on the taxonomic framework both to interpret what it means, and even to revise the initial analysis in terms of evolutionary theorizing. It may even result in lumping together what may be separate species in a static taxonomy.

My argument is that much of the difficulty in resolving the species problem lies not in some heritage of classificatory "typology," "Platonism," or "essentialism" as opposed to "population thinking" has been claimed by those influenced by discussions of Ernst Mayr. Nor is it demanding some new metaphysics of individuals or a radical revision of naming and reference theory (Ghiselin, 1997). Instead, the resolution of this issue requires the acknowledgement of the importance of dynamic human intentionality in science itself. Depending on varying interests, the thematizations of the natural world can have different, and even mutually-exclusive intents. There also may be practical issues of importance that decide the utility of "class" or "entity" conceptualizations of species in different contexts. Definitions of species on morphological grounds, for example, those making sharp distinctions based on constant morphological characters, may be more important for defining "endangered" species, than the holistic and historical meaning of species as historical lineages that may be of interest to evolutionary theory (Agapow et al., 2004).

To this point, the criticism can be raised that I have done nothing more than argue for a kind of pragmatic epistemological stance which simply relativizes the question of species to different human interests and does not require the apparatus of Husserl. I have myself previously endorsed such a solution as a practical option in line with

[21] I find Michael Polanyi's distinction of "focal" and "subsidiary" awareness a useful way to understand these relations between the objects of directed attention and the accepted background assumptions of the life-world in a way that Husserl does not explore (Polanyi, 1964, Chap. 4).

the arguments of Philip Kitcher and J. J. Gibson (Sloan, 2002 in Auxier and Hahn, 2002). But this is to position intentionality only at the level of the natural attitude of scientific inquiry.

Here is where the second order of phenomenological analysis becomes relevant. This is the intentionality involved when I assert "evolutionary theory is true, and the history of the species of *Bedotia* displays one of the ways it has worked in time." Or to speak of the past, I might claim that "the fossil record of Madagascar and Africa reveals evidence of the joining of these two bodies in the Cretaceous, after which they split." In these cases, a theoretical judgement is being made concerning the relations of objects and truth claims are being offered concerning the meaning of these relationships. The intentionality then becomes "categorial."[22] As Robert Sokolowski puts this:

> What happens in categorial intentions is that the things we perceive become elevated into the space of reasons, the domain of logic, argument, and rational thinking. Categorial experience is the transit point leading from perception to intelligence, where language and syntax come into play. (Sokolowski, 2000, pp. 94–95)

To move to this level of analysis is to move to consideration of the "transcendental" rather than "empirical" ego. This is not to introduce something mysterious, nor to inject a ghostly self into the picture. It is instead the recognition that the "self" can both be an object of study on one hand, and at the same time the dynamic agent of rationality behind such objectification. When one has introduced judgments of the truth of propositions, the assessment of the validity of a theoretical construct, or a self-reflection on the process of rationality itself, one is involving this transcendental dimension of reflection. In the specific case under examination, it may indeed be the preferred scientific judgement that the concern with biological species as "chunks of a temporal nexus," as "gene pools," or as "interbreeding populations," rather than collections of empirically discrete individual organisms arranged in static classes, satisfies a higher theoretical goal and is justifiably advocated as the preferred scientific stance. But this conclusion is not one made *within* the domain of scientific inquiry itself. It is to step outside this framework to make truth claims about what is being asserted.

The analogy with a game of chess might be helpful here.[23] Within the game there are discrete pieces, specific intentional strategies, and objects of specific moves that can be identified and analyzed in relation to a specified goal—the capture of the opponent's king. These are immanent to the specific game and take place within the larger domain of the rules of game of chess. But we can also step outside the game to describe the game itself, its origins, and the history of the development of specific allowable moves such as castling. In this case we have adopted a "transcendental" perspective in the sense intended by Husserl.

Within the normal inquiry of natural science, we are effectively within the chess game. Here my move is my scientific intention to describe a new species and characterize it as distinct from a related species. I am operating within the rules and

[22] For complexities in this see Siewert (2016), Sect. 3.

[23] I am developing on this analogy with revisions as employed by Sokolowski (2000, pp. 118–119).

specified procedures of the *International Code of Zoological Nomenclature.* But I can also step outside this framework and critique the assumptions of these rules. Or I can make claims about the evolutionary relationships of a group of species within a defined genus that have no meaning within the *Code*, and I may then theorize on the relationships of this group in history to similar freshwater fish on the mainland of Africa from which I may claim they have descended. I may even reflect on the issue of theory construction in general and the relations of theory to empirical evidence.

In adopting this transcendental perspective of phenomenology, my initial classificatory conception of species as a "set" or class, defined by essential properties according to the *International Code*, has not disappeared or been discredited by my subsequent theoretical reflections. My original classification is how I distinguished my rook from a pawn in the first place. But I now interpret these "species" relationally and draw lines of historical connection between them. I see them as connected horizontally by reproduction, and vertically by materially-connected ancestor and descendant lineages acted upon by environmental changes and other causes of change summarized in the concept of natural selection. Empirical evidence (fossils) might even be summoned to support such diachronic claims. But the further implicit or explicit claim that these are true conclusions, and that the reasons given should be compelling to others and accepted by a scientific community, involves stepping outside the game to make a transcendental judgment of truth or falsity.

What this analysis underlines again is the place of the knower in the knowing act itself.[24] It also shows the difficulty facing replacement ontologies which oppose scientific and common experiential realities. These arise commonly in scientific accounts when assertions of scientific realism are then offered in some way as replacements for the primordial givenness of the life-world itself. We then have the puzzle over Eddington's two tables. And if such replacement ontologies are turned upon the human subject, as we see when one encounters the strong claims of eliminative materialism, strong AI, genetic determinism, and reductive sociobiology, the status of the knowing agent making such claims becomes ultimately paradoxical.

12.5 Conclusion

In closing I should address possible misunderstandings of my argument and suggest a direction in which I would wish to expand these reflections. The central point I emphasize is the importance of a rich conception of intentionality in scientific knowing, and the ever-present nature of human subjectivity which cannot be eliminated in this scientific inquiry. I have also suggested that this realization can assist us in dealing with a long-standing scientific controversy that seems otherwise unresolvable. Within the world defined by the "natural" attitude, multiple different scientific

[24] I see this to be the point that Michael Polanyi, with an imperfect and often inadequate vocabulary, was trying to articulate with his concept of "personal" knowledge.

thematizations are possible in response to different human interests or intentionalities. We also see the necessity of stepping outside this natural framework to another level of reflection if we are to make truth claims, assess various intentionalities, and argue theoretically. And to do this involves recognition of the existence of the rational knowing agent in our scientific inquiry that is not reducible to the two-dimensional domain of the natural attitude.

To expand my arguments more adequately, I would certainly need to consider important arguments and positions that depart from the central insights I have drawn generally from the Phenomenological tradition as articulated by Husserl and his subsequent interpreters. Husserl's original project has been repeatedly criticized for its tendencies toward subjective Idealism and solipsism, and its divorce from concrete lived experience (Gadamer, 1972). The solution to this, as advocated by Gadamer and others, has commonly involved an appeal for a necessary integration of Husserl's insights with those of his student Heidegger and his "existentialist" phenomenology.

My own conclusion, to the contrary, is that this is better pursued in the direction of the non-Heideggerian "realist" and "embodied" insights that ultimately owe some debt to Aristotelianism in some form.[25] This has been developed by the tradition of "philosophical anthropology," a program of inquiry that while long known in German circles, is only slowly making its way into Anglophone philosophical discussions.[26] While emerging from some of the same historical and social context as that which framed Husserl's philosophical program, this anthropological tradition opposed the move of Husserl toward transcendental idealism.[27] This alternative line of inquiry, developed particularly by Max Scheler, Helmuth Plessner, and Arnold Gehlen, breaks with the abstract cognitive approach of Husserl that has led many to criticize his project as leading in the direction of subjective Idealism. As Joachim Fischer has characterized this difference between "philosophical" anthropology and Phenomenology, "[...] perception begins not with the efforts of subjectivity but "'elsewhere,' 'indirectly,' at the factual existence of life [...]; the focus is external, on the living world" (Fischer, 2009, p. 155). This does not substantially alter my argument concerning the importance of intentionality in empirical science, the possibility of stepping outside this to a "transcendental" perspective, nor the recognition of human rationality as something standing apart from its specific intentionalities, all of which I feel are warranted by rigorous phenomenological analysis. But it does mean a deeper connection with concrete human historical and biological existence without making the jump into Heideggerian existentialism.

Where this might be possible is through an exploration of the relation of the phenomenological perspectives to their embodiment that takes account of the unique

---

[25] In drawing together some traditional Aristotelian and Scholastic insights with the project of German philosophical anthropology, I am indebted to the important dissertation of Dylan Belton (2020). I also acknowledge my long-term debt to the many writings of Marjorie Grene (Sloan, 2007).

[26] Honenberger (2015a); See also Fischer (2009); Moss (2015 in Honenberger 2015b, 2020), and originally by Grene (1965).

[27] See Spiegelberg (1965), I, Chap. 5; (Bernstein, 2019 in Plessner 2019 [1928]; Philipse, 1995 in Smith & Smith 1995).

relation of the human being to the rest of nature by the fact of the unusual positionality of the human being in the organic world. This theme, emphasized by Plessner's concept of "eccentric positionality," draws our attention to the full implications of the upright posture assumed by human beings in the Paleolithic for the development of language, culture, and self-reflection, and the new layers of intentionality itself that accompany this transition (Dunbar, 2011 in Auletta et al., 2011; Krüger, 2015, in Honenberger 2015b; Moss, 2015 in Honenberger 2015b, 2020; Plessner, 2019, Chap. 7). If this perspective is taken, it seems possible to connect the intentionality of consciousness and the transcendental standpoint developed in detail by Husserl to a more biologically-grounded intentionality that can be understood to have developed through a historical process of evolution. To do this would not, however, be easy, and to be true to Husserl's project, it would need to avoid the reductionism that Husserl interpreted as "psychologism."[28]

Whatever are the possibilities in such a project, I suggest that we can find in a return to Husserl's insights a clear rationale for the re-positioning of scientific inquiry in relation to the given nature of consciousness and self-reflection. For most of ordinary life, including that of our detailed scientific inquiry, this can be ignored or rendered subsidiary, simply standing as the constant background to the "natural" attitude we assume in these inquiries. Recognition of this framing of human inquiry within this context becomes critically important, however, when the scientific gaze is turned, as is increasingly being done at present, upon human beings, upon conscious life, upon the motivations for behaviour, rendering the subjective individual an "object" of scientific attention. When this scientific gaze becomes simply a Frankenstein monster that renders us problematic to ourselves, one that opens the door to increasingly sophisticated forms of technological control, to a denial of ethical responsibility, to the subjection of human knowing to explanatory claims that threaten the status of scientific and philosophical rationality itself, it is important to return to some of the rich philosophical wellsprings that philosophers like Husserl and his successors and interlocutors have offered us from a Continental perspective.

Jean Gayon, I acknowledge, did not pursue these issues in this way. But his constant inquiry and his deep learning and insights, revealed to me in so many discussions, was an inspiration to explore in different ways his constant interest in keeping alive a dialogue between traditions of inquiry. I am pleased to dedicate this essay to his memory. Although he was only able to see a draft of these reflections in 2017, his remarks were very important for my decision to pursue these further.

Acknowledgements For this revision I am indebted to comments by David Depew, Phillip Honenberger, Vittorio Hösle, Lenny Moss, Dylan Belton, Karl Ameriks, M. Katherine Tillman, and members of the Philosophy of Biology Circle at the University Texas, hosted by Sahotra Sarkar. Responsibility for interpretations is my own.

---

[28] In his June 10, 1931 Berlin lecture, Husserl directly criticized "philosophical anthropology" as "psychologistic," and specifically identified Wilhelm Dilthey as its main proponent, without mention of Scheler, Plessner and others (Husserl, 1941, trans. McCormick and Elliston, 1981, pp. 315–323). On the Husserl-Dilthey relationship see Tillman (1976).

References

Agapow, P. M., Binida-Emonds, O. R., Crandall, K. A., Gittleman, J. L., et al. (2004). The impact of species concepts on biodiversity studies. *Quarterly Review of Biology, 79*, 161–179.

Ameriks, K. (1977). Husserl's realism. *The Philosophical Review, 86*, 498–519.

Auletta, G., Leclerc, M., & Martinez, R. (Eds.). (2011). *Biological evolution: Facts and theories.* Gregorian & Biblical Press.

Auxier, R., & Hahn, L. E. (Eds.). (2002). *The philosophy of Marjorie Grene.* Open Court Press.

Beckermann, A. (1992). Introduction: Reductive and nonreductive physicalism. In A. Beckermann, H. Flohr, & J. Kim (Eds.), *Emergence or reduction? Essays on the prospects of nonreductive physicalism* (pp. 1–22). de Gruyter Press.

Belton, D. (2020). *The sojourning animal: An anthropology of detachment and attunement.* Unpublished Ph.D. Dissertation, Department of Theology, University of Notre Dame.

Bernstein, J. M. (2019). Introducing Helmuth Plessner's philosophical anthropology. In H. Plessner (2019 [1928]), *Levels of organic life and the human. An introduction to philosophical anthropology* (pp. xxxvii–lxv, Trans. M. Hyatt). Fordham University Press.

Besoli, S. (2012). The primacy of functioning intentionality in Edmund Husserl. In A. Salice (Ed.), *Intentionality: Historical and systematic perspectives* (pp. 175–202). Philosophia Verlag.

Bianco, G., Van de Vijver, G., & Wolfe, C. T. (Eds.). (2023). *Canguilhem and continental philosophy of biology.* Springer.

Bickle, J. (2003). *Philosophy and neuroscience: A ruthlessly reductive account.* Kluwer Academic Press.

Brenner, A., & Gayon, J. (Eds.). (2009). *French studies in the philosophy of science: Contemporary research in France.* Springer.

Byers, D. (2002). *Intentionality and transcendence: Closure and openness in Husserl's phenomenology.* University of Wisconsin Press.

Cairns, D., Embree, L., Kersten, F., & Zaner, R. (2001). Husserl's intentionality. *Journal of the British Society for Phenomenology, 32*, 116–124.

Craver, C., & Darden, L. (2013). *In search of mechanisms: Discoveries across the life sciences.* University of Chicago Press.

Crick, F. (1994). *The astonishing hypothesis: The scientific search for the soul.* Simon and Schuster.

Darwin, C. (1964 [1859]). *The origin of species by means of natural selection.* Facsimile of 1st ed. Harvard University Press.

Dennett, D. (1987). *The intentional stance.* MIT Press.

Dunbar, R. (2011). How humans came to be so different from monkeys and apes. In G. Auletta, M. Leclerc, & R. Martinez (Eds.), *Biological evolution: Facts and theories* (pp. 275–289). Gregorian & Biblical Press.

Dupré, J. (1999). On the impossibility of a monistic account of species. In R. A. Wilson (Ed.), *Species: new interdisciplinary essays* (pp. 3–22). MIT Press.

Eddington, A. S. (1929). *The nature of the physical world.* Cambridge University Press.

Ereshefsky, M. (Ed.). (1992). *The units of evolution: Essays on the nature of species.* MIT Press.

Ereshefsky, M. (2017). Species. In E. N. Zalta (Ed.), *The Stanford encyclopedia of philosophy* (Fall 2017 ed.), https://plato.stanford.edu/archives/spr2017/entries/species/

Fischer, J. (2009). Exploring the core identity of philosophical anthropology through the works of Max Scheler, Helmuth Plessner, and Arnold Gehlen. *IRIS, 1*(1), 153–170.

Friedman, M. (2010). Science, history, and transcendental subjectivity in Husserl's *Crisis*. In D. Hyder & H.-J. Rheinberger (Eds.), *Science and the life-world: Essays on Husserl's "Crisis of European sciences"* (pp. 100–115). Stanford University Press.

Gadamer, H. G. (1972). The science of the life world. *Analecta Husserliana, 2*, 173–185.

Gayon, J. (1992a). L'espèce sans la forme. In J. Gayon & J.-J. Wunenburger (Eds.), *Les figures de la forme. Philosophie biologique et théorie esthétique* (pp. 49–61). Éditions L'Harmattan.

Gayon, J. (1992b). L'individualité de l'espèce: une thèse transformiste? In J. Gayon (Ed.), *Buffon 1988. Actes du colloque international du bicentenaire de la mort de Buffon, Paris-Montbard-Dijon* (pp. 475–489). Vrin.
Gayon, J. (Ed.). (1992c). *Buffon 1988: Actes du colloque international du bicentenaire de la mort de Buffon, Paris-Montbard-Dijon.* Vrin.
Gayon, J. (2008). Les espèces et les taxons monophylétiques sont-ils des individus? In P. Ludwig & T. Pradeu (Eds.), *L'individu: perspectives contemporaines* (pp. 128–150). Vrin.
Gayon, J. (2009). Philosophy of biology: An historico-critical characterization. In A. Brenner & J. Gayon (Eds.), *French studies in philosophy of science. Contemporary perspectives* (pp. 201–212). Springer.
Gayon, J. (2014). Espèce, biologie. In *Encyclopaedia universalis*, http://www.universalis.fr/encyclopedie/espece-biologie/
Gayon, J. & Burian, R. (Eds.). (2007). *Conceptions de la science: hier, aujourd'hui, demain: hommage à Marjorie Grene.* Ousia.
Gayon, J., & Wunenburger, J.-J. (Eds.). (1992). *Les figures de la forme. Philosophie biologique et théorie esthétique.* Éditions L'Harmattan.
Ghiselin, M. (1974). A radical solution to the species problem. *Systematic Zoology, 23*, 536–544.
Ghiselin, M. (1997). *Metaphysics and the origin of species.* State University of New York Press.
Grene, M. (1965). *Approaches to a philosophical biology.* Basic Books.
Grene, M. (1974). *The knower and the known.* University of California Press.
Gutting, G. (Ed.). (2005). *Continental philosophy of science.* Blackwell.
Hardy, L. (2014). *Nature's suit: Husserl's phenomenological philosophy of the physical sciences.* Ohio University Press.
Heidegger, M. (1962). *Being and time* (7th ed., J. Macquarrie & E. Robinson, Trans.). SMC Press.
Hempel, C. (1966). *Philosophy of natural science.* Princeton University Press.
Honenberger, P. (2015a). Grene and Hull on types and typological thinking in biology. *Studies in History and Philosophy of Biological and Biomedical Sciences, 50*, 13–25.
Honenberger, P. (Ed.). (2015b). *Naturalism and philosophical anthropology: Between transcendental and empirical perspectives.* Palgrave Macmillan.
Honenberger, P. (2021). Natural artificiality, niche construction, and the content-open mediation of human behavior. *Biology and Philosophy, 36,* 1–25.
Honenberger, P. (2023). All knowing is orientation: Marjorie Grene's ecological epistemology. In G. Bianco, G. van de Vijver & C. T. Wolfe (Eds.), *Canguilhem and continental philosophy of biology* (pp. 39–60). Springer.
Hull, D. (1965). The effect of essentialism on taxonomy: two thousand years of stasis. *British journal for the philosophy of science, 15*, 314–326, *16,* 1–18.
Hull, D. (1999). On the plurality of species: Questioning the party line. In R. A. Wilson (Ed.), *Species: New interdisciplinary essays* (pp. 23–48). MIT Press.
Husserl, E. (1970a [1913]). *Logical investigations* (2nd ed., J. Findlay, Trans.). Humanities Press.
Husserl, E. (1970b). *The crisis of European sciences and transcendental phenomenology* (D. Carr, Trans.). Northwestern University Press.
Husserl, E. (1972 [1913]). *Ideas: A general introduction to pure phenomenology* (W. R. Gibson, Trans.). Collier Books.
Husserl, E. (1999 [1931]). *Cartesian meditations* (D. Cairns, Trans.). Springer.
Husserl, E. (1941). Phänomenologie und Anthropologie. *Philosophy and phenomenological research, 2,* 1–14. Translated in P. McCormick & F. Elliston (Eds.). (1981). *Husserl: shorter works* (pp. 315–323). University of Notre Dame Press.
Hyder, D. & Rheinberger, H.-J. (Eds.). (2010). *Science and the life-world: Essays on Husserl's "Crisis of European sciences".* Stanford University Press.
Jacob, P. (2019), Intentionality. In E. N. Zalta (Ed.), *The Stanford encyclopedia of philosophy* (Winter 2014 Edition). https://plato.stanford.edu/archives/win2014/entries/intentionality/
Jonas, H. (1966). *The phenomenon of life.* University of Chicago Press.

Kant, I. (2013, [1777]). Of the different human races. In J. M. Mikkelsen (Ed.), *Kant and the concept of race. Late eighteenth-century writings* (pp. 55–71). State University of New York.
Kitcher, P. (1984). Species. *Philosophy of Science, 51*, 308–333.
Kitcher, P. (1987). Ghostly whispers: Mayr, Ghiselin, and the "philosophers" on the ontological status of species. *Biology and Philosophy, 2*, 184–192.
Kitcher, P. (1989). Some puzzles about species. In M. Ruse (Ed.), *What the philosophy of biology is: Essays dedicated to David Hull* (pp. 183–208). Springer.
Kockelmans, J. (1966). *Phenomenology and physical science*. Duquesne University Press.
Krüger, H.-P. (2015). Intentionality and mentality as explanans and explanandum: Michael Tomasello's research program from the perspective of philosophical anthropology (pp. 183–218). In P. Honenberger (Ed.), *Naturalism and philosophical anthropology: Nature, life, and the human between transcendental and empirical perspectives* (pp. 183–218). Palgrave Macmillan.
Ludwig, P., & Pradeu, T. (Eds.). (2008). *L'individu: perspectives contemporaines*. Vrin.
Machamer, P., Darden, L., & Craver, C. (2000). Thinking about mechanisms. *Philosophy of Science, 67*, 1–25.
McCormick, P., & Elliston, F. (Eds.). (1981). *Husserl: Shorter works*. University of Notre Dame Press.
Meincke, A. S., & Dupré, J. (Eds.). (2020). *Biological identity: Perspectives from metaphysics and the philosophy of biology*. Routledge.
Merlin, F., & Huneman, P. (Eds.). (2018). *Philosophie, histoire, biologie: mélanges offerts à Jean Gayon*. Éditions Matériologiques.
Méthot, P.-O. (2023) Analytic and continental approaches to biology and philosophy: David Hull and Marjorie Grene on 'What Philosophy of biology is not'. In G. Bianco, G. Van de Vijver, & C. T. Wolfe (Eds.), *Canguilhem and Continental philosophy of biology* (pp. 13–38). Springer.
Mikkelsen, J. M. (Ed.). (2013). *Kant and the concept of race: Late eighteenth-century writings* (J. M. Mikkelsen, Trans.). State University of New York Press.
Moss, L. (2015). The hybrid hominin: A renewed point of departure for philosophical anthropology. In P. Honenberger (Ed.), *Naturalism and philosophical anthropology* (pp. 171–182). Palgrave Macmillan.
Moss, L. (2020). Normativity, system-integration, natural detachment and the hybrid hominin. *Phenomenology and the Cognitive Sciences, 20*, 21–37.
Nagel, T. (2012). *Mind and cosmos: Why the materialist neo-Darwinian conception of nature is almost certainly false*. Oxford University Press.
Natanson, M. (1969). Philosophy and psychiatry. In E. Straus, M. Natanson, & H. Ely (Eds.), *Psychiatry and philosophy* (pp. 85–110). Springer.
Newman, S. (2020). The origins and evolution of animal identity. In A.S. Meincke & J. Dupré (Eds.), *Biological identity: perspectives from metaphysics and the philosophy of biology* (pp. 128–148). Routledge.
Nye, M.-J. (2014). *Michael Polanyi and his generation: Origins of the social construction of science*. University of Chicago Press.
Oderberg, D. (2007). *Real essentialism*. Routledge.
Philipse, H. (1995). Transcendental idealism. In B. Smith & D. W. Smith (Eds.), *The Cambridge companion to Husserl* (pp. 239–322). Cambridge University Press.
Pigliuccci, M. (2003). Species as family resemblance concepts: The (dis)solution to the species problem? *BioEssays, 25*(6), 596–602.
Plessner, H. (2019 [1928]). *Levels of organic life and the human: an introduction to philosophical anthropology* (Hyatt, Trans.). Fordham University Press.
Polanyi, M. (1964). *Personal knowledge: Towards a post-critical philosophy* (Rev. ed.). Harper.
Reydon, T. A. C. (2005). On the nature of the species problem and the four meanings of 'species.' *Studies in History and Philosophy of Biological and Biomedical Sciences, 36*, 135–158.
Rheinberger, H.-J. (1997). Cytoplasmic Particles in Brussels (Jean Brachet, Hubert Chantrenne, Raymond Jeener) and at Rockefeller (Albert Claude), 1935–1955. *History and Philosophy of the Life Sciences, 19*(1), 47–67.

Rheinberger, H.-J. (2010). *An epistemology of the concrete: Twentieth-century histories of life*. Duke University Press.
Richards, R. A. (2010). *The species problem: A philosophical analysis*. Cambridge University Press.
Roger, J. & Fischer, J.-L. (Eds.). (1987). *Histoire du concept d'espèce dans les sciences de la vie*. Fondation Singer-Polignac
Ruse, M. (Ed.). (1989). *What the philosophy of biology is: Essays dedicated to David Hull*. Springer.
Ruse, M., & Richards, R. J. (Eds.). (2009). *The Cambridge companion to the "Origin of species."* Cambridge University Press.
Salice, A. (Ed.). (2012a). *Intentionality: Historical and systematic perspectives*. Philosophia Verlag.
Salice, A. (2012b). *Phänomenologische Variationen*. Intention and fulfillment in early Phenomenology. In A. Salice (Ed.), *Intentionality: Historical and systematic perspectives* (pp. 203–242). Philosophia Verlag.
Searle, J. (1983). *Intentionality: An essay in the philosophy of mind*. Cambridge University Press.
Searle, J. (2012). Foreword. In A. Salice (Ed.), *Intentionality: Historical and systematic perspectives* (pp. 9–22). Philosophia Verlag.
Siewert, C. (2016). Consciousness and intentionality. In E. N. Zalta (Ed.), *The Stanford encyclopedia of philosophy* (Spring 2017 Edition). https://plato.stanford.edu/archives/spr2017/entries/consciousness-intentionality.
Sloan, P. (1987). From logical universals to historical individuals: Buffon's conception of biological species In J. Roger & J.-L. Fischer (Eds.), *Histoire du concept d'espèce dans les sciences de la vie* (pp. 97–136). Fondation Singer-Polignac.
Sloan, P. (2002). Reflections on the species problem: What Marjorie Grene can teach us about a perennial issue. In R. E. Auxier & L. W. Hahn (Eds.), *The Philosophy of Marjorie Grene* (pp. 225–255). Open Court.
Sloan, P. (2007). Télélogie et forme: un réexamen. In J. Gayon & R. Burian (Eds.), *Conceptions de la science: hier, aujourd'hui, demain: hommage à Marjorie Grene* (pp. 343–67). Ousia.
Sloan, P. (2009). Originating species: Darwin on the species problem. In M. Ruse & R. J. Richards (Eds.), *The Cambridge companion to the "Origin of species"* (pp. 67–86). Cambridge University Press.
Sloan, P. (2013). The species problem and history. *Studies in History and Philosophy of Biological and Biomedical Sciences, 44*, 237–241.
Sloan, P. (2015). Questioning the zoological gaze. In P. R. Sloan, G. McKenny, & K. Eggleson (Eds.), *Darwin in the twenty-first century: Nature, humanity, God* (pp. 232–266). University of Notre Dame Press.
Sloan, P. (2018). Locating the human in the biological world: A way into the species problem. In F. Merlin & P. Huneman (Eds.), *Philosophie, histoire, biologie: mélanges offerts à Jean Gayon* (pp. 219–241). Éditions Matériologiques.
Sloan, P. R., McKenny, G., & Eggleson, K. (Eds.). (2015). *Darwin in the twenty-first century: Nature, humanity, God*. University of Notre Dame Press.
Smith, D. R., & McIntyre, R. (1982). *Husserl and intentionality: A study of mind, meaning, and language*. Reidel.
Smith, B., & Smith, D. W. (Eds.). (1995). *The Cambridge companion to Husserl*. Cambridge University Press.
Sokolowski, R. (2000). *Introduction to phenomenology*. Cambridge University Press.
Sparks, J. (2001). *Bedotia masoala*: A new species of atherinoid rainbowfish (Teleostei: Bedotiiade) from the Masoala peninsula, northeastern Madagascar. *Copeia, 2001*(2), 482–489.
Spiegelberg, H. (1965). *The phenomenological movement: a historical introduction* (2nd ed., 2 Vols.). The Hague.
Stamos, D. (2003). *The species problem: Biological species, ontology, and the metaphysics of biology*. Lexington Books.
Straus, E., Natanson, M., & Ely, H. (Eds.). (1969). *Psychiatry and philosophy*. Springer.
Tillman, M. K. (1976). Dilthey and Husserl. *Journal of the British Society for Phenomenology, 7*, 123–130.

Tymieniecka, A.-T. (Ed.). (1999). *Life; Scientific philosophy: phenomenology of life and the sciences of life* (2 vols.). Kluwer Academic Press.
Wilkins, J. S. (2009). *Species: A history of the idea.* University of California Press.
Wilson, R. A. (Ed.). (1999). *Species: New interdisciplinary essays.* MIT Press.
Winsor, M. (2006). The creation of the essentialism story: An exercise in metahistory. *History and Philosophy of the Life Sciences, 28*, 149–174.

Chapter 13
Vitalism and the Construction of Biology: A Historico-Epistemological Reflection

Charles T. Wolfe

Abstract What is the historical epistemology of the life sciences? In what way does it differ from historico-philosophical reflection on "foundational" or "conceptual" issues in the sciences tout court? This is a question to which Jean Gayon and his mentor Georges Canguilhem devoted a considerable amount of effort, yielding somewhat different answers, as I will try to show. One obvious difference, as P.-O. Méthot has shown, is Gayon's appropriation of anglophone philosophy of biology; another is Canguilhem's way of presenting his work as restricted to contextualized, historicist claims while in fact it is shot through with strong normative (and at times metaphysical) claims on Life, normativity, and value. In this essay I reflect on how to conduct historical epistemologies of the life sciences, focusing on two interrelated cases I have worked on in the past: vitalism and the constitution of biology.

Keywords Vitalism · Historical epistemology · Biology · Canguilhem · Gayon

13.1 Introduction

Jean Gayon was instrumental in the formation of my thinking about what one could call—and he did—the historical epistemology of the life sciences. This is a rather non-standard intellectual object, as most forms of history and philosophy of science (HPS, HP&S, and their social, practice-based, or otherwise 'instrumental' twists) do not focus on the life sciences, and conversely, most forms of the history of medicine or the history of biology are rather allergic to 'philosophy' or theoretical considerations (Gayon, 1998, 2009, 2010). Indeed, Anglophone historical epistemology (unlike the older *épistémologie historique*) seems to deliberately downplay philosophical language and themes in favor of knowledge-gathering practices, histories of knowledge, and so on. This is quite different from what we find in the work of Gayon's mentor, Georges Canguilhem, who could write about how Van Helmont or Willis were "vitalist physicians" (an odd characterization of Willis, but that is neither

C. T. Wolfe (✉)
Université Toulouse 2 Jean-Jaurès, Toulouse, France
e-mail: ctwolfe@gmail.com

P.-O. Méthot (ed.), *Philosophy, History and Biology: Essays in Honour of Jean Gayon*, History, Philosophy and Theory of the Life Sciences 30,
https://doi.org/10.1007/978-3-031-28157-0_13

here nor there) that the history of philosophy could not ignore (Canguilhem, 1965, p. 89).

So, what is the historical epistemology of the life sciences, then? It is not strictly defined, whether as a method or as a set of texts, but one could imagine it seeking to analyze discursive (and/or scientific) constellations in which Kant and Claude Bernard, or Leibniz and Théophile de Bordeu, are given equal precedence as to their vocabulary, intellectual context, and influence outside of their official 'field' (Métraux, 2004). In what follows I seek to contribute to this way of approaching the conceptual construction of the life sciences, by focusing on the interrelation between (emerging) vitalism and (emerging) biology, in the mid- to late eighteenth century. I say 'emerging' because both vitalism and biology seem to emerge and crystallize terminologically around the same time, between the 1770s and the 1790s (Wolfe, 2019a). To be fair to Gayon, though, one should note that he sometimes insisted that his work did *not* belong to the genre of historical epistemology (being much more influenced by the Anglophone genre of philosophy of biology[1])—although one should bear in mind Canguilhem's proclivity for deflecting such considerations, asserting sometimes that he was not a philosopher but a historian of science, or conversely that he was not a proper historian of science, but a philosopher (or simply an academic, that is, a teacher rather than the possessor of a personal philosophical system or *œuvre*).[2]

The problem of the status of vitalism in what used to be called, in Canguilhem and Jacques Roger's time, "biological thought" but also, sometimes, "biological philosophy" (Gayon, 1995) is well known. 'Vitalism' was, and still is typically used to refer to the doctrine, or type of doctrines, lying at the very limit (or margins) of reasonable scientific practice (Bechtel & Richardson, 1998). Francis Crick, Jacques Monod, and other luminaries have denounced vitalism in these terms, often referring to Hans Driesch's notion of entelechy, or more confusingly, to this notion *and* to Bergson's far more abstract notion of *élan vital* (for reappraisals of these notions, see Chen, 2019 and Donohue and Wolfe, eds., forthcoming). I do not wish in the following to study the reception history of concepts of vitalism, but such a study would doubtless show marked differences in (a) the degree of hostility and (b) the exact meanings given to the term, between Anglophone usage on the one hand, and the German and French usages on the other hand (in which varieties of anti-reductionism have been more prevalent). To return to the denunciation of vitalism, one might say this point boils down to the claim that it is a case of scientific *fraud*: the vitalist (on this view) insists that a scientific study of living nature should include mysterious vital forces or principles, whether these are derived from the study of embryology, cell theory, organismic biology, etc. The vitalist then is a kind of crank (Crick, 1966).

---

[1] In his study of Gayon's contribution to *both* historical epistemology *and* the philosophy of biology, P.-O. Méthot notes Gayon's effort to distinguish himself from the 'French' tradition (Méthot, 2018), but also notes subsequently that Gayon's research "retained the imprint of French epistemology" (Ibid., p. 33). See also chapters by Méthot and Loison, this volume.

[2] Canguilhem, "La médecine et son histoire" (1972 interview by F. Proust in *Tonus*), in Canguilhem (2018, p. 564).

Now, I have tried on a number of occasions (Wolfe, 2011, 2015, 2019a, 2020b; Wolfe & Terada, 2008; Wolfe & Normandin, 2013) to dispel or otherwise critique this common intuition, notably by seeking to *historicize* the issue, that is, to show—here is the 'historical' part of 'historical epistemology'—that there are different forms of vitalism, some of which may be committed to vital forces, and others not. Sometimes I sought to bring these forms under the heading of a broad conceptual distinction—the more 'epistemological' side of historical epistemology—between *substantival* and *functional* vitalism. The former posits the existence of vital forces as substantially real, like tables and chairs, while the latter seeks to grasp functional properties of organization (a.k.a. 'organism') without giving these a metaphysical foundation. At other times I sought to present this in terms of a distinction between *metaphysical* and *non-metaphysical* vitalism. Indeed, Bichat and Bernard are quite explicit in their way of reproaching, say, Montpellier vitalism for being too metaphysical, not experimental enough, too enamored with philosophical concepts, etc. (Bechtel, 2013)—and a century before Bernard, Albrecht von Haller, when defending his concept of the irritability of muscular fibre, had criticized his predecessor Francis Glisson in just such terms (Giglioni, 2008; Wolfe, 2020b).

Here, I return to this idea of historicizing vitalism, with a particular focus on how vitalism suitably understood relates to the constitution of biology as a science—a question which has recently been studied with other focal points, that is, other narratives of constitution, that do not focus on vitalism, but on more Leibnizian, Kantian and *naturphilosophisch* lines of development.[3] Indeed, vitalism itself has (re)emerged in recent historiography as a more complex and protean entity, without its impact on the development of biology having been reassessed per se.[4]

13.2 Defining Vitalism, and the Particular Case of Montpellier Vitalism

What is vitalism? Who is, or was, a vitalist? I have tried several times in other places to answer this deceptively simple question (most recently Wolfe, 2020b). While the term is often used loosely to apply to all kinds of theories of 'animate matter' or 'ensouled matter', I suggest that the term be restricted to theories in which the difference between 'life' and 'non-life' (living matter and non-living matter, living bodies and dead bodies, bodies and machines, biology and physics, etc.) is crucial, however this difference is spelled out. Now, it so happens that this stipulative definition applies quite well to the doctrines of the 'Montpellier School', that is,

[3] Duchesneau (2018), Zammito (2018), Bognon-Küss and Wolfe (2019a, 2019b), Wolfe (2019a), Wolfe (2019b).

[4] Cimino and Duchesneau (1997), Rey (2000), Williams (2003), Wolfe and Terada (2008), Wolfe (2011, 2017, 2019a, 2019b, 2020b), Wolfe and Normandin (2013), Steigerwald (2014), Gissis (2014). Gayon's reflection on Enlightenment vitalism is limited to Gayon (1994). On vitalism in the twentieth century, especially in the context of Logical Empiricism, see Gayon (2010) and Chen (2019).

the school associated with the Montpellier Faculty of Medicine, especially in the mid-to late eighteenth century—which is where the term is first used (in fact as a self-description), notably by Charles-Louis Dumas, Dean of the Montpellier Faculty in the 1790s (Rey, 2000; Williams, 2003; Wolfe, 2019a). Vitalism in this context is strongly tied to the study of the systemic functions of the organism, with a notable emphasis on functions such as organic sensibility (i.e., sensitivity).

This is not the place to delve into the specifics of the framework of Montpellier vitalism—its tension with the more mechanistic (at least in rhetoric) but also more experimentally oriented 'Paris School', its relation to earlier ideas and impact on nineteenth-century developments, and the similarity or divergences between different individual actors therein (thus elsewhere I've contrasted Barthez to Bordeu and Ménuret, including as concerns their degrees of commitment to a kind of substantival metaphysical vitalism). I wish instead to try, within various limits and constraints, to reconstruct a 'biological concept' articulated within this vitalist framework, or differently put, to see to what extent this vitalist theorizing contributes to, plays a role in, or (more minimally) is part of a shared tacit concern with the articulation of biology in the late eighteenth century.

The noted *dix-huitièmiste* Reill proposes the following definition of what he calls Enlightenment vitalism: its proponents, according to Reill,

> sought to discover the secrets of living nature, defined as a dynamic entity in which active forces worked in harmony within a complex set of interrelations. They posited the existence of self-activating forces and 'principles' in living matter, envisioning a living thing as a teeming interaction of drives, sympathies, and elective affinities, united in what they called 'organized bodies'. (Reill, 2010, p. 65)

I disagree with Reill as regards the positing of "self-activating forces or principles"—I have found literally no such invocations in the Montpellier vitalists, and indeed Théophile de Bordeu, one of the more prominent figures in this school, specifically expresses a kind of polite skepticism about the positing of an ontologically specific vital force that would explain why living bodies behaved as they did.

In his major work, the 1751 *Recherches anatomiques sur la position et la fonction des glandes*, Bordeu, after some professions of ontological neutrality ("whatever the case may be"), continues in a more affirmative way: "all living parts are directed by an ever-vigilant self-preserving force" (*Recherches anatomiques*, § CVIII, in Bordeu, 1818, vol. 1, p. 163). Here one might think: there is the typical vitalist appeal to a mysterious vital force! But significantly, he continues in rather deflationary (or at least agnostic) tones: "does this force belong, in certain respects, to the essence of a part of matter, or is it a necessary attribute of its combinations?... I can only suggest a *way* of conceiving things, *metaphorical* expressions, comparisons" (Ibid.).

More useful in Reill's definition is his emphasis on "organized bodies," indeed an anchoring concept (along with its cognates 'organization' and, later, 'organism') in both this species of vitalism *and* what will become the science of biology, at least in the programmatic statements for a 'biology' written in the later eighteenth century by Kielmeyer, Treviranus, Lamarck and others in sometimes more Kantian, sometimes more Newtonian frameworks (Zammito, 2018). Reill continues by stressing that in vitalism, "relation, *rapport*, *Verwandschaft* [affinity], and interconnection replaced mechanical aggregation as one of the defining principles of matter. By emphasizing the centrality of interconnection, Enlightenment vitalists modified the concept of cause and effect. In the world of living nature, each part of an "organized body" was both cause and effect of the other parts" (Reill, 2010, p. 66; for more on whole-part relations in Montpellier vitalism, see Wolfe, 2019b).

Today, after a century-odd output of writings from theoretical biologists and their kin, we expect 'organismic' appeals to be explicitly opposed to 'reductionist' theories. And clearly such oppositions did not have to wait for molecular biology, the Century of the Gene or even Jacques Loeb's reductive biochemistry project (Allen, 2005; Creager, 2017). Reactions to Cartesian animal-machine models and later, to La Mettrie's *homme-machine*, clearly involved the denunciation of the reduction of life to its other (whether this was understood as dead matter, 'mere' matter or the machine: Huneman & Wolfe, 2017). But interestingly—and this is often overlooked as commentators somehow complacently repeat the opposition between mechanism and vitalism—the vitalist concept of organization is not *opposed* to the machine concept (this is one of the few points on which I differ from Rey's landmark 2000 study of Montpellier vitalism). Rather, the vitalist concept is explicitly *structural*; it is more of a 'meta-mechanism' or an 'expanded mechanism' than it is an *anti*-mechanism.

13.3 Mechanism and Vitalism: Circles and Swarms

Since vitalism is often associated with the idea that the difference between a living body and a dead body (or a machine, or a bundle of atoms, or matter per se) lies in the presence of a kind of animating 'force' in the former, it is worth emphasizing how much, in texts of the representative Montpellier vitalists, we find an emphasis on *structure*, which in its simplest form, I have elsewhere referred to as "expanded mechanism" (Wolfe, 2019a).

Granted, vitalist attempts to model living systems—or better, to clarify the conditions under which such modelling could occur—can also be reactive with regard to machine paradigms. As Bordeu wrote in his *Recherches sur l'histoire de la médecine*, reflecting on his studies: "Spare us, once and for all, all these tiny fibres, pressures, globules, thick substances, sharp angles, lymph, hammers and all the rest of the equipment from mechanical workshops with which [earlier doctors] filled the living body—they were the playthings of our fathers" (Bordeu, 1818, II, p. 670).

But primarily, and more significantly, Montpellier vitalists understand organization, not as the mysterious manifestation of a hidden hegemonic entity (like the soul), but in structural terms. Thus Jean-Joseph Ménuret de Chambaud, another key member of this school, although less-known and underappreciated until recently,[5]

[5] Ménuret (1739–1815) published mainly under the name Jean-Jacques, although his given name was Jean-Joseph, for reasons that remain unknown; his birth date is usually wrongly given as 1733.

when he criticizes mechanism, does so to note that mechanism did not justice to the complexity of what he called, using a technical term of the period, the "animal economy" (which in this context is essentially a precursor word for "organism"). Indeed, mechanism did not do justice to the uniquely "organic structure" of the animal economy:

> They [*sc.* the mechanists] believed that movement merely obeyed the ordinary laws that apply to all *inorganic* machines, treated the human body geometrically, and rigorously calculated all the various degrees of force required for particular actions, or how much was lost, etc.— but all these calculations, which obviously varied tremendously, shed no light on the *animal economy*. They did not even pay attention to the *organic* structure of the human body which is the source of its main properties. (Ménuret, 1765a, p. 364b).s

Structure itself is not the object of an analysis in these texts, but my point here is (a) that it is simply a more complex type of organization than that of the machine, (b) without being its opposite. Indeed, Ménuret asserts that "everything leads us to believe that the human body is like the other machines which art can assemble and disassemble, displaying their tiniest springs" (Ibid., p. 362b). Ménuret also spoke of the "human machine" as a chemically propelled complex system in equilibrium (Ménuret, 1765b, p. 435b): thus *neither* a simple machine like a pulley or a funnel, *nor* an organism understood in a sense ontologically absolutely separate from machines. Diderot expresses this nicely, although he leaves open just how fundamental, or irreconcilable the difference is: "Difference between the clamp of a wooden or iron set of pliers, and that of pliers made of flesh or two fingers. The wooden clamp does not feel, the flesh clamp feels; the wooden clamp does not suffer, the flesh clamp suffers; the wooden clamp cannot be tickled" (Diderot, *Éléments de physiologie*, in Diderot, 1975, XVII, p. 335, 499).

Organisation was indeed a technical term in the eighteenth century for 'physiological structure'; the *Encyclopédie* defines it as the "arrangement des parties qui constituent les corps animés" (Diderot & D'Alembert, 1751–1780, p. IX, 629b); it is comparable, and related to, terms such as 'organized bodies' and 'animal economy'. It is an organism concept. As such, it actually belongs to the core of the conceptual dimension of the constitution of biology.

The organism concept is crucial in the constitution of biology, as I will discuss in the following section. The term 'organism' itself appears much earlier than 'biology', as Tobias Cheung has shown (Cheung, 2000, 2014; see also Toepfer, 2009), but it was a rare term for most of the eighteenth century. Claims for the need of a specific science of living entities in the later decades of the century often explicitly pointed to the unique form of organization of these entities. For instance, when Caspar Friedrich Wolff sought to redefine embryology, he stated that one of the main goals of his "theory of generation" is not just to defend epigenesis as an account of the formation of the embryo, but also more broadly, to define a science of the causes and effects specific to the formation of organisms ("organized bodies," Wolff, 1966 [1764], 36f.). What is specific to the embryo and more generally the organism is the presence of a

---

Between late 1758 and 1761, when he was aged 19–22, he contributed a staggering number of medical entries to the *Encyclopédie*, which have only recently been identified as his (Rey 2000).

force Wolff calls *vis essentialis*. Other authors do not define the specificity in terms of a force but strictly of organization.

This specific mode of organization, in the early modern and Enlightenment periods, is often described with the metaphor of the circle, specifically, the idea of the "*circle of action*," referring back to the old Hippocratic idea that "*everything concurs, consents and conspires together in the body,* in relation to the animal economy" (Leclerc, 1696, III, p. 250). This circularity is primarily mean to call attention to the non-linear character of the relation between parts in the organism (animal economy): it is very difficult therein to determine what is a cause, and what is an effect. Hence this type of organization is also described, in more detail, using the metaphor of the bee-swarm. Bordeu writes,

> Might I make use of a comparison which, however rough, may be useful?
>
> I compare the living body, in order to properly assess the particular action of each part, to a swarm of bees which cluster together, and hang from a tree like a bunch of grapes; I find the image [...] that one of the lower organs was an *animal in animali*, to be quite helpful. Each part is, so to speak, not quite an animal, but a kind of independent machine which contributes in its way to the general life of the body.
>
> Hence, following the comparison to a bee swarm, it is a whole stuck to a tree branch, by means of the action of many bees which must act in concert to hold on; some others become attached to the initial ones, and so on; all concur in forming a fairly solid body, yet each one has a particular action, apart from the others [...]. (Bordeu, *Recherches*, 1751, § CXXV; Bordeu, 1818, I, p. 187)

There is a lot to be said about this presence of the bee-swarm as a core metaphor for organism (specifically, organismic unity[6]) in Montpellier vitalism (for more discussion, see Kleiman-Lafon & Wolfe, 2021). I simply wish to emphasize that, however much Montpellier vitalism was a 'medical affair', it nevertheless is strongly committed to a concept of organism (which, as noted above, is constitutive of the emergence of biology), focusing on the type of relations within the organism. Notably, just as the swarm is composed of individual bees, each of whom is alive and thus 'has a life of its own', but the swarm, like a superorganism, has an identity of its own. Similarly, the vitalist conception of the organism is that it is composed of organs which are like individual lives, and the organs themselves are synergic combinations of even smaller lives: "the life of each organ of the animated body is not a mere unitary life (*vie simple*) but in truth, the product of as many *particular lives* as there are *living molecules* entering into the composition of this organ" (Grimaud, 1776, p. 12).[7]

In short, the vitalist concept of organization is a *structural* concept. Granted, it makes use of a 'thick' concept of Life, never fully defined (indeed, Bordeu, Ménuret and others are careful not to make potentially essentialist statements on the definition of Life itself) but at work in concepts such as the relational and synergic model of

[6] Diderot, a great reader of the vitalists, writes, "All our organs [...] are only distinct animals which the law of continuity holds together in a general state of sympathy, a single unity, a single identity" (*Rêve de D'Alembert*, in Diderot, 1975, XVII, p. 122).

[7] This work, an *Essai sur l'irritabilité* signed "Mr 'D.G.," is actually by J.C.M.G. de Grimaud, Chair of Anatomy in Montpellier in the late decades of the vitalist school.

the 'one Life composed of many lives', as cited above. But it does not distinguish the organization of living organisms from the states of inert or aggregate matter, in terms of the presence of a vital force (or consciousness, or self-notice that the bee-swarm is not a colony governed by one individual). If there is teleology here, it is not of the top-down kind (Wolfe, 2019b). As such, this concept marks a point of intersection with the history of biology, to which I now turn.

13.4 The Constitution of Biology[8]

Despite the appearance of more careful specialized scholarship over the past few decades (see for overview Toepfer, 2011), it is still standard to trace the usage of the term 'biology' back to Treviranus and Lamarck around 1800: Treviranus' *Biologie, oder Philosophie der lebenden Natur, für Naturforscher und Aerzte* appeared in 6 volumes starting in 1802; Lamarck's *Hydrogéologie, ou Recherches sur l'influence qu'ont les eaux sur la surface du globe terrestre* in 1801–1802 (in fact, Lamarck planned a work entitled *Biologie, ou Considérations sur la nature, les facultés, les développements et l'origine des corps vivants*, which he never completed).[9] But, as Jacques Roger observed nicely, "biology did not just appear suddenly at the end of the eighteenth century like Athena born from the head of Zeus" (Roger, 1980, 1995, p. 210). Earlier, the term 'biology' was used in the context of German *Naturphilosophie*, in Theodor Georg August Roose, Karl Friedrich Burdach, and Carl Christian Erhard Schmid, in the 1790s,[10] and Peter McLaughlin has pointed to its usage in a treatise of natural philosophy by a Danzig professor named Michael Christoph Hanov, the 4-volume *Philosophia naturalis sive physica dogmatica*, of which volume 3, published in Halle in 1766, treats "geology, biology and botany."[11]

Yet the exact meaning of the term 'biology' remains unstable until the end of the century (Linnaeus had already used it a generation earlier, but to mean something

[8] I am reprising here analyses which first appeared in Wolfe (2019a; notably the "Excursus" on biology, pp. 201–210) and Bognon-Kuss and Wolfe (2019a, 2019b). On the problem of life (the theory of life, the ontology of life, the concept of life) in relation to "life science" in the early modern period (that is, leading up to the period under discussion here) cf. Wolfe (2020a).

[9] A part of this manuscript exists, but did not circulate; it was only published in the twentieth century by Pierre-Paul Grassé (Lamarck 1944). Grassé considers the text to be from 1812–1814 but Pietro Corsi has shown more recently that *Biologie* represents older material, earlier than his 1802 book (Corsi 2006, 38n.).

[10] Roose (1797), Introduction, Schmid (1798–1801), Burdach (1800; Burdach stands out in this context for being a physician). On the history and development of biology in those years (1795–1802), see Barsanti (1995), Caron (1988), Corsi (2006), Kanz (2002), Toepfer (2011), Gayon (2014) and Wolfe (2019a).

[11] McLaughlin (2002). The full title of Hanov's work—he uses the term in the text of volume 4 rather than 3—is *Philosophiæ naturalis sive physicæ dogmaticæ tomus I continens physicam generalem, cœlestem et æthereum. Tanquam continuationem systematis philosophici Christiani L.B. de Wolff* (Halle, 1762). And, as was often the case (in Linnaeus and others), 'biology' is not used in Hanov's text as we would understand it. For further analysis of Hanov, see Duchesneau (2018).

close to 'biography', a usage which Kai Torsten Kanz has traced back as far as 1660: Gayon, 2014; Kanz, 2002), and conversely, if one approaches the issue from the standpoint of the existence of a new science of life and how to designate it, the terminology was quite variable, including 'zoonomy', 'general zoology', 'biology', 'physiology', 'bionomy', 'biogeography' and 'general natural history', a term which was used until the mid-nineteenth century, e.g. by Isidore Geoffroy Saint-Hilaire.[12] The term 'life science' or 'science of life' (*Lebenswissenschaft*, *Wissenschaft des Lebens*) itself appears in the title of an 1800 work by Christoph Meiners (Meiners, 1800), but in an antiquated sense, which refers to morals—the 'art of living', as it were.

It is thus widely agreed that biology as a science of the functioning and development of living bodies emerged in the beginning of the nineteenth century, integrating methodological or empirical advances in various disciplines, namely physiology, embryology, comparative anatomy, natural history and medicine. The fact that the word 'biology' was simultaneously and independently coined by several authors from different national and disciplinary backgrounds (Bichat 1800, Hanov 1766, Lamarck 1809, Treviranus, 1802–1822) is commonly seen as a testimony of this epistemic emergence (Barsanti, 1995; McLaughlin, 2002; Wolfe, 2019a). Even though scientists had, of course, been dealing with living phenomena prior to this, what is striking here is the constitution of a unified framework for developing research focused on "vital" phenomena.

How and on what basis or bases did biology emerge as a unified science of life? I will not seek to answer that question here, but one can distinguish between several types of possible answers, of which I'll discuss three.

First, histories of 'generation', later to be renamed developmental biology, tend to focus on the phenomenal novelty of this set of experiences—of the embryo, the egg, the polyp, but also, monstrous births, hybridization and the like.[13] The phenomena of generation disturbed a number of tacit metaphysical items of consensus, as well as the privileged mathematical and mechanical models that had emerged in the Galilean, Cartesian and Newtonian intellectual cultures. It is perhaps with such phenomena in mind that Diderot wrote, in the mid-eighteenth century (in the *Pensées sur l'interprétation de la nature*), that a "great revolution in the sciences" is coming, which he associates notably with "the history of nature," i.e., natural history, a field which he understands in a broad sense very much as a proto-biology; he adds that

[12] At times the name for this projected science is no longer one we recognize, e.g. "Zoonomia" (Schmid, Erasmus Darwin and others) or "Bionomia" (Hanov, and later Comte, as noted in Gayon, 2014); at other times, it is a familiar name but with an expanded meaning, like Kielmeyer's "allgemeine Zoologie" (by which he meant a "physics of organized beings"—in other words, a biology: Lenoir, 1982, 50f.), or Claude Bernard's usage of "physiology" to mean the science of life (Bernard, 1879, I, 1ère leçon, 3). Cf. Kanz (2002, p. 11). Auguste Comte, in his lectures, discusses all of these terms (not as candidates for the name of the science but as describing different parts of that science), as described in Gayon (2014, 14n.).

[13] See e.g., Roger (1963), Goldberg (2022), and on monsters, Gayon (2004).

within a few more years there will no longer be any great mathematicians.[14] That is, the phenomena of life will induce a "revolution in the sciences," away from mathematics.

Second, a more taxonomic, classificatory impetus, itself also associated with a program for natural history, but emphasizing a more systematic dimension, which is understood as bringing about a kind of *coupure* between a former, more qualitative 'protobiology' and a new, systematic 'biology'. Thus Maupertuis, in his *Lettre sur le progrès des sciences* (1752), speaks explicitly of turning natural history into a "real science," i.e., no longer just offering "agreeable tableaus for us to contemplate," but rather, presenting "the general processes of Nature, in its production and preservation."[15] Such processes would amount to biological laws: natural history understood as a general theory of life—hence, a biology—would study living beings in terms of their organization and their functions. We find a similar kind of *coupure* in Kant's opposition between a mere "description of Nature" (*Naturbeschreibung*), which he associates with Linnaeus, and the "history of nature" (*Geschichte der Natur*), which he presents as new (Sloan, 2006, pp. 633–636; Zammito, 2018, p. 176f.). Linnaeus himself emphasized that classification into genera and species was the highest point that science (presumably, life science) could reach (Linnaeus, 1751, § 290).

Third, we are often told that biology is a science of organisms; that the concept of organism is an (irreducible and constitutive) building-block for the science of biology. The mechanical philosophy—Boyle, Hobbes, Descartes but also iatromechanism as a research program in physiology and medicine, with figures such as Giorgio Baglivi and Hermann Boerhaave (Clericuzio & Conforti, 2020; Wolfe, 2020a)—rested on the notion that natural phenomena result from interactions between material particles governed by the laws of mechanics. This enabled the formulation of laws of motion and the invention of particular mechanisms. It is hard to grasp the specific nature of the organization of the 'organized body' or organism within such a framework (although less hard than is often said: Hutchins et al., 2016; Shapiro, 2003). Some two centuries later, Claude Bernard stated that scientifically, "we are right to treat the organism like a machine," but he immediately introduced a somewhat puzzling distinction: we are "mistaken when we treat it as a *mechanical machine*, fixed and unchanging"; rather, "the organism is *an organic machine*, which possesses a flexible mechanism, thanks to special organic processes which exist therein; but these do not transgress any of the general laws of mechanics, physics or chemistry" (notebook fragment, cit. in Canguilhem, 1989, p. 552). Bernard points to an important dimension

[14] Diderot, *Pensées sur l'interprétation de la nature* (1753/1754), § 4, reprinted in Diderot (1975). On natural history as 'biology' in Diderot, see Wolfe (2009), Wolfe and Shank (2019); for a similar argument regarding Buffon, see Roger (1979). Roger comments that the second volume of Buffon's *Histoire naturelle* (1749) contains, under the heading "History of Animals," chapters which fully belong to what today would be called "general biology" (1979, 49); Gayon observes that "natural history dealt with a number of the objects that biology later took over" (Gayon, 2014, 17)—a more charitable view than in Gayon (1998). For an opposite view, defending an axiomatic and mathematical background for the emergence of biology, via an emphasis on Christian Wolff, see van den Berg and Demarest (2020). For a more anatomically focused narrative, see Schmitt (2006).

[15] Maupertuis, *Lettre sur le progrès des sciences* (1752), § 13, in Maupertuis (1756, vol. II, 386).

in the constitution of biology, namely chemistry (the specific chemistry of organic beings[16]), but I quote him simply to 'bookend' the coming-together of an organism concept, as something that is neither *merely* mechanical, nor *transcendent* with regard to ordinary mechanical laws. Recall Ménuret's (vitalist) insistence that mechanism was not false but insufficient, as it did "not even pay attention to the organic structure of the human body which is the source of its main properties" (Ménuret, 1765a, 364b).

Obviously, all of these need to be combined with some version of a 'systematic' claim, that is, a claim for the specific systematic unity of this new science. Thus, to take one example among many, Treviranus claimed to be inaugurating a science which was independent of medicine and natural history, the object of which was to study life in its various manifestations and discover its laws. The Introduction to his *Biologie* includes a chapter entitled "Object and importance of biology," in which he presents this new science as studying "the various forms and manifestations of life," its conditions and laws, and "which causes govern its action" (Treviranus, 1802, vol. I, p. 4).[17] In his propaedeutic to medicine, Burdach suggests that "biology" be the name for a very synthetic (syncretistic) part of medicine: that which studies the morphological, physiological and psychological features of human beings (Burdach, 1800, p. 62). As Gayon comments (with reference notably to Lamarck and Treviranus, but the point applies to several less famous individuals, some of whom are studied in Zammito, 2018), "all of these authors held that the proper ambition of biology was to discover laws that could account for the fundamental unity of living beings" (Gayon, 2014, p. 17).

At this point a familiar objection comes to mind: haven't I been guilty of (unpardonable) anachronism? This is Foucault's well-known objection to the rather internalist, idealistic and indeed Whiggish histories of biology prevalent in the 1960s, which he (and Canguilhem, and Roger) struck decisive blows against. Foucault claimed influentially in that "Life did not exist" before the emergence of biology as a science bearing that name, in the nineteenth century, a claim which I think should

---

[16] A separate paper would be needed to show how this chemical focus (a) allows of a vitalist inflection and, importantly (b) need not amount to focusing on vital *forces*. In Blumenbach for instance, the study of biological functions is explicitly tied to a strong concept of vital force: his *Bildungstrieb* is truly a *Lebenskraft* which, alone, enables the explanation of how physical and chemical laws subserve the laws of organization in the course of embryogenesis (Blumenbach 1787, § 42–48; cf. Duchesneau 2011). For instance, if we examine the points of contact between nascent biology and slightly more emerged chemistry—around what will come to be called organic chemistry (Holmes 1974)—with an eye to the conceptual conditions of the emergence of biology, it then appears that this emergence had less to do with the affirmation of irreducible vital forces, and more with an increased focus on new, chemical conceptions of self-organizing processes (Bognon-Küss, 2019). I also do not engage here with the debate surrounding the 'Lenoir Thesis', i.e., the extent to which Kant influenced Blumenbach and other's Kant-betraying programs for a biology (see Lenoir, 1982 and the extensive and convincing critique in Zammito (2012), although he acknowledges that what may have been a misunderstanding of Kant was "influential... in the constitution of biology as a discipline in the succeeding decades": Zammito (2018, p. 236).

[17] Joan Steigerwald has noted that Treviranus's *Biologie* "provides an ambiguous prototype for a science of biology," however, because Treviranus established neither a new school or discipline nor created an institutional or technical basis for empirical inquiry into life (Steigerwald, 2014, p. 105).

be taken with a considerable grain of salt (Foucault also makes the less controversial claim that early modern natural history could not constitute itself as biology, although I have given some indications above that argue against this claim). Recall the two overlapping passages in *Les mots et les choses* (*The Order of Things*). First, the category of Life is the "most anachronistic" of categories one could apply to the systems of knowledge of the eighteenth century. Thus.

> Historians want to write histories of biology in the eighteenth century, but they do not realize that biology did not exist then, and that the pattern of knowledge that has been familiar to us for a hundred and fifty years is not valid for a previous period. And that, if biology was unknown, there was a very simple reason for it: that life itself did not exist. All that existed was living beings, which were viewed through the grid of knowledge constituted by natural history. (Foucault, 1966, p. 139)

He repeats this more or less identically later on in the book: "early modern natural history could not constitute itself as biology. For until the end of the eighteenth century, life does not exist—only living beings" (Foucault, 1966, 173). In fact, it seems to me that, contra Foucault, the obsession with Life and the growing recognition that a specific science will be required to study it, is a striking eighteenth-century development. Further, as Jacques Roger observed, if there was no biology, there could also be no biologists (Roger, 1980, p. 258). Yet there were many naturalists, philosophers, natural philosophers who were consistently and carefully focused on Life and the properties of living beings. My point is not that there was a ready-made, self-standing biology while Foucault says there was none, but minimally (a), that there were projects for a unified science of life, (b) there were practitioners, as Roger observes, that we don't know what to do with if we impose the oddly systematic, epochal categorization of Foucault.[18]

Foucault adds an interesting criterion much further on: before Lamarck, "the dividing line between the living and the non-living is never a decisive problem" (Foucault, 1966, pp. 244–245). Indeed, a question that remained implicit in my brief 'digest' of the emergence(s) of biology, which has a more speculative—or at least, dangerously conceptual—ring to it, is the extent to which any of these, or all these approaches taken as a unified whole, depend on an ontology of life? Treviranus, for instance, pointed out that if botany and zoology are treated as parts of biology, they appear "in a new light," no longer mere chunks of nomenclature, because they take a step beyond traditional classifications: "medicine, physiology and pathology will all be renewed"; he also quotes Stahl's insistence on how "one first has to know what life is—what we commonly call 'life'."[19] And here we return, of course, to the

[18] For an excellent, less tendentious discussion of the shifting meanings of 'biology' and its predecessors, 'physiology' and 'natural history', and an analysis of the relation between 'philosophy' and these terms, see Gayon (1998) and Gayon (2014). For the newer view that the eighteenth century *was* significantly concerned with 'vital' matters, see Reill (2005), Zammito (2018), Bognon-Küss and Wolfe (2019a, 2019b), Wolfe (2019a), and the reflections in Huneman (2019). Zammito criticizes Foucault's claim that there was no historicized sense of Nature in the eighteenth century, as well (Zammito, 2018, p. 173).

[19] "Ante omnia itaque scire convenit, quid sit illud quod vulgata appellatione vita dicitur" (Stahl, *Theoria medica vera*, cit. Treviranus 1802, vol. I, 11n.).

vitalist terrain—in which, recall, the concern with the divide between the living and the non-living is crucial (40–50 years prior to when Foucault would allow for talk of Life and biology).

However, without opting for the (too-neat) radical solution of Foucault, and stipulating: no biology before actual biology, it would not be desirable either to insist that such ontologies of Life either are life science or are precondition thereof. Notice that when I sought to connect the vitalist concept of organism (animal economy, organized body, etc.) to the emergence of biology, it was not a return to insisting on 'who' deserves to be at the centre of the emergence of biology, or which of these projects is genuinely original, marking a 'discontinuity' or a 'break'.[20] Rather, I wonder to what extent this emergence can be related to the new vitalist focus on organization, in the decades immediately prior to Treviranus and Lamarck's usage of 'biology'.

13.5 Conclusion: Vitalism and 'Biology'

The set of doctrines associated with the Montpellier Faculty of Medicine in the mid-to-late eighteenth century may seem like a surprising choice for an attempt at a historico-conceptual reflection on the emergence of biology as a systematic discipline concerned (at least in that phase of its existence) with the nature of Life—one could say (and Jean Gayon did say, once[21]) that this concerns *medical* rather than *biological* vitalism. The latter is typically associated with the focus on embryology found in authors like Driesch, in which vitalist claims refer to, or are based on, evidence concerning the specific forces at work in generation. Yet if we reflect on notions such as the "animal economy" in Montpellier vitalism (Wolfe, 2017, 2019a; Wolfe & Terada, 2008) we can see that they are intended as models of organization. Not as theories of health or disease, or even models of the body as we might have encountered them in debates between iatromechanists and iatrochemists in the previous century (Clericuzio & Conforti, 2020), but models of organization, in other words, a core notion of biological reflection.

Does vitalism participate in the constitution of biology? I have given some indications as to how this may be understood. Yet simply asserting such participation is a perilous type of claim because it seems guilty of the elementary mistake of 'idealism': how could a theoretical constellation give rise to a scientific discipline in its materiality? Indeed, several practitioners but also polemicists at the time accused the Montpellier vitalist school of being too philosophical, too metaphysical, too much of a theory and insufficiently experimental (Wolfe, 2013). One answer could be to treat both as sociological objects and study e.g., Montpellier and Paris medicine as

[20] Some put Lamarck in this position, because he considered life as a set of laws resulting from specific kinds of organization, while others consider that Treviranus should occupy this position, or that the emergence of a discipline should not be traced back to individual factors.

[21] Discussion after my presentation on Canguilhem and vitalism at the conference *Vie, vivant, vital: vitalisme* (Université de Montpellier, June 2009).

'networks'. That is not my approach, and it does not help answer the question I posed (unless again, biology is itself reduced to a network of power and influence, patronage, center and peripheries…). A different difficulty is that 'vitalism' and 'biology' are themselves nexuses of different influences, vocabularies, and traditions. The story I have told here would look very different if focused on Lamarck (Gissis, 2019), Blumenbach (Duchesneau, 2011; Steigerwald, 2006) or Kant (Huneman, 2008). I focused on vitalism's contribution to an 'organizational' concept, itself deeply … biological. Which is not to say that biology in its emerging phases 'was vitalist' or (a weaker claim) that claims for the autonomy of biology (whether ontological or methodological) are de facto vitalist (although the more mechanism-friendly, structural, and less metaphysical sense of vitalism I've argued for is indeed quite compatible with that narrative).

As Jacques Roger suggested quite reasonably, "Now that vitalism is no longer a threat to the modern biochemical interpretation of life, it should be possible to study its long history calmly, and even to define the positive role it played in certain stages in the history of biological research" (Roger, 1980, 1995, p. 210). Vitalism is not just one theory among others that can be refuted or eliminated in the course of the history of the life sciences (like, say, preformationism), because it was fundamentally concerned with the definition of life (if not by offering a strict one, then by articulating careful criteria for what such a definition should look like, or should not look like). Vitalism is also a component in struggles for definition of an experimental life science that involve demands for the autonomy of such a science. Vitalism thus understood is part of the struggles for legitimacy (terminological and other) in the self-definition of the biomedical sciences. I'd like to think these reflections on the historical epistemology of vitalism and the constitution of biology would have interested Jean Gayon and been the basis for future discussions.

References

Allen, G. E. (2005). Mechanism, vitalism and organicism in late nineteenth and twentieth-century biology: The importance of historical context. *Studies in the History and Philosophy of Biological and Biomedical Sciences, 36*, 261–283.

Barsanti, G. (1995). La naissance de la biologie: observations, théories, métaphysiques en France 1740–1810. In C. Blanckaert, J.-L. Fischer, & R. Rey (Eds.), *Nature, histoire, société: essais en hommage à J. Roger* (pp. 197–227). Klincksieck.

Bechtel, W. (2013). Addressing the vitalist's challenge to mechanistic science: Dynamic mechanistic explanation. In S. Normandin & C. T. Wolfe (Eds.), *Vitalism and the scientific image in post-enlightenment life science, 1800–2010* (pp. 345–370). Springer.

Bechtel, W., & Richardson, R. C. (1998). Vitalism. In *Routledge encyclopedia of philosophy* (pp. 639–643). Routledge.

Bernard, C. (1879). *Leçons de physiologie opératoire*. J.-B. Baillière et fils.

Blumenbach, J. F. (1787). *Institutiones physiologicæ*. J.C. Dieterich.

Bognon-Küss, C. (2019). Between biology and chemistry in the 18th century: How nutrition shapes vital organization. *History and Philosophy of the Life Sciences, 41*, 1–46.

Bognon-Küss, C., & Wolfe, C. T. (2019a). The idea of 'philosophy of biology before biology': A methodological provocation. In C. Bognon-Küss & C. T. Wolfe (Eds.), *Philosophy of biology before biology* (pp. 4–23). Routledge.
Bognon-Küss, C., & Wolfe, C. T. (Eds.). (2019b). *Philosophy of biology before biology*. Routledge.
Bordeu, T. de. (1818). *Œuvres complètes*, précédées d'une Notice sur sa vie et ses ouvrages par M. le Chevalier de Richerand, 2 vols. Caille et Ravier.
Burdach, K. F. (1800). *Propädeutik zum Studium der gesammten Heilkunst*. Breitkopf & Härtel.
Canguilhem, G. (1965). *La connaissance de la vie* (Revised ed.). Vrin [1952].
Canguilhem, G. (1989). Vie. *Encyclopedia universalis* (1st ed. (19680, Vol. 23, pp. 546–553).
Canguilhem, G. (2018). *Œuvres complètes,* tome V*: Histoire des sciences, philosophie biologique et commémorations (1966–1995)*. Textes présentés et annotés par C. Limoges. Vrin.
Caron, J. (1988). 'Biology' in the life sciences: A historiographical contribution. *History of Science, 26*, 223–268.
Chen, B. (2019). Revisiting the logical empiricist criticisms of vitalism. *Transversal: International Journal for the Historiography of Science, 7*, 25–40.
Cheung, T. (2000). *Die Organisation des Lebendigen. Die Entstehung des Biologischen Organismusbegriffs bei Cuvier, Leibniz und Kant*. Campus-Verlag.
Cheung, T. (2014). *Organismen: Agenten zwischen Innen- und Außenwelten 1780–1860*. Transcript.
Cimino, F., Duchesneau. F. (Eds.) (1997). *Vitalisms from Haller to the cell theory*. Olschki.
Clericuzio, A., & Conforti, M. (2020). Iatrochemistry and iatromechanism in the early modern era. In D. Jalobeanu & C. T. Wolfe (Eds.), *Encyclopedia of early modern philosophy and the Sciences* (p. xx). Springer.
Corsi, P., et al. (2006). Biologie. In P. Corsi & J. Gayon (Eds.), *Lamarck, Philosophe de la nature* (pp. 37–64). PUF.
Creager, A. N. H. (2017). A chemical reaction to the historiography of biology. *Ambix, 64*(4), 343–359.
Crick, F. (1966). *Of molecules and men*. University of Washington Press.
Diderot, D., & D'Alembert, J. R. (Eds.) (1751–1780). Encyclopédie ou Dictionnaire raisonné des sciences, des arts et des métiers..., 36 vols. Briasson
Donohue, C., & Wolfe, C. T. (Eds.). (forthcoming). *Vitalism and the contemporary life sciences*. Springer.
Duchesneau, F. (2011). Blumenbach et la théorie des forces vitales. In P. Nouvel (Ed.), *Repenser le vitalisme* (pp. 73–88). PUF.
Duchesneau, F. (2018). *Organisme et corps organique de Leibniz à Kant*. Vrin.
Foucault, M. (1966). *Les mots et les choses*. Gallimard.
Gayon, J. (1994). Le vitalisme entre vie et mort dans la pensée des Lumières. In *Philosopher sans frontières: hommage à Joachim Kopper* (pp. 99–119). Centre Gaston Bachelard.
Gayon, J. (1995). La philosophie biologique dans l'œuvre de Jacques Roger. Postface à Jacques Roger, *Pour une histoire des sciences à part entière*, texte établi et présenté par Claude Blanckaert (pp. 459–471). Albin Michel.
Gayon, J. (1998). La philosophie et la biologie. In J.-F. Mattéi (Ed.), Encyclopédie philosophique universelle, vol. IV: Le Discours philosophique (pp. 2152–2171). PUF.
Gayon, J. (2004). La banalisation biologique du monstre. In J.-C. Beaune (Ed.), *La vie et la mort des monstres* (pp. 93–97). Champ Vallon.
Gayon, J. (2009). Philosophy of biology: An historico-critical characterization. In A. Brenner & J. Gayon (Eds.), *French studies in philosophy of science. Contemporary research in France* (pp. 201–212). Springer.
Gayon, J. (2010). Vitalisme et philosophie de la biologie. *REPHA—Revue étudiante de philosophie analytique, 2*, 7–18. Also in P. Nouvel (Ed.). (2011). *Repenser le vitalisme* (pp. 15–31). PUF.
Gayon, J. (2014). Biologie et philosophie de la biologie: paradigmes. In T. Hoquet & F. Merlin (Eds.), *Précis de philosophie de la biologie* (pp. 11–24). Vuibert.
Giglioni, G. (2008). Whatever happened to Francis Glisson? Albrecht Haller and the fate of eighteenth-century irritability. *Science in Context, 21*, 465–493.

Gissis, S. B. (2014). The continuing vitality of the problématique of vitalism? *Studies in History and Philosophy of Biological and Biomedical Sciences, 47*, 196–200.
Gissis, S. B. (2019). 'Biologie': Lamarck's endeavor of a science of living entities. In C. Bognon-Küss & C. T. Wolfe (Eds.), *Philosophy of biology before biology* (pp. 170–187). Routledge.
Goldberg, B. (2022). Generation theory. In D. Jalobeanu & C. T. Wolfe (Eds.) *Encyclopedia of early modern philosophy and the sciences* (pp. xx). Springer.
Grimaud, J. C. M. G. de [published under initials D.G.]. (1776). *Essai sur l'irritabilité.* Bonnet frères.
Holmes, F. L. (1974). *Claude Bernard and animal chemistry. The emergence of a scientist.* Harvard University Press.
Huneman, P. (2008). *Métaphysique et biologie. Kant et la constitution du concept d'organisme.* Kimé.
Huneman, P. (2019). Philosophy after philosophy of biology before biology. In C. Bognon-Küss & C. T. Wolfe (Eds.), *Philosophy of biology before biology* (pp. 198–212). Routledge.
Huneman, P., & Wolfe, C. T. (2017). Man-machines and embodiment: From Cartesian physiology to Claude Bernard's 'living machine.' In J. E. H. Smith (Ed.), *Embodiment (Oxford Philosophical Concepts)* (pp. 257–297). Oxford University Press.
Hutchins, B., Basse Eriksen, C., & Wolfe, C. T. (2016). The embodied Descartes: Contemporary readings of *L'Homme*. In D. Antoine-Mahut & S. Gaukroger (Eds.), *Descartes' Treatise on Man and its reception* (pp. 287–304). Springer.
Kanz, K. T. (2002). Von der BIOLOGIA zur Biologie. Zur Begriffsentwicklung und Disziplingenese vom 17. bis zum 20. Jahrhundert. In U. Hoßfeld & T. Junker (Eds.), *Die Entstehung biologischer Disziplinen II. Beiträge zur 10. Jahrestagung der DGGTB in Berlin 2001* (Verhandlungen zur Geschichte und Theorie der Biologie, 9) (pp. 9–30). VWB.
Kleiman-Lafon, S., & Wolfe, C. T. (2021). Unsystematic vitality: From early modern beeswarms to contemporary swarm intelligence. In P. Fratzl, W. Schäffner, M. Friedman, & K. Krauthausen (Eds.), *Active materials* (pp. 261–297). De Gruyter.
de Lamarck, J.-B. (1944). *Biologie ou Considérations sur la nature, les facultés, les développemens et l'origine des corps vivans* (1800), ed. P.-P Grassé. *La Revue Scientifique, 82*, 267–276.
Le Clerc, D. (1696). *Histoire de la médecine.* Chouet et Ritter.
Lenoir, T. (1982). *The strategy of life. Teleology and mechanics in nineteenth-century German biology.* University of Chicago Press.
Linnaeus, C. (1751). *Philosophia botanica.* Godofr. Kiesewetter.
McLaughlin, P. (2002). Naming biology. *Journal of the History of Biology, 35*, 1–4.
Meiners, C. (1800). *Allgemeine kritische Geschichte der älteren und neueren Ethik oder Lebenswissenschaft nebst einer Untersuchung der Fragen: Gibt es dann auch wirklich eine Wissenschaft des Lebens? Wie sollte ihr Inhalt, wie ihre Methode beschaffen seyn*? (Vol. 1). Johann Christian Dieterich.
Maupertuis, P.-L. M. de. (1756). *Œuvres* (2 Vols.). Bruyset.
Ménuret de Chambaud, J.-J. (1765a). Œconomie Animale (Médecine). In D. Diderot & J. R. D'Alembert (Eds.), *Encyclopédie ou Dictionnaire raisonné des sciences, des arts et des métiers* (Vol. XI, pp. 360–366). Briasson.
Ménuret de Chambaud, J.-J. (1765b). Spasme (Médec. Patholog.). In D. Diderot & J. R. D'Alembert (Eds.), *Encyclopédie ou Dictionnaire raisonné des sciences, des arts et des métiers* (Vol. XV, pp. 434–438). Briasson.
Méthot, P.-O. (2018). De l'épistémologie historique à la philosophie de la biologie: le double héritage philosophique de Jean Gayon. In F. Merlin & P. Huneman (Eds.), *Philosophie, histoire, biologie: Mélanges offerts à Jean Gayon* (pp. 15–50). Matériologiques.
Métraux, A. (2004). 1064332 atomes et un cercle de vie. Quelques réflexions à peine philosophiques. *Multitudes, 16*(2), 41–47.
Reill, P. H. (2005). *Vitalizing nature in the Enlightenment.* University of California Press.

Reill, P. H. (2010). Eighteenth-century uses of vitalism in constructing the human sciences. In D. R. Alexander & R. L. Numbers (Eds.), *Biology and ideology from Descartes to Dawkins* (pp. 61–87). University of Chicago Press.
Rey, R. (2000). *Naissance et développement du vitalisme en France de la seconde moitié du 18e siècle à la fin du Premier Empire*. Voltaire Foundation
Roger, J. (1963). *Les sciences de la vie dans la pensée française du XVIIIe siècle. La génération des animaux de Descartes à l'Encyclopédie*. Armand Colin.
Roger, J. (1979). Chimie et biologie: des molécules organiques de Buffon à la physico-chimie de Lamarck. *History and Philosophy of the Life Sciences, 1*, 41-64.
Roger, J. (1980). The living world. In G. S. Rousseau & R. Porter (Eds.), *The ferment of knowledge. Studies in the historiography of eighteenth-century science* (pp. 255–283). Cambridge University Press.
Roger, J. (1995). Le monde vivant (1980). In C. Blanckaert (Ed.), *Pour une histoire des sciences à part entière* (pp. 192–224). Albin Michel.
Roose, T. G. A. (1797). *Grundzüge der Lehre von der Lebenskraft*. Christian Friedrich Thomas.
Schmid, C. C. E. (1798–1801). *Physiologie philosophisch bearbeitet* (3 Vols). Akademische Buchhandlung.
Schmitt, S. (2006). *Aux origines de la biologie moderne: l'anatomie comparée d'Aristote à la théorie de l'évolution*. Belin.
Shapiro, L. (2003). The health of the body-machine? Or seventeenth century mechanism and the concept of health. *Perspectives on Science, 11*(4), 421–442.
Sloan, P. R. (2006). Kant on the history of nature: The ambiguous heritage of the critical philosophy for natural history. *Studies in History and Philosophy of Biological and Biomedical Sciences, 37*, 627–648.
Steigerwald, J. (2014). Treviranus' biology: generation, degeneration and the boundaries of life. In S. Lettow (Ed.), *Gender, race, and reproduction. Philosophy and the early life sciences in context* (pp. 105–127). SUNY Press.
Steigerwald, J. (Ed.). (2006). *Kantian teleology and the biological sciences*, Special Issue of *Studies in History and Philosophy of the Biological and Biomedical Sciences, 37*(4).
Toepfer, G. (2009)."Organisation" und "Organismus"—von der Gliederung zur Lebendigkeit—und zurück? Die Karriere einer Wortfamilie seit dem 17. Jahrhundert. In M. Eggers & M. Rothe (Eds.), *Wissenschaftsgeschichte als Begriffsgeschichte. Terminologische Umbrüche im Entstehungsprozess der modernen Wissenschaften* (pp. 83–106). Transcript.
Toepfer, G. (2011). Biologie. In G. Toepfer (Ed.), *Historisches Wörterbuch der Biologie. Geschichte und Theorie der biologischen Grundbegriffe* (Vol. 1, pp. 254–295). J.B. Metzler Verlag.
Treviranus, G. R. (1802–1822). *Biologie oder die Philosophie der lebenden Natur*. J.F. Röwer.
van den Berg, H., & Demarest, B. (2020). Axiomatic natural philosophy and the emergence of biology as a science. *Journal of the History of Biology, 53*, 379–422.
Williams, E. A. (2003). *A cultural history of medical vitalism in Enlightenment Montpellier*. Ashgate.
Wolfe, C. T. (2009). 'Cabinet d'histoire naturelle', or: The interplay of nature and artifice in Diderot's naturalism. *Perspectives on Science, 17*(1), 58–77.
Wolfe, C. T. (2011). From substantival to functional vitalism and beyond, or from Stahlian animas to Canguilhemian attitudes. *Eidos, 14*, 212–235.
Wolfe, C. T. (2013). Vitalism and the resistance to experimentation on life in the eighteenth century. *Journal of the History of Biology, 46*, 255–282.
Wolfe, C. T. (2015). Il fascino discreto del vitalismo settecentesco e le sue riproposizioni. Scienze e filosofia da Copernico a DarwinIn P. Pecere (Ed.), *Il libro della natura* (Vol. 1, pp. 273–299). Carocci.
Wolfe, C. T. (2017). Models of organic organization in Montpellier vitalism. *Early Science and Medicine, 22*, 229–252.
Wolfe, C. T. (2019a). *La philosophie de la biologie avant la biologie: une histoire du vitalisme*. Classiques Garnier.

Wolfe C. T. (2019b). Il problema del tutto e delle parti: il caso del vitalismo di Montpellier. In G. D'Anna, F. Piro, M. Sanna, & F. Toto (Eds.), *Morfologie del rapporto parti-tutto* (pp. 257–270). Mimesis

Wolfe, C. T. (2020a). Life as concept and as science in early modern thought. In D. Jalobeanu & C. T. Wolfe (Eds.), *Encyclopedia of early modern philosophy and the sciences*. Springer.

Wolfe, C. T. (2020b). Vitalism in early modern medical and philosophical thought. In D. Jalobeanu & C. T. Wolfe (Eds.), *Encyclopedia of early modern philosophy and the sciences*. Springer.

Wolfe, C. T., Normandin, S. (Eds.) (2013). *Vitalism and the scientific image in post-Enlightenment life science* (pp. 1800–2010). Springer

Wolfe, C. T., & Shank, J. B. (2019). Diderot. *Stanford Encyclopedia of Philosophy.* https://plato.stanford.edu/entries/diderot/

Wolfe, C. T., Terada, M. (2008). The animal economy as object and program in montpellier vitalism. *Science in Context, 21*(4), 537–579.

Wolff, C. F. (1966) [1764]. *Theorie von der Generation; Theoria generationis*. Georg Olms.

Zammito, J. (2012). The Lenoir thesis revisited: Blumenbach and Kant. *Studies in History and Philosophy of the Biological and Biomedical Sciences, 43*(1), 120–132.

Zammito, J. (2018). *The gestation of German biology. Philosophy and physiology from Stahl to Schelling*. University of Chicago Press.

Chapter 14
Lamarck: Philosopher of Nature

Stéphane Tirard

Abstract This paper presents the chapter on the history of the inheritance of acquired characters that Jean Gayon published in 2006 in a book entitled *Lamarck, Philosophe de la nature*. Gayon's detailed analysis of this notion clarifies the oft-misunderstood role of the French naturalist Jean-Baptiste Lamarck in its elaboration and demonstrates how several authors contributed to it during the first half of the nineteenth century. Beyond the specific case of the inheritance of acquired characters, Gayon's article is an important contribution to the history of heredity in general. Above all, it confirms Gayon's strong interest for the study of the history of concepts, which is probably one of the most important aspects of his oeuvre.

Keywords Lamarck · Heredity of acquired traits · Transformism · Jean Gayon

14.1 Introduction

Lamarck, Philosophe de la nature (Corsi et al., 2006)[1] aims to reassess Lamarck's work from around 1800, a period during which the naturalist's thought underwent a profound transformation that would lead to his transformist thesis. The work is divided into four parts. The first chapter, written by Gabriel Gohau, focuses on two little-studied books by Lamarck, *Mémoires de physique et d'histoire naturelle* (1807) and *Système analytique des connaissances positives de l'homme* (1820), and proposes

[1] This book is the result of reflections that its four authors have had during four years of seminars (*Lamarck seminar* from 1998 to 2000 and *Seminar on Lamarckism* from 2000 to 2002) organized by the REHSEIS team (Paris 7), then with the Mahler Centre (University of Paris 1), the IHPST (University of Paris 1) and the Centre François Viète (University of Nantes). All translations are mine.

Translated from Tirard (2018).

S. Tirard (✉)
François Viète Centre for Epistemology and History of Science and Technology, Nantes Université, 2 rue de la Houssinière, 44 322 Nantes, France
e-mail: Stephane.tirard@univ-nantes.fr

P.-O. Méthot (ed.), *Philosophy, History and Biology: Essays in Honour of Jean Gayon*, History, Philosophy and Theory of the Life Sciences 30,
https://doi.org/10.1007/978-3-031-28157-0_14

a comparison of Lamarck's "philosophy" before and after 1800. Pietro Corsi, for his part, analyzed Lamarck's "biology" project, that is to say, his "specific physics of life", which was ultimately a failure. In the third part, I studied the notion of "spontaneous generation" as used by Lamarck, who made it one of the crucial aspects of his evolutionary thinking. Finally, in the fourth chapter, Jean Gayon focused on the issue of the "inheritance of acquired characters" that Lamarck called the "transmission of acquired modifications" and which, as is well known, lay at the center of his transformist perspective. Gayon is interested in the emergence of this notion, wrongly attributed to Lamarck, which has a long history, the complexity of which can only be grasped by exploring the eighteenth and nineteenth centuries.

During the conversations the four of us had during the seminars in which the idea of a common book arose, we learned that Jean Gayon had long contemplated such an investigation into the concept of the "inheritance of acquired characters". His chapter, which belongs to a body of work in the history and philosophy of the sciences, focusing especially on evolutionary biology, is one of the few texts he devoted to Lamarckism. Apart from the work that interests us here, this subject seems to be present only in one title of his bibliography, a collaboration with Richard M. Burian and Laurent Loison, published in 2017: "*The Contribution—and Collapse—of Lamarckian Heredity in Pastorian Microbiology: 1. Lysogeny, 1900–1960*" (Burian et al., 2017).

In the present chapter, I will summarize and discuss Jean Gayon's analysis of the "inheritance of acquired characters", the aim of which is to break with the banality of an expression long presented as closely—if not exclusively—linked to the history of Lamarckism. Gayon's analysis is organized into three stages. The first is a study of the use of the term "heredity" in the biological sense, i.e., "natural heredity", which became a widespread notion at the turn of the eighteenth century. The second part, focusing on Lamarck himself, features a very precise examination of his vocabulary and his formulations of the idea that would later be called the "heredity of acquired traits". Finally, the third part identifies the historical factors, outside of Lamarck's framework, which also participated in the genesis of the expression and its later banalization.

14.2 Gayon's Analysis of the Concept of Inheritance of Acquired Characters

The chapter opens with some clarifications on the choice of words composing what Jean Gayon calls the "verbal expression of a content of thought" (*expression verbale d'un contenu de pensée*). Regarding the word "heredity", he reminds us that, as a noun, it was not used in connection with transmission by generation before the nineteenth century; before this period, physicians and naturalists only used the adjective "hereditary". Regarding the word "characteristics", it turns out that naturalists (those who adopted the Lamarckian worldview between 1810 and 1840) spoke of "acquired modifications" rather than "acquired characteristics". Finally, as to the word "acquired" itself, it does not seem to raise specific issues; it was commonly used by all the authors considered in Gayon's research.

Before appearing in biological discourses, heredity was a technical term in law and medicine. In the legal field, heredity referred to property and especially to privileges. The French Revolution, for example, proclaimed the end of the heredity of privileges. In addition, physicians long had an interest in hereditary diseases (as an adjective). The transition from the adjective to the substantive came later and was noticeable in Prosper Lucas's 1847 treatise on "natural heredity" (Lucas, 1847). Jean Gayon took this opportunity to highlight the work of Carlos Lopez Beltrán (1992), who has shown that this transition took place among French medical doctors who no longer restricted the word's sphere of reference to pathological phenomena or anomalies but applied it to a wide variety of biological phenomena. This transition can therefore be described as a "de-medicalization" or "de-pathologisation" of the concept of heredity. Finally, it is important to remember that the term heredity also entered the terminology of natural history through discussions of human variations and races.

This first part of Jean Gayon's text is therefore intended as a clarification of the vocabulary of heredity and does not deal with the inheritance of acquired characteristics in the strict sense. On this subject, he follows Conway Zirkle (1946) in stressing that although the idea of acquired traits was very much present in nineteenth century medicine, natural history, anthropology, and theology (e.g., the transmission of Adam's original sin), the expression itself did not occur before this period.

The second part of Gayon's text offers an analysis of some remarks of Lamarck, who himself never used either the word "heredity" or the word "hereditary". Lamarck also did not speak of "acquired characters" but of "new faculties that have been acquired" or of "acquired changes". In his chapter, Jean Gayon systematically reviews Lamarck's most important texts, starting with the *Opening Address* of his 1801 Museum course, where he gives a brief glimpse of his transformist ideas:

> Lamarck indicates here three categories of factors which, together, he believes are sufficient to modify a species. Firstly, there must be variations in the "surrounding environment" and/or internal variations. Secondly, the faculties modified by these variations must become stronger with use. Finally, the concomitant modifications of the organic parts concerned must be transmitted and propagated by generation. (Gayon, 2006, p. 116)

This first examination of Lamarck's work led Jean Gayon to the following central claim of his chapter:

> The process Lamarck mentions may well have an implicit notion of heredity, but it is minimal. It is simply postulated that the nature and state of the modified parts "are preserved (...) by generation". One could almost say that this is the purest concept one can imagine of "heredity of acquired traits" (...) free of any theoretical content traditionally attached to the vocabulary of heredity. (Gayon, 2006, p. 116)

Lamarck thus used the notion of inheritance of acquired characters in what Gayon calls a "framework vision" (*vision cadre*) (Gayon also spoke of a "theoretical world") of organic change that would be found also in the works of neo-Lamarckian biologists. This framework is based on a transformist view of species, an anti-preformism, a causal privilege accorded to circumstances, a gradual and cumulative view of organic changes, and a role for individual actions. It is, therefore, within this global framework that Lamarck first articulated the process that would later be renamed "heredity of acquired traits". However, if a certain vision of heredity is indeed present, "it is absorbed in a global explanation of change by 'circumstances', the various elements of which are interlinked" (Gayon, 2006, p. 118).

Gayon's analysis then turns to the most well-known texts in which Lamarck formalized his thinking: *Recherches sur l'organisation des corps vivants* (1802), *Philosophie zoologique* (1809), and *Histoire naturelle des animaux sans vertèbres* (1815–1822). Some important elements should be highlighted. In the *Recherches*, Lamarck shows how changes—i.e., the shape of the body, the number and condition of the organs and faculties—can be acquired through habit and how these changes make it possible to rethink the chain of beings. In addition, the naturalist indicates that any "acquired change" must have been acquired in both parents. Two important aspects that appear later in his *Philosophie zoologique* should be clarified at the outset. The first is that Lamarck extends his thinking to plants, noting that in plants "there are no actions and, therefore, no *habits as* such" (Lamarck, 1809, p. 223). Jean Gayon notes the importance of this point by specifying that:

> This consideration of plants enables Lamarck to mobilize his most convincing examples, borrowed both from human experience of domestic plant cultivation and from the knowledge that botanists draw from their collections. In the case of cultivated plants, the changes are indeed rapid and well-documented. (Gayon, 2006, p. 123)

The second point, which derives from the first and which is of particular relevance to our subject, is the formulation in *Philosophie zoologique* of two laws in which Lamarck did not use the terms habit, action, and movement, which would restrict the scope of his analysis to animals.

> First Law.
>
> In any animal which has not reached the end of its development, the more frequent and sustained use of any organ gradually fortifies that organ, develops it, enlarges it, and gives it a power proportionate to the duration of its use; while the constant lack of use of that organ imperceptibly afflicts it, deteriorates it, progressively diminishes its faculties, and eventually causes it to disappear.
>
> Second Law
>
> All that nature has caused individuals to acquire or lose by the influence of circumstances in which their race has long been exposed, and, consequently, by the predominant influence of any one organ, or by that of a constant defect in the use of any one part; it preserves it by generation to new individuals from it, provided that the changes acquired are common to both sexes, or to those who produced these new individuals. (Lamarck, 1809, p. 235)

Taken together, these two laws would become known later as the "law of use and disuse" and would be one of the keystones of Lamarck's "framework vision". The question of heredity is present, even though Lamarck did not use the words "heredity" and "hereditary".

Later, in *Histoire naturelle des animaux sans vertèbres* (1815–1822), Lamarck formulated four laws which, according to him, "concern organisation, and which govern all the acts that take place within it by the forces of life" (Lamarck, 1815, p. 181). Let us recall them by way of comparison:

> *First law:* Life, by its own forces, tends continually to increase the volume of the whole body that possesses it, and to extend the dimensions of its parts up to a limit which it determines.
>
> *Second law:* The production of a new organ in an animal body results from a new want supervening, which continues to make itself felt, and from a new movement that this want gives birth to, and continues.
>
> *Third law:* The development of the organs and their power of action are constantly in ratio of the employment of these organs.
>
> *Fourth law:* All that has been acquired, laid down or changed, in the organisation of the individuals during the course of their life, is conserved by their generation and transmitted to the new individuals proceeding from those which have undergone those changes. (Lamarck, 1815, pp. 181–182)

Lamarck, however, seems to have simplified these laws: when he commented on them, he took up, as Jean Gayon noted, the earlier 1809 formulations and spoke of a law of "development" and a law of "transmission". Moreover, the fourth law is characterized by a vocabulary that is still close to the vocabulary of heredity. Lamarck went from using the words "conservation" and "propagation" to that of the "transmission" of "all that has been acquired, laid down or changed, in the organisation of the individuals during the course of their life". However, Lamarck kept a focus on the modification of species rather than on the phenomena of heredity.

In the last part of his chapter, Jean Gayon argues that the notion of acquired characters and Lamarckian thought do not neatly overlap. More specifically, he shows that if certain historical developments of this notion did pass through Lamarck's works, others took different paths. Thus, the fact that Lamarck was an essential link in this history must be put into perspective. Two lineages must be distinguished: one French and the other British. The first one includes the names of Charles-Georges Leroy, Nicolas de Condorcet, Pierre Jean Georges Cabanis, Frédéric Cuvier, François-Désiré Roulin, Auguste Comte, and Pierre Flourens. This review, which is impossible to summarize here without distorting it, begins with Condorcet and his *Esquisse d'un tableau historique des progrès de l'esprit humain* (1794). In this work, Condorcet reflects on the possibility of an indefinite perfectibility of the human species and claims that "the organic perfectibility or degeneration of races in plants, in animals, can be considered as one of the general laws of nature" (Condorcet, 1970, p. 236 [1794]). No proof of Leroy's (1768) influence on Condorcet exists, although Condorcet did claim in his *Lettres sur les animaux* that *animals,* as Jean Gayon points out, "were capable of perfectibility not only individual, but also specific, and based on the transmission to their offspring of the behaviours learned in the course of their lives" (Gayon, 2006, p. 135). As to Condorcet, he believed that the physical and moral faculties of the human species were perfectible. It is not possible to know whether Lamarck had read Condorcet, but the latter's work was well-known and widely available.

For his part, Cabanis stated in *Les rapports du physique et du moral de l'homme* (1802) that habits acquired during a lifetime can be transmitted to the next generation and he formulated "the hypothesis of the emergence of elementary forms of life from inorganic elements, as well as the derivation of animal species from each other" (Gayon, 2006, p. 137). The dates of publication suggest that Lamarck and Cabanis had not read each other's work before developing their respective arguments; the latter did not use the word "heredity" either, and did not intend to develop a theory of the modification of species.

Cuvier's *Recherches sur les caractères ostéologiques qui distinguent les principales races du chien domestique* (1811) is particularly noteworthy. This work introduces the first use of the word "hereditary" in the context of natural history: "The breed usually perpetuates itself and the same is true for most of the traits we have examined: new proofs that accidental changes always end up being hereditary" (Cuvier, 1811, p. 353).

Jean Gayon then turns to the contribution of Roulin, who did use the adjective "hereditary", and the examples provided in *Recherches sur quelques changemens observés dans les animaux domestiques transportés de l'ancien dans le nouveau continent* (1829). According to Geoffroy Saint-Hilaire, domestic animals showed the transmission of certain acquired habits. Hereditary transmission was also admitted by Comte (although he criticized Lamarckian ideas) in the forty-second lesson of his *Cours de philosophie positive* in 1835. For his part, Flourens considered that "[t]he issue of the heredity of acquired modifications is one of the most important and wide-ranging issues of general physiology" (Flourens, 1841, p. 112). This in turn led Gayon to claim that: "Flourens's expression of 'heredity of acquired modifications' preceded that of 'heredity of acquired traits'. If this is the case, it would have been the decisive milestone in what led from a conception of evolution (Lamarck) to a specific conception of heredity" (Gayon, 2006, p. 147). The last French author mentioned by Gayon is Prosper Lucas, whose *Traité philosophique et physiologique de l'hérédité naturelle* (1847) provided a general overview of all the main conceptions of heredity. It marked the already mentioned "de-pathologisation" of heredity. But Lucas was a fixist and was opposed to Lamarckian transformism.

The other historical lineage is British; it includes the works of James Cowles Prichard and Charles Lyell. According to Gayon, Prichard used hereditary language from a "non-modificationist" perspective. Trained in medicine, anthropology, and natural history, Prichard stated in 1813: "We may be allowed to conclude with a considerable degree of confidence that no acquired varieties of constitution become hereditary or in any manner affect race" (1813, p. 203). According to him, in all species and in the human species in particular, variation could not be indefinite. However, his views on this matter changed over time, and so in 1836 he opted for the inheritance of acquired characters, while remaining in favour of the fixed nature of species.

As for Lyell, the second volume of the *Principles of Geology* (1832) includes a critique of the Lamarckian hypothesis. He used the word "hereditary" but rejected the claim that the changes could go beyond the species limit.

Jean Gayon's conclusion of the section devoted to Prichard and Lyell is the high point of the chapter; he demonstrates there how the heredity of acquired modifications and a large part of Lamarckian biology proved to be compatible with fixist thought. Here it is *in extenso*:

> Lyell and Prichard therefore share a common strategy. Both admit the Lamarckian argument of the transmission of acquired skills through generation. But they believe that the changes thus produced do not go beyond a certain point, which is characteristic of each species. These two authors show in an exemplary manner how part of the Lamarckian doctrine of the influence of circumstances (the second law of use and non-use, or law of transmission by generation), a doctrine that was originally fundamentally transformist, could be taken up and accredited in a non-transformist context. We have characterised this theoretical context as 'subtle fixism'. We could just as well have said 'limited transformism'. This, I believe, is the main reason for the later success of the expression 'heredity of acquired characters'. Certainly, this formula became the rallying sign of the neo-Lamarckian transformists of the late nineteenth century. But, as it was constructed, in an area of interaction between naturalistic and medical discourse, it was not necessarily associated with unlimited transformism. (2006, p. 157)

Thus, if "the inheritance of acquired characters" is indeed the "verbal expression of a content of thought", Jean Gayon's brilliant analysis opens-up a window into its complex and plural dynamics. The "inheritance of acquired characters", which was present for a long time in biological thought without being named, even by Lamarck himself, was in fact common to both fixism and transformism and would become a very important part of the nineteenth century life sciences. To underline this point one last time, let us borrow from Jean Gayon this quotation from Flourens: "The issue of the heredity of acquired modifications is one of the most important and wide-ranging issues of general physiology" (Flourens, 1841, p. 112).

14.3 Conclusion

Beyond the historical and philosophical insights, this chapter by Jean Gayon illustrates the nature of his formidable method of analysis, which is at the heart of his many contributions to the field of historical epistemology. His examination of the biological concepts in this study (as in all his works) is based on a detailed reconstruction of his subject matter that alternates between explanations and moments of synthesis that mark historiography. But the text we have just examined also testifies to a taste and probably a passion for words, and among the words of science those that designate concepts, one of Jean Gayon's goals being to reveal the meaning and the history of this meaning. This is certainly an indication of a fruitful meeting point between philosophy and history of science.

References

Beltrán, C. B. (1992). *Human heredity 1750–1870; The Construction of a domain, PhD Dissertation*. University of London.

Burian, R. M., Gayon, J., & Loison, L. (2017). The contribution—and collapse—of Lamarckian heredity in Pastorian microbiology: 1. Lysogeny, 1900–1960. *Journal of the History of Biology, 50*(1), 5–52.

Cuvier, F. (1811). Recherches sur les caractères ostéologiques qui distinguent les principales races du chien domestique. *Annales du Museum d'histoire naturelle* (Vol. 18, pp. 333–353). G. Dufour.
Corsi, P., Gayon, J., Gohau, G., & Tirard, S. (2006). *Lamarck. Philosophe de la nature*. Préface par Armand de Ricqlès. Presses Universitaires de France.
de Condorcet, J. A. N. (1970). *Esquisse d'un tableau historique des progrès de l'esprit humain [1793-1794]*. Vrin.
Flourens, P. (1841). *Résumé analytique des observations de Frédéric Cuvier sur l'instinct et l'intelligence des animaux*. C. Pitois.
Gayon, J. (2006). Hérédité des caractères acquis. In P. Corsi, J. Gayon, G. Gohau, & S. Tirard (Eds.), *Lamarck. Philosophe de la nature* (pp. 105–163). Presses Universitaires de France.
Lamarck, J.-B. (1802). *Recherches sur l'organisation des corps vivants*. Maillard.
Lamarck, J.-B. (1807). *Mémoires de physique et d'histoire naturelle*. Muséum d'histoire naturelle.
Lamarck, J.-B. (1809). *Philosophie zoologique*. Dentu.
Lamarck J.-B. (1815). *Histoire naturelle des animaux sans vertèbres, t.1*. Déterville.
Lamarck, J.-B. (1820). *Système analytique des connaissances positives de l'homme*. Déterville.
Lucas, P. (1847). *Traité philosophique et physiologique de l'hérédité naturelle dans les états de santé et de maladie du système nerveux, avec l'application méthodique des lois de la procréation au traitement général des affections dont elle est le principe*. J.-B. Baillière.
Lyell, C. (1830–1833). *Principles of geology, being an attempt to explain the former changes of the earth's surface, by reference to causes now in operation* (3 Vols.). John Murray.
Prichard, J. C. (1813). *Researches into the physical history of man*. John and Arthur Arch.
Roller, R.F.-D. (1829). Recherches sur quelques changements observés dans les animaux domestiques transportées de l'ancien dans le nouveau continent. *Annales Des Sciences Naturelles, 16*, 16–34.
Tirard, S. (2018). Lamarck, philosophe de la nature. In F. Merlin & P. Huneman (Eds.), *Philosophie, histoire, biologie: mélanges offerts à Jean Gayon* (pp. 283–292). Éditions Matériologiques.
Zirkle, C. (1946). The early history of the idea of the inheritance of acquired characters and of pangenesis. *Transactions of the American Philosophical Society, 35*, 91–150.

Chapter 15
Between Science and Ideology: The Case of Clémence Royer

Gérard Chazal

Abstract Clémence Royer (1830–1902) was the first translator of Darwin's *On the Origin of Species* into French. She prefaced her translation with a long text and included several notes. She defended the theory of evolution against creationism, but she developed a view of organic evolution that was more ideological than scientific. This paper examines Royer's changes from the original text, which led Darwin to request another French translation of his work. Overall, Royer's work falls squarely within the scope of the history of "*Après Darwin*" that Jean Gayon has meticulously analyzed in his work.

Keywords Darwinism · Eugenics · Evolution · Progress · Selection · Sorting · Anti-christianity

15.1 Introduction

The relationship between philosophy and the sciences has undergone many changes. Some authors have even thought that philosophy is only a particular form of ideology, in opposition to which science must impose its own approach, just as the patient search for truth, even provisional and partial, must be opposed to the peremptory affirmation of a belief lacking solid foundations. The boundary between scientific hypothesis, possibly awaiting proof, and ideological speculation, is not always easy to draw. And, even if, a posteriori, we can clearly see what is scientific and what is ideological, the place of philosophy has yet to be determined. Sometimes we see philosophy claiming to provide the framework for science, but in so doing also offering a dubious interpretation of what science is. We have also seen philosophy trying to understand the scientific approach to draw from it values that one can hope to generalise, i.e., one envisages extending an empirically determined law outside the

Translated from Chazal (2018).

G. Chazal (✉)
Université de Bourgogne, Dijon, France
e-mail: gerard.chazal@orange.fr

P.-O. Méthot (ed.), *Philosophy, History and Biology: Essays in Honour of Jean Gayon*, History, Philosophy and Theory of the Life Sciences 30,
https://doi.org/10.1007/978-3-031-28157-0_15

domain of its initial application. But while we can well admit what some philosophers have called "nomadic concepts" (Stengers, 1987), such extensions remain hazardous. Thus, it has not been proven that what can be of value in the scientific approach to biology can be of value in the organisation and life of a society, even if the transition from the biological to the social can often be tempting, for it presents what may appear to be explanations, though they may turn out to be nothing more than illusions. Conversely, if democracy is the least bad of all political systems, as Churchill put it, and remains a real value for many, it is not relevant in science, as it would be absurd to try to settle a scientific controversy by means of a vote.

15.2 Science and Ideology in Biology

Many of us of in my generation have wondered one day or another, on the margins of our philosophical studies or in the later stages of a career, whether at a colloquium or seminar, a meeting, or after a talk, whether there is a place for philosophy between science and ideology and whether the second is not the fatal deviation that constantly threatens the first. Science seemed to us so perfectly defined, its fields so clearly delineated that, by comparison, it seemed very risky to try to answer the question of what philosophy was, what its role was. On the other hand, insofar as many philosophers linked their practice of philosophy to intellectual commitments whose foundations did not always seem well assured, it was tempting to assimilate philosophy to ideology in the sense that Marx gave to this term in *The German Ideology* (Marx & Engels, 1976 [1932]). Was anyone who wanted to be a philosopher finally destined to be at home nowhere and destined only to wander through marshlands, where any search for solid ground seems futile? Even the field of rationality could seem uncertain when, among those who claimed to be philosophers, voices were raised in praise of the most irrational thoughts. Moreover, by instrumentalizing science, certain philosophies can sometimes produce counter-values that take precedence over all truth and in so doing erase it.

One of the most paradigmatic examples of such a shift in the scientific approach, where the status of statements becomes uncertain and it becomes difficult to distinguish between scientific hypothesis and ideological deviation, arose in the wake of Darwin (Gayon, 1998 [1992]). Here, as Jean Gayon's work has shown, mainly in *Darwin et l'après-Darwin*, philosophy played an important role in disentangling from the theses of social Darwinism and eugenics what follows logically from Darwinian premises and what was but an ideological reconstruction based as much on certain works of the Paris Anthropological Society, founded by Broca in 1859 (the same year in which *On the Origin of Species* was published), and the philosophy of Herbert Spencer, as on Darwin's own hypotheses. Here, I would like to try to illustrate this point with the case of Clémence Royer (1830–1902), who first translated Darwin's work into French, but who mixed in so many considerations wrongly deduced from the hypothesis of natural selection that Darwin asked for a new translation of his work.

In opposition to the creationist theses that had long prevailed, the science of the eighteenth and especially the nineteenth centuries gradually established the idea of the transformation of species, thus making man an animal among other animals, the result of the blind work of natural selection. What we are witnessing, therefore, is a long process that progressively substitutes acquired knowledge for imposed beliefs. Diderot had suspected a possible evolution of life through the emergence of new species; Buffon had considered the transformist hypothesis before finally rejecting it; Lamarck developed it, basing the evolution of the different organs of living beings on the "law" of use or disuse, and entrusting the direction of this evolution to the environment. Finally, it was Darwin who proposed the most relevant hypothesis with his theory of evolution based on the existence of small individual variations on which natural selection could operate. If a variation is favourable, in each environment, individuals who carry it will be able to live longer and produce more offspring than those without it. As this variation is hereditary, it will be the cause of a gradual evolution, for over time there will be more and more individuals carrying this variation within the species, until it eventually becomes the norm, having previously been only the exception. For a variation to be selected it must therefore provide those who benefit from it with an advantage as regards survival and/or reproduction. Hence, in Darwin's case, there exists a double selection: natural selection and sexual selection. This scientific formulation of the theory of evolution, as first established by Darwin, is well known and may seem clear. However, it was and still is violently attacked by most creationist ideologues, though it was also twisted to serve certain purposes which, in a covert and insidious manner, partly denied it and thus deformed it.

15.3 Evolution by Natural Selection

The Darwinian hypothesis was based on the one hand, on a considerable mass of observations accumulated over the years, particularly during Darwin's voyage aboard the *Beagle*, and, on the other hand, on the example of artificial selection practised systematically from the nineteenth century onwards by European livestock farmers and probably observable empirically since the beginnings of agriculture and animal husbandry. I will not discuss here the value of the argument of artificial selection, to which it was objected that, once it ceased, the population returned to its initial state of diversity, or the fact that it introduced an element of teleology that does not exist in natural selection. What is certain, however, is that Darwin did not derive from the facts of evolution the idea of an intention or purpose that guided it. For him, there is no will among individuals to survive and reproduce to the detriment of other individuals of the same species, as would appear to be suggested by the concept of the "struggle for life" popularised essentially by Spencer. Hereditary variations are a matter of chance; natural and sexual selection work on these random data, even if the underlying mechanisms of heredity were unknown at the time. Darwin sought to overcome this ignorance by means of the pangenesis hypothesis, which remained a simple hypothesis formulated at the end of the book, *Variation of Animals and Plants under the Action of Domestication* (1868).

I mention this aspect of Darwinian theory because Royer criticized it and substituted for it another hypothesis, "dynamogenesis" (Royer, 1877).[1] Thus Darwin, in the extension of physical science as it had developed since Descartes, expunged any idea of finality. Species derive from each other and from common ancestors according to a mechanism that does not presuppose any project to build a natural and hierarchical chain of beings. The question of man remains. Darwin did not address it in *On the Origin of Species* (1859), his first work, but it is the subject of his last work, *The Descent of Man* (1871). Seeing man as part of the evolutionary continuum, Darwin argued that with human beings there arises, in addition to the social instincts we share with many other species, language, culture, civilisation, morality, which in a way cancel out the crude natural selection at work in the world of other living beings. Of course, all these characteristics of the human species are the result of natural selection, and yet at the same time they are opposed to the continuation of this selection by allowing for the development of empathy and compassion—perhaps simply because the appearance of consciousness enabled humans not only to recognise other members of their own species but also to "put themselves in the place" of others. Patrick Tort has developed this reading of Darwin's latest work, finding in it the materialist foundation of an ethics without transcendence arising through what he calls "the reversive effect of evolution" (Tort, 2000, 2002, 2005a, 2005b).

Whatever the merit of Tort's reading of Darwin, it must be recognised that Darwin's work remained in need of proof, left a certain number of problems unresolved (those concerning the mechanisms at the origin of small variations as well as those of their heredity), and above all, as a result, lent itself to interpretations that went beyond the known scientific facts and thus brought into play philosophical or ideological commitments. This field was thoroughly explored by Gayon (1998 [1992]). After Darwin, a space for research and interpretation was opened in which social Darwinism and eugenics could lay claim to the status of scientific theory, attempting to base an anthropology on Darwinian theories. After summarising Darwin's theory, I would like, by way of example, to make a few remarks about what social Darwinism was, basing my claims mainly on the work of Clémence Royer, whose work exemplifies the far frontiers of science and its interpretations.

15.4 Translating Darwin

The first French translation of *On the Origin of Species*, in 1862, was made by Royer, who preceded it with a long preface and some notes. Darwin was so reluctant to accept this translation that he requested certain corrections, which were made in subsequent editions (in 1866, 1870, and 1882 by Royer) and eventually requested a new translation by Jean-Jacques Moulinié. Royer attributed Darwin's reservations

[1] All translations are by the author.

about this translation to her criticism of the pangenesis hypothesis. In a note to her book, *Deux hypothèses sur l'hérédité*, she wrote:

> In the preface to the third edition of my translation of *On the Origin of Species* (Darwin, 1870), I had already published a critique of the theory of pangenesis and announced my dynamic theory, which I then set about putting to paper. Public events have so far delayed its publication. The divergence of my views on this subject with M. Ch. Darwin explains his desire to see published in France another translation of his main work by a less independent disciple of the *magister dixit*. (Royer, 1877, p. 18)

If it is true that Darwin's pangenesis is only a hypothesis which in 1868 aimed to explain the variations on which selection operates, it is also true that it was not until the following century that the mechanisms of heredity were effectively clarified. However, Royer's criticism has its weaknesses, and her hypothesis proved to be just as inadequate as that of Darwin. One can thus criticize Royer first for wrongly comparing the "cellular gemmules" to Pasteur's germs, which she rejects: "Darwin's pangenesis is, like Pasteur's *panspermia*, a hypothesis that lacks a basis, i.e., a first, real, observed and self-explanatory fact" (Royer, 1877, p. 6). Then, after having reproached Darwin for the immaterial character of his gemmules, she writes: "The heredity of organic characters cannot be due to a transmission of matter" (Royer, 1877, p. 18). Recognizing only matter and force, with form resulting from the application of the latter to the former, Royer is obliged to replace heredity by "germs", i.e., particles, with heredity by force. And it is thus by force that life will be defined: "In any case, life appears to us as an essentially dynamic phenomenon, as a force in motion" (Royer, 1877, p. 26). The hypothesis of "dynamogenesis" does not appear, however, to be any more scientifically founded than that of pangenesis. Moreover, this criticism of pangenesis was put forward only after the reservations formulated by Darwin himself.

In fact, one might think that Darwin's dissatisfaction with Royer's translation rests on three aspects of this work which not only reveal differences between the two thinkers but even stem from opposing visions. Royer's translation, preface, and notes amount to a genuine interpretation of Darwin that may appear to some as an ideological instrumentalization of Darwinism. This view is confirmed by Royer's, 1870 work entitled *Origine de l'homme et des sociétés* (Royer, 1870).

The first reason for reticence is probably the anti-religious use that Royer intends to make of the work and of Darwinian theory. Admittedly, the Darwinian theory of a transformation of species is de facto opposed to the idea of a single moment of divine creation of all the species we know today. There is indeed a certain incompatibility between the creationist conceptions of most religions and the theses of Darwin. The fierce opposition that Darwin's work stirred up in religious circles is a sure sign of this. But in *On the Origin of Species*, Darwin did not emphasize this opposition because to do so would have been to go beyond the strictly scientific approach of accounting for the facts and to enter a theological and philosophical debate in which Darwin at the time clearly did not wish to get involved. However, not only does Royer engage in this debate, but the whole first part of her preface is a kind of history of reason in the struggle against religions, from Egypt and Greece to Christianity, a history that is not dissimilar to the succession of theological, metaphysical, and positive ages found in Auguste Comte. Royer enlists the work of Darwin in her fight as an "atheist freethinker". Obviously, Christianity is the main target in this affair:

> The germ of this religion was the apostolic Christolatry[2] of the Pauls and the Johns, so different from the master's doctrine. Propagated by an ignorant, dominating, and corrupt priesthood, it spread like an obscure veil over all intellects and applied the brake of faith to the legitimate curiosity of human genius, at the very moment when civilised Rome was collapsing in the face of the barbarian invaders. (Royer in Darwin, 1870, p. XXX)

Still in a manner reminiscent of Comte, she lists the geniuses of human thought, who together laid down the steppingstones required to leave religious thought behind: "But the hour of emancipation sounded, nevertheless. While Copernicus, Columbus, Galileo, Kepler, and Newton revealed the true system of the world, Bacon, Descartes, Leibniz, and Locke opened new roads before the spirit" (Royer in Darwin, 1870, p. XXXI). It is therefore in opposition to religion and all creationist theories that Darwin is supposed to have provided an answer to the old question of "the origin and nature of organic forms" (Royer in Darwin, 1870, p. XXXV). She thus makes of Darwin's thought a sort of nominalism, opposed to the realism of species with its links to creationism and the fixity of species (*fixisme*):

> Now, it [Darwinism] is precisely the doctrine of nominalism and the most absolute individualism, it is the absence of any idea or ideal prototype, it is also the tendency of natural freedom to make specific characteristics diverge almost constantly, by varying and individualising forms, that M. Charles Darwin has demonstrated today in his beautiful book *On the Origin of Species*. (Royer in Darwin, 1870, p. XXXVII)

Once again Royer draws anti-religious philosophical conclusions from Darwin's work that may be considered legitimate, but which go beyond the strictly scientific approach of the Darwinian project of 1859, and it is understandable that Darwin could have felt instrumentalized in a fight that was not of his own choosing.

The second reason for Darwin's reticence is the confusion Royer maintains between certain terms and so also between the notions they represent. Thus, notions of evolution and progress, selection, and sorting, reintroduce a form of finalism into Darwin's writings. And probably of most importance of all, the very way she translated Darwin's title in the first edition of 1862 could only elicit certain disavowals from Darwin himself. Indeed, the English title, *On the Origins of Species by Means of Natural Selection or the Preservation of Favoured Races in the Struggle for Life* became, *De l'origine des espèces ou des lois du progrès chez les êtres organisés* (*On the Origin of Species or the Laws of Progress in Organized Beings).* First, the word "selection" has vanished; and in the main body of the text, she translated the English word "selection" by "élection" (election). Whereas the term "selection" implies a posteriori *that* a new trait gives a certain advantage or disadvantage to the individual who bears it in the competition for survival and reproduction, the term "election" implies the idea of a priori criteria which would define what is favourable or unfavourable and as a function of which nature would either conserve or eliminate individuals. This amounts to replacing a natural mechanism, as described by

[2] In French, "*christolâtrie*" is a neologism coined by Royer.

Darwin, with a kind of judgement made by nature regarding the value of a new trait. Further, by introducing the notion of progress, this title gives the idea of an evolution of species from the lesser to the greater, from imperfect forms to more perfect forms; it is, in other words, the introduction of finalism, which is completely absent from Darwin's theories, and a return to Lamarck.

Of course, Royer's finalism is not based on the idea of transcendence, on some divine will, but to speak of progress in the organisation of living beings is to accept, as Darwin never did, the idea that evolution is heading somewhere, towards a more perfect state, an "omega point", as Teilhard de Chardin will later say, reintroducing the notion of transcendence. It was at Darwin's request that, from the 1866 edition onwards, the title was changed to, *De l'origine des espèces par sélection naturelle ou des lois de transformation des êtres organisés (On the Origin of Species by Natural Selection or the Laws of Transformation of Organised Beings*), where the idea of "laws" does not imply that transformation is oriented towards the better. The word "selection" was reintroduced in place of "election", though Royer claimed that she was giving in to a misguided use of the term. However, this notion of progress in evolution, for which selection would be the means, also leads to the vision of a hierarchical order in the living world. This is clearly stated in the opening words of the book, *Origine de l'homme et des sociétés*: "Man is the terminal and superior summit of the great tree of life" (Royer, 1870, p. 3).

Lastly, while in *On the Origin of Species* Darwin does not address the question of man, as he did in his last work, Royer addresses this question in her preface and even devotes an entire work to the topic (Royer, 1870), in which, taking up theses that belong more to Spencer than to Darwin, she develops an approach which would later become social Darwinism. In fact, she takes up the philosophy promoted by Spencer in the Anglo-Saxon world, which extended the principle of natural selection, which had in the meantime become the "struggle for life", to the organisation of human societies. Differences between individuals concerning strength, health, and of course intelligence, are assimilated to the variations of Darwinian theory and selection must somehow separate them out, favouring those individuals who exhibit superiority. The consequence is that the human species can only pursue its development and triumph in the direction of "progress" if this selection is allowed to play out freely. It therefore becomes harmful for humanity to thwart the "natural" selection of individuals by helping the weak to survive and reproduce, thus passing on their "weakness" to the following generations. Worse still, by promoting the development of the weakest elements of society through state intervention as a regulator in economic and social matters, we risk causing the degeneration of the human species—somewhat like how a wild animal species that we begin to feed and protect would gradually lose its ability to survive on its own. The result is an extreme form of economic liberalism in which free competition plays the role of natural selection.

15.5 From Darwin to Social-Darwinism

This interpretation, or rather this extension of the theory of evolution into the social and economic field, is formulated in the preface to the first edition of the translation of *On the Origin of Species* and then systematically developed in the *Origine de l'homme et des sociétés*. From the very beginning of this latter work, Royer tells us that it is a question of drawing on the latest discoveries of science to combat the philosophy of Rousseau, thus clearly indicating the philosophical character of her thesis. The goal is to defend social inequalities—against the Rousseau of *Le discours sur l'origine de l'inégalité parmi les hommes*—which are assimilated to the Darwinian variations that cause the evolution of species. Whereas for Rousseau, man in the state of nature, free from any comparison of individuals with one another, experiences a kind of equality "by default", inequalities are introduced through life in society, which thus becomes a site of appropriation, jealousy, and ultimately of all the misfortune of men. For Royer, conversely, nature is profoundly inegalitarian, and it is these inequalities that become the motor of evolution towards a higher degree of perfection. Thus, with a clear view to evoking Rousseau, the last chapter of the book is entitled, "On the Origin of Inequalities". However, Royer distinguishes natural inequalities, seen as factors of progress in the species, and which have enabled us to pass from the primitive anthropoid to modern man, from abusive inequalities which can hinder this evolutionary development of humanity.

> It will now be easy for us to resolve the related question of the origin of social inequalities, their necessary causes, and their fatal exaggerations throughout history – exaggerations which have caused these inequalities to deviate and so often move away from their original purpose, which was none other than social utility, the necessity of the species. (Royer, 1870, p. 538)

Social inequalities are also natural since they inherit their necessity from nature, and it is only in their exaggerated form that they escape the order of nature. They also have "a goal", which is tantamount to introducing a kind of immanent finality to the order of living beings, a finality that leads them to a kind of perfection: the white Western man.

> It follows from all the above that man, both physical and mental, being the product of successive variations of previous animal species, is the result of individual, ethnic and specific inequalities, which have gradually constituted him as a species, race or individual. (Royer, 1870, pp. 438–439)

Evolution necessarily led to man. One thus slides from variations to justified inequalities. Along the way, Royer develops a certain hierarchy of the living world, a type of great chain of being, of which man occupies the apex. And the finalism thus introduced is a finalism without God, reminiscent, as already mentioned, of Lamarck and the legacy of acquired characteristics. Primates are thus only a first draft of man, a more or less failed attempt, a steppingstone on the long journey towards man. She writes:

> All living anthropoids are therefore only very imperfect drafts (*ébauches*) of human organisation, failed attempts that have only succeeded in perpetuating themselves in regions where the heat of the climate somehow diminishes the vital competition between species, while at the same time making life abundant and easy for those who can bear it. [...] If the original rootstock from which humanity emerged prevailed over these poorly endowed emulators and perpetuated itself as it progressed, it is because from the outset, through a series of fortunate and rapid variations, it immediately benefited from a more harmonious organisation. (Royer, 1870, p. 160)

The very terms used by Royer are not lacking in ambiguity and suggest a finalist vision of evolution. Thus, the term "draft" (*ébauche*) suggests a kind of craftsman, such that the different species are in some way prototypes of the final object, which is man. This vision is of course quite foreign to Darwinian thought and takes us beyond strictly scientific description. Similarly, the idea that human organisation would be the most harmonious in the series of living beings amounts to a form of anthropocentrism that a more objective vision would have to abandon. The survival of the great apes, which are nothing more than evolutionary "failures" in relation to the human form, can only be explained by the existence of an environment, climate, and abundance of resources, which reduces natural selection and allows them to persist when they should really have been eliminated, on account of their disharmonious bodies:

> For primitive man, the first truly anthropoid primate, to have first and constantly defeated his original rivals in the struggle for life, he must never have had the vaulted spine of the gorilla, or its descending lumbar spinal apophyses, or the short legs of the chimpanzee, or the long arms of the gibbon, or the misshapen foot of the orangutan. (Royer, 1870, p. 160)

The "ugliness" or the "deformity" of the great apes stems obviously from the point of view of man as perfect form, in which Royer remains trapped. For her, the beautiful, well-formed, and harmonious is what she appreciates as such. With our reduced hair cover, our clumsiness when climbing, our upright gait, we may appear ugly and deformed in the eyes of chimpanzees. Let us add also that the idea of "competition between species" for survival belongs more to Wallace's theory than to Darwin's. Natural selection has replaced the Creator God in establishing the great chain of beings. For Royer, selection operates sometimes within a species, sometimes between individuals able to survive, and sometimes between competing species for the sharing of resources.

The human species is thus perfectible, if selection phenomena or rather the "struggle for life" is at play within and between human societies:

> And in the same way, when the first truly anthropomorphic animal appeared, it lived surrounded by many varieties and species having like it, but perhaps a little less like it, the virtual tendency to take on the human form and the ability to perfect itself through a series of successive variations and advances. If the human race ultimately seems to be if not the only perfectible one, then at least the most perfectible and perhaps the only one perfectible indefinitely, it is because it has inherited this tendency to progress, to perfect itself, from the whole series of its ancestors, themselves undoubtedly more variable than their fellows and contemporaries. (Royer, 1870, pp. 137–138)

Here we see a term appearing once again that is more Lamarckian than Darwinian: tendency to progress. In fact, this tendency minimises the randomness of the variations that appear over the course of a generation, whether because they depend directly on this "virtual tendency" and so are no longer random, or because selection itself is the expression of this tendency to progress. A distortion of Darwinian thought clearly arises in this idea of inheritable perfectibility. This is a notion completely foreign to Darwin, for whom only variations that appear at random in a generation are inheritable, and not this ill-defined trait of perfectibility and a tendency to progress. After all, one can obviously ask oneself which criteria define greater and lesser degrees of perfection. One can indeed consider that what makes a living form perfect is its very existence, its permanence in time, and therefore in a certain way its adaptation to the conditions of life at a given time and place. But then there will be as many perfect forms as there are environments. Or, as the above quotation suggests, what is perfect is man and one falls into a kind of anthropocentrism. Evolution becomes evolution towards and for man.

Royer will also draw from this vision an argument in favour of the racist theses of racial inequality, for this perfectibility is differential. Social organisation and morality are only of value insofar as they allow one group to prevail over another in the struggle for life and for control of the environment. For Royer's idea of progress is closely linked to the idea of struggle and elimination of the defeated; progress presupposes the elimination of the weakest:

> And if each variety in the course of progress had not succeeded in exterminating the inferior varieties with which it was struggling, not only would the human species never have become what it is today, but it would not even have succeeded in separating its specific type clearly enough from that of the anthropoids from which it descended and to which it was still so close, so as to become forever distinct from them and without any possible mixing or return to its brutal atavistic forms. (Royer, 1870, pp. 246–247)

Consequently,

> [i]t is easy to imagine that the varieties, families or tribes who combined the greatest intrepidity and the most ruthless cruelty towards the rival tribes with whom they fought for vital resources, with the most developed social customs and moral feelings, were those who constantly came out on top in this endlessly secular struggle; for they alone were able to oppose a sufficiently compact bundle of united forces to the more fragmented forces of their enemies to prevail over them and, through their destruction, to expand their place in the sun, that is to say, their domain of undisputed domination over the surface of the earth and over the animal species. (Royer, 1870, pp. 247–248)

These developments accord with the criticism put forward at the beginning of *L'origine de l'homme et des sociétés* of Rousseau's theses in *Le discours sur l'origine des inégalités*:

> We must therefore forever relegate to the status of a myth the hypothesis of the Golden Age, of a primitive state of innocence and harmlessness of the human race, during which it would have been endowed only with gentle and peaceful instincts and in which, having in common with us only our most noble feelings, it would have had none of our violent passions. (Royer, 1870, p. 248)

Going further, any concession made to the weak with a view to protecting them thus becomes a sin against the law of progress and evolution, and Royer, in the last chapter of her work, attacks with great virulence not only the socialist theorists of equality, especially Proudhon, but also Christianity, which is suspected of developing an ideology of protecting the weak that hinders the march of humanity towards progress and happiness. For "ultimately, there is no legal inequality that does not derive from a factual inequality; no social inequality that should not have, and initially did not have, its starting point in a natural inequality. This is the law we dare to formulate" (Royer, 1870, p. 545). Social inequalities are therefore only a reflection of the natural inequalities that make possible evolution, and therefore also progress, by eliminating the weakest. However, social inequalities sometimes tend to outlive the natural inequalities that initially caused them and thus to block evolution; they then become bad. For Royer, the fundamental moral value is therefore no longer equality but equity—what she calls equality of freedom, according to which each individual is free to develop themselves and to fight for their survival–, which rewards each according to their merit. The laws protecting the weak, the sick, the destitute, the useless... are thus inequitable, since they favour the survival of those who hinder by their very existence and possible reproduction evolution towards progress. Her 1870 work thus ends with these words:

> It is time, if it is not late, to demonstrate to our masses (*nos foules*), still Christian by their instincts, but whose intelligence is in search of true beliefs and certain moral rules, that the law of equity and the path to happiness for all is the equality of freedom and progress through inequality, which, from the animal that gave birth to man, and then from man also, can in the future give birth to the divine race that will govern the earth with justice, joy, and peace. (Royer, 1870, p. 587)

The demonstration is aimed at "our masses" and not the whole of humanity today, where by "our masses" Royer means white European populations or white populations of European origin. For the selection applied to humanity hierarchizes groups after having hierarchized species:

> That is to say, from the Mincopies of the Andaman Islands, the Maori of New Zealand, the Tasmanians of Van-Deimen, the Hottentots and Bushmen of southern Africa, the inhabitants of Tierra del Fuego or the Eskimos, to the first bimanous anthropoid with thirty-two teeth and thirty-two vertebrae, walked upright on his own two feet and only occasionally climbed trees, there is an infinitely smaller distance to our European peoples than from these insignificant hordes. It can even be said, without fear of error, that from an intellectual point of view, a Mincopy, a Bushmen, a Papuan or even a Lapp is more closely related not only to the monkey, but to the kangaroo, than to Descartes, Newton, Goethe, or Lavoisier. (Royer, 1870, pp. 542–543)

The races are therefore fundamentally unequal. This fully justifies the colonial enterprise and the right of Europeans to the riches of Africa, Asia, and the planet. It provides a justification that she believes to be based on evolutionary theories of the racist theories common in her time. We may recall that in 1903, in a work for children, Paul Bert distinguishes four races, European whites, Asian yellows, African blacks, and American reds, adding: "Only it must be known that whites, being more intelligent, more hard-working, more courageous than the others, have invaded the whole world" (Bert, 1903, p. 18). There will therefore arise a natural repugnance, born of this selection, to interbreeding between races and which could homogenise populations and erase the inequalities necessary for "progress":

> The repugnance to blood mixing first manifested itself in the superior races and in females even more than in males. Today it is a universal fact that, while crossbreeding occurs between white and inferior races, union, unless it is the result of violence, occurs between white and Negro, Indian or Australian; and it is only exceptionally that we find examples of crossbreeding between white women and men of other races. (Royer, 1870, pp. 540–541)

What an odd type of feminist thinking! For Royer, this tendency is mainly brought about by women, since she sees such crossbreeding mainly between white men and women of colour, and rarely in the other direction, forgetting, on the one hand, that the situation of European women in the nineteenth century hardly allowed them to have sexual adventures, let alone marriage with men of colour, and that they could only become pregnant by a man of colour through violence, and, on the other hand, that in the case of the white man having sexual intercourse with a woman of colour, it was also, very often, through an act of violence by the master towards the slave.

15.6 Conclusion

Thus, it is clear how Darwinism, from the outset, undergoes a kind of overcoming or ideological distortion towards sociological and political positions that are in no way scientifically justified. We can also understand Darwin's reticence in the face of the translation and comments made by Royer, and we understand them all the better since the latest work on the *Descent of Man*, which, unlike the *Origin of Species*, deals with the question of man, in no way justifies Royer's theses, which, before the term was proposed by Galton, were part of what was to become social Darwinism, and which was later invoked as justification for eugenic theories and practices. On the contrary, Darwin sees empathy and sociability as the results of natural and sexual selection. However, it would be a little too easy to condemn Royer by accusing her of eugenics, unbridled liberalism, and notorious racism. Our judgment would necessarily be that of humans of the beginning of the twenty-first century and not of the end of the nineteenth century. What the story of Royer shows, however, is the way in which science can generate, at the limits of its own domain, theses that may initially appear to stem from its achievements before we realise, through the very development of knowledge, that they are part of an ideological construction, sometimes closely linked to a political and social history. For Social Darwinism is a posteriori closely linked to the situation of the working classes in the booming capitalist industrial society and to colonial expansion.

These debates, at the frontiers of science and ideology, which took place after Darwin and which we have just illustrated by the case of Royer, in some respects show what the critical role of philosophy can be.

References

Bert, P. (1903). *La deuxième année d'enseignement scientifique*. Armand Colin.
Chazal, G. (2018). Entre science et idéologie: le cas Clémence Royer. In F. Merlin & P. Huneman (Eds.), *Philosophie, histoire, biologie: mélanges offerts à Jean Gayon* (pp. 165–181). Éditions Matériologiques.
Darwin, C. (1870). *De l'origine des espèces par sélection naturelle ou des lois de transformation des êtres organisés, troisième édition, traduction de Clémence Royer, avec préfaces et notes du traducteur*. Guillaumin et Masson.
Gayon, J. (1992). *Darwin et l'après-Darwin: une histoire de l'hypothèse de sélection naturelle*. Kimé.
Gayon, J. (1998) [1992]. *Darwinism's struggle for survival: heredity and the hypothesis of natural selection*. Cambridge University Press.
Marx, K. & Engels F. (1976) [1932]. *The German ideology*. Progress Publishers.
Royer, C. (1870). *Origine de l'homme et des sociétés*. Masson et Guillaumin.
Royer, C. (1877). *Deux hypothèses sur l'hérédité*. Ernest Leroux éditeur.
Stengers, I. (Ed.). (1987). *D'une science à l'autre. Des concepts nomades*. Le Seuil.
Tort, P. (2000). *Darwin et la science de l'évolution*. Gallimard.
Tort, P. (2002). *La seconde révolution darwinienne—Biologie évolutive et théorie de la civilisation*. Kimé.
Tort, P. (2005a). *Darwin et le darwinisme*. Presses universitaires de France.
Tort, P. (2005b). *Darwin et la philosophie: religion, morale, matérialisme*. Kimé.

Chapter 16
Is Darwin Resistible? Mill on Design and Natural Selection

Denis Forest

Abstract How is it possible that in *Three Essays on Religion* Stuart Mill gives more credit to the design argument than to the Darwinian explanation of biological adaptations? Three main types of reasons are presented. The first derives from the theory of knowledge contained in *A system of logic* and Mill's understanding of both Darwin and Design: making use of the method of agreement, the design argument is an instance of genuine inductive reasoning, which for Mill is basically sound, while Darwin follows the "hypothetical method" that Mill disvalues. Moreover, the whole Darwinian strategy, inherited from Whewell's view of consilience, is foreign to Mill's view of science. The second type of reasons derives from Mill's metaphysics: within his phenomenalist framework, Mill is not committed to a naturalistic ontology, and he has no reason to favor the elimination of the supernatural. Finally, fruitfulness is not a key epistemic value for him, and Mill's reservations on Darwin mirror the limitations of his own epistemic norms.

Keywords Design argument · Adaptation · Darwinism · Hypothetical method · Consilience · Metaphysics · Logic

16.1 Introduction: On the Reasons to Acknowledge the Power of Natural Selection and the Problem of Mill's Reluctance

I shall begin with a personal memory. The first time I had an opportunity to hear Jean Gayon giving a presentation was in the early 1990s. I had never taken one of his courses, and although I had bought the French edition of *Darwinism's Struggle for Survival* (1998 [1992]) I had yet to sit down and read it. I still recall his talk vividly. The type of geometric beauty it possessed could elicit either fascination or

Translated from Forest (2018).

D. Forest (✉)
IHPST, Université Paris 1 Panthéon-Sorbonne, Paris, France
e-mail: denis.forest@univ-paris1.fr

P.-O. Méthot (ed.), *Philosophy, History and Biology: Essays in Honour of Jean Gayon*, History, Philosophy and Theory of the Life Sciences 30,
https://doi.org/10.1007/978-3-031-28157-0_16

shock and revulsion. In fact, it was as though Gayon were presenting a model that one could adopt or reject, a certain idea of what philosophy and history of science might be, at a very high level. As luck would have it, the subject matter of the lecture was Darwin and, more specifically, one passage in the introduction to *The Variation of Animals and Plants under Domestication* (Darwin, 1868) a crucial text Gayon had considered in a variety of contexts since his dissertation (see Gayon, 2011). In addition to being historically informative, the talk had also a much broader interest: the topic was the justification of a scientific theory, and the relationship between science and philosophy of science. Gayon showed that there are two levels of justification of natural selection as an explanatory principle in Darwin: first of all, "arguments establishing the plausibility of natural selection as a general process existing in nature" (Gayon, 2011, p. 60) and, secondly, the hypothetico-deductive strategy that consists in showing that natural selection explains and unifies "a large array of 'independent classes of facts'" (Ibid.)[1]

In the present chapter, I shall consider from a different angle the question of one's reasons to reject or to accept Darwin's theory, and deal with one of Darwin's contemporaries, the philosopher John Stuart Mill, and his *Essays on Religion,* a text written very late in his career.[2] This text is disconcerting, especially because when Mill examines the design argument, his appraisal of Darwin's theory (considered as a plausible alternative to design) is the opposite of what might be expected of him. In this collection of essays, published posthumously in 1874, Mill writes about the argument from design that "we now at last reach an argument of a scientific character, which does not shrink from scientific tests, but claims to be judged by the established canons of Induction. The design argument is wholly grounded on experience" (Mill X [1969], p. 446). And he concludes: "(...) in the present state of our knowledge, the adaptations in Nature afford a large balance of probability in favor of creation [of organisms] by intelligence [rather than in favor of evolution by natural selection]" (Ibid., p. 450).

This conclusion is disconcerting, first because of the Humean background. Even before Kant's examination of the physico-theological proof, Hume had criticized the design argument as a very weak form of analogical reasoning: see the second Part of the *Dialogues concerning Natural Religion* in which Philo, who is usually Hume's spokesman in the *Dialogues,* expresses the following views:

> [...] But can you think, *Cleanthes*, that your usual phlegm and philosophy have been preserved in so wide a step as you have taken, when you compared to the universe houses,

---

[1] This second strategy is inherited from Whewell's philosophy (Ruse, 1975).

[2] The *Three Essays on Religion,* published for the first time posthumously in 1874, includes "Nature", "Utility of Religion", and "Theism". "Nature" offers a critique of the idea that conformity to nature is a criterion of what is morally good, and a rejection of the idea of divine omnipotence. "Utility of religion" concentrates on the relations between religion and morality. The third essay, "Theism", is the one that directly examines the truth of religion and the proofs of God, including the design argument. This third essay, thought to date from 1868 and 1870 (Reader, 2002, p. 167), was written long after the other two.

In the text, I refer to the *Collected Works of John Stuart Mill*, specifying the volume with a Roman numeral followed by the publication date of the volume in square brackets.

> ships, furniture, machines, and, from their similarity in some circumstances, inferred a similarity in their causes? Thought, design, intelligence, such as we discover in men and other animals, is no more than one of the springs and principles of the universe [...] But can a conclusion, with any propriety, be transferred from parts to the whole? (Hume, 1995, p. 58)

In this passage, Philo-Hume is telling us: the universe is not like a machine; intelligence is only one of its many principles; and inferring that the cause of *some things* is the cause of everything is arbitrary, and amounts, in his own terms, to a "partiality in our own favor". How could the author of *A System of Logic* be less demanding than Hume on such a crucial point?

This judgment is disconcerting, in the second place, if considered in reference to the whole contents of the *Three Essays on Religion.* Mill, in his examination of rational theology in these *Essays,* strongly sets himself apart from it:

1. He rejects the necessity of a First Cause of the universe (Mill X [1969], p. 436): neither objects nor the whole of objects need to be caused, but only events;
2. He accepts a form of diachronic emergence; that is, the possibility that new properties or new capacities might appear in the course of the history of the universe, and specifically, intelligence may be one such property or capacity. Therefore, for Mill, intelligence does not necessarily come first (Mill X [1969], p. 440);
3. In direct relation to the bitter feud he had with Henry Longueville Mansel,[3] he rejects on moral grounds the thesis of divine omnipotence and the idea of a providential government of the universe (Ibid., p. 389).

Lastly, Mill's initial reaction to *On the Origin of Species* was not at all unfavorable. A letter to Alexander Bain, dated April 11, 1860, contains the earliest written reference Mill ever made to the book, that had been published in 1859[4]:

> [Darwin's book] far surpasses my expectation. Though he cannot be said to have proved the truth of his doctrine, he does seem to have proved that it *may* be true, which I take to be as great a triumph as knowledge & ingenuity could possibly achieve on such a question. Certainly, nothing can be at first sight more entirely unplausible than his theory & yet after beginning by thinking it impossible, one arrives at something like an actual belief in it, & one certainly does not relapse into complete disbelief. (cited in Hull, 2000, p. 52)[5]

Mill's opinion of the book is clearly positive, even though the way he formulates it already reflects the ambivalent attitude that would later emerge in his published works, *A System of Logic* and the *Essays on Religion.*

A question then arises: why, at the time when he was writing the *Essays,* does Mill maintain that the explanation by design of biological adaptations is preferable to the Darwinian one? Why didn't the well-known antagonism between Whewell and Mill (Snyder, 2006) lead Mill to adopt a position that would once again permit him to

---

[3] See Reader (2002).

[4] See also the letter of Fawcett of July 16, 1861, where he reports to Darwin that "he [Mill] considers that your reasoning throughout *[On the Origin of Species*] is in the most exact accordance with the strict principles of logic" (Snyder, 2006, p. 194).

[5] To be compared to Whewell's own reaction, quoted by Snyder (2006, p. 195).

distance himself from Whewell, who had clearly declared his opposition to any form of transmutation of species? (Whewell, 2001)? I suggest that the explanation for Mill's attitude towards the choice between Darwin and design can be found in Mill's fundamental positions in two fields of philosophy to which he made a significant contribution: theory of knowledge and metaphysics.[6]

16.2 The Millian Analysis of the Design Argument

According to Mill, the design argument is not based on "mere analogy" (Mill X [1969], p. 447) as it is believed to be. Its best reconstruction involves no comparison, along Paley's lines, between such artifacts as watches on the one hand and organisms on the other hand. The design argument is an inductive argument, based on the first of the four "Methods of experimental enquiry" described in the *System of Logic,* the Method of Agreement (Mill VII [1973], pp. 388–390). The Method of Agreement consists in "comparing different instances of a phenomenon" and identifying the circumstances that invariably precede or follow this phenomenon, depending on whether one is seeking to identify its cause or its effect. The process can be represented using the following diagram:

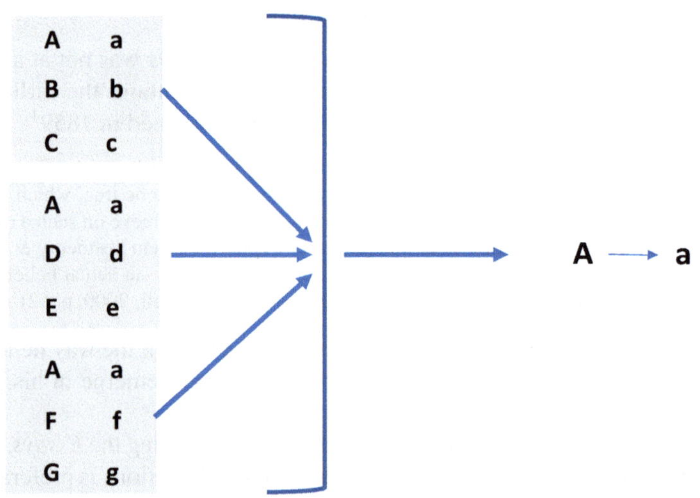

In cases where the cause is sought, the method of agreement will permit us to identify an "invariable antecedent." Nevertheless, by Mill's own admission, the method of agreement presents certain flaws: the invariable antecedent to a phenomenon might *not* be its cause, even in the limited sense of a proximal event which "completes the cause" (Mill VII [1973], p. 389). The method of agreement fails to be conclusive

[6] I analyze a third one, moral philosophy, in Forest (2017).

when we cannot produce or reproduce the effect by manipulating its assumed cause and are reduced to mere observation. This method, although imperfect, can nevertheless guide us in identifying a genuine cause. Here is how we can reconstruct Mill's application of the method of agreement to the structure of the eye in the third *Essay*.

1. The parts of the eye (designated by the sets of lower-case letters abc / ade / afg) all share a single distinctive property ('distinctive' meaning that the property is not shared with other parts of the organism): contributing to sight.
2. Having a beginning in time, these parts must be the result of a shared cause or combination of causes.

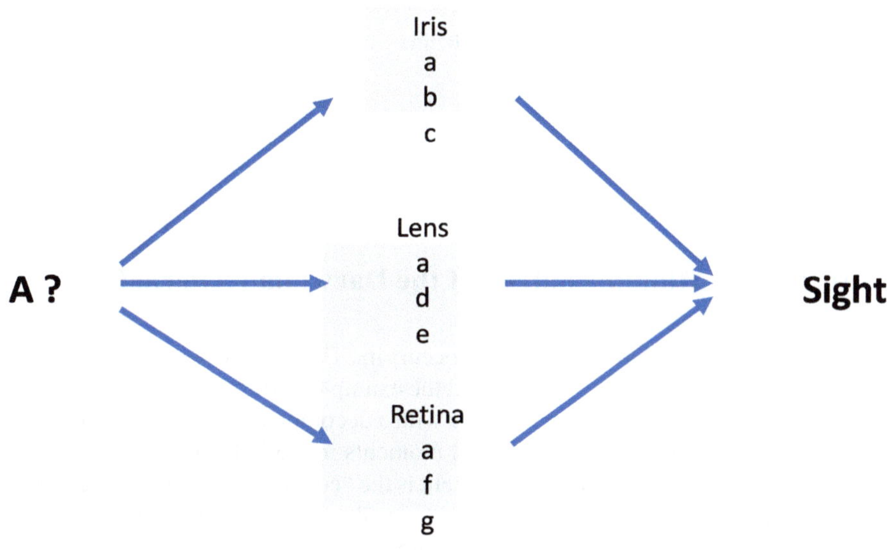

3. There must be some connection between the remarkable fact of sight and the cause (A) which brought all these elements together. According to Mill, the design argument consists in inferring the ultimate cause of the parts of the eye from their shared effect.

In his own words: "Sight, being a fact not precedent but subsequent to the putting together of the organic structure of the eye, can only be connected with the production of that structure in the character of a final, not an efficient cause: that is, it is not Sight itself but an antecedent Idea of it that must be the efficient cause. But this at once marks the origin as proceeding from an intelligent will" (Mill X [1969], p. 448).

In the present context, the point of the Method of agreement is to start with the instantiation of a single property (contributing to sight) by several entities (the parts of the eye) and to infer from this the nature of their common cause.

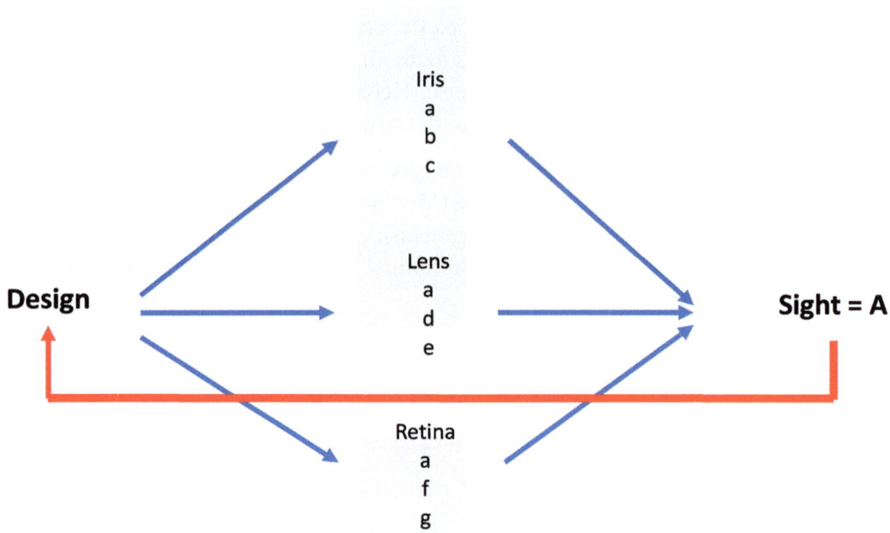

16.3 The Millian Analysis of the Darwinian Alternative

This is where the reference to Darwin occurs in "Theism". As we know, in Chapter VI of *Origin of Species*, Darwin also used the example of the structure of the eye. Mill's reference to Darwin goes hand in hand with a decomposition of the design argument. For Mill, there are indeed two pivotal moments in the inductive reasoning involved in the design argument. The first moment is the "connection" between the recurrence of a property shared by a set of effects (contributing to sight), on the one hand, and the cause that these effects have in common, on the other.

The second moment consists in identifying the nature of this cause, by resorting to a form of causality by intention. But Mill adds immediately that "I regret to say, however, that this latter half of the argument is not as inexpungable as the former." An alternative way of conceiving the common cause does exist: Creative forethought is not absolutely the only link by which the origin of the wonderful mechanism of the eye may be connected with the fact of sight. There is another connecting link on which attention has been greatly fixed by recent speculations. [...] This is the principle of the "survival of the fittest" (Mill X [1969], pp. 448–449). In this case, the distal cause of sight is not design, but the patience with which nature has recruited variations favorable to this effect. And Mill concludes: "The theory if admitted would be in no way whatever inconsistent with Creation. But it must be acknowledged that it would greatly attenuate the evidence for it" (Ibid., p. 450).

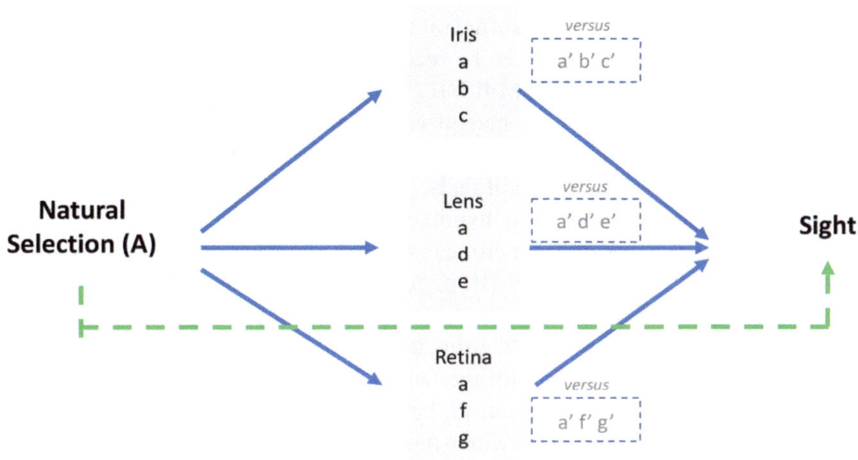

16.4 Rules of Induction and Rules of Hypothesis

Why does Mill, with all his intellectual esteem for Darwin, maintain that "the adaptations in Nature afford a large balance of probability in favor of a creation [of organisms] by intelligence"? To answer that question, one must return to the comment on Darwin added by Mill to *A System of Logic* in 1862 (Mill VII [1973], p. 498). Mill's evaluation of Darwin's theory in that development is a consequence of the general theoretical framework of the book, where he offers a distinction between the "rules of Induction" (Book III, Chapter VIII) and the domain of "Hypotheses" (Book III, Chapter XIV). Mill proposes the following definition of a hypothesis:

> An hypothesis is any supposition which we make (either without actual evidence, or on evidence avowedly insufficient) in order to endeavor to deduce from it conclusions in accordance with facts which are known to be real; under the idea that if the conclusions to which the hypothesis leads are known truths, the hypothesis itself either must be, or at least is likely to be true. (Mill VII [1973], p. 490)[7]

The Hypothetical method is for Mill a truncated version of what he calls the Deductive method, which consists normally in three steps: induction, ratiocination, and verification. Ratiocination is the act of computing what the consequences of the hypothesis are in any given case while verification is the act of comparing the result of ratiocination with the actual phenomenon. This means that the Hypothetical method is the association of ratiocination and verification with no inductive antecedent to support the hypothesis itself. In some cases, when ratiocination is sufficient to prove that the phenomena could not be accounted for by another hypothesis than the one under consideration, it will have according to Mill the value of a true induction.

---

[7] On the status of hypotheses in Mill's view of science, see Snyder (2006, pp. 121–125).

But in other cases, it will not be sufficient that a hypothesis is able to "account for all the known phenomena" for it to be received as true, because other hypotheses could account for these as well (Mill VII [1973], p. 500). For Mill, the Darwinian hypothesis belongs to the second category: in the context of the design argument, he sees no reason to assume that no other hypothesis could account for biological adaptations and other related natural facts.

Mill is happy to concede that hypotheses have an important heuristic value (Skorupski, 1989, p. 197), that sometimes inductive enquiries are mistakenly considered as hypothetical (Mill VII [1973], p. 505) and that, from an historical point of view, "nearly everything that is now theory was once hypothesis" (Ibid., p. 496). Moreover, some hypotheses are "legitimate", and legitimacy is associated by Mill in the case of natural selection with the fact that the causal explanation offered by Darwin involves "not only a *vera causa*, but one proved to be capable of producing effects of the same kind with those which the hypothesis ascribes to it" (Ibid., p. 499). When Mill speaks of a *vera causa* in this passage, he seems close to Herschel's conception of it (something whose causal power can be empirically verified) (Snyder, 2006, p. 201). And he is apparently referring to artificial selection, which offers evidence that it is possible to produce effects of the same kind as those ascribed to natural selection by the Darwinian hypothesis. This means that for him the problem raised by the Darwinian view of nature is less the existence of *some* form of natural selection than the extent of its power: in his own terms, "the question of possibility is entirely one of degree" (Mill VII [1973], p. 499). Mill does not think that the combination of the model of artificial selection and empirical premises (like the limitation of resources and the facts of variation and inheritance) have the value of a true inductive procedure (Gayon, 2011, p. 64). And as he has no sympathy for what Whewell calls the consilience of inductions (the ability to explain classes of independent facts as a criterion of truth), in his view the whole Darwinian strategy fails. The rules of hypothesis "have seldom been so completely fulfilled", but these rules are in themselves foreign to the regime of proof. To say that Darwin has offered "an unimpeachable example of a legitimate hypothesis" (Mill VII [1973], p. 499) is in fact to deliver a scathing judgment on the intrinsic value of Darwin's theory.

16.5 The Role of Metaphysics in the Millian Evaluation of Darwin

To understand Mill's verdict in the *Essays on Religion*, the reference to his theory of knowledge is not sufficient. There are two other important fields that come into play: metaphysics and ethics. The former deserves our attention here, because it directly concerns the question of the legitimacy of an explanation that relies upon a supranatural cause.[8]

[8] On the moral dimension of the problem, see Forest (2017).

Nowhere does Mill assert that one reason for preferring the Darwinian explanation of biological adaptations would be that it does not require to take in consideration a supranatural entity. In the *Essays*, the reference to the "supernatural" (Mill X [1969], p. 429) does not automatically disqualify the design argument. Worse: in itself, it is not even sufficient to weaken it.

To make sense of Mill's position, let us recall his phenomenalism. According to the famous statement in *An Examination of Sir William Hamilton's Philosophy*, matter is, properly understood, nothing but the "permanent possibility of sensation" (Mill IX [1979], p. 183). In the phenomenalist framework, then, organisms, like any material entity, are simply clusters of such permanent possibilities, and the question of their ultimate nature is beyond the scope of our scientific investigation. We can also safely assume that Mill had remained immutably faithful to his phenomenalism at the time he was writing the *Essays on Religion*, as suggested by his discussion of the relationship between thought and the material brain in the third *Essay*, the one where he also discusses Darwin and design (Mill X [1969], pp. 461–462).

Accordingly, the limitations of Mill's naturalistic commitment must be considered. As we know, Mill is a monist as far as scientific methodology is concerned. In his *Logic*, he models his conception of the "moral sciences", the ones we currently refer to as the social or human sciences, after the natural sciences (Book VI of his *Logic* in Mill VIII [1973]). Mill maintains that in the future, the only way for the moral sciences to go beyond what he disdainfully calls "empiricism" to acquire a genuine scientific character is by adopting the methods of the natural sciences. Yet his naturalism is of a purely syntactical, not of a semantic type. The causal laws to which science leads are laws of succession, not laws of constitution. For this reason, Mill sees no need to supplement the natural sciences with a naturalistic metaphysics, and his phenomenalism forbids him from rejecting out of hand design and causality by intention to solve the problem of the mutual convenience between biological structures and functions.

16.6 Conclusion

From *A System of Logic* to the *Three Essays on Religion*, it is not Mill's positions on Darwin that undergo a change, but the perspective he adopts in two different contexts. In his *Logic,* natural selection is a more valid hypothesis than Cartesian vortices are. In the *Essays on Religion*, however, design as the product of an inferior form of inductive reasoning (the method of agreement being for him the weakest of all inductive procedures) is still preferable to one of the most rigorous applications of the hypothetical method.

Certainly Mill, writing at a time where the fate of Darwinism was impossible to predict, can be credited for assessing the merits of Darwinism in relation to a state of knowledge at a given point in history rather than in any absolute manner. Hence, we could say that his position is open to the future of its revision; in formulating his reservations, Mill would only be asserting, with some reason, that in 1870 the explanatory power of the Darwinian program is not yet known. Or, to use his words "he [Darwin] has opened a path of inquiry full of promise, the results of which none can foresee" (Mill VII [1973], p. 499).

Yet Mill's lukewarm reaction to Darwin may also invite us to question the epistemic standards advocated by the *System of Logic*. Mill proceeds as if good natural science was only a matter of empirical confirmation. As a result, in his appraisal of Darwin, he considers what Laudan calls the "context of acceptance", leaving aside Laudan's "context of pursuit" (Laudan, 1977). If we take this second context in consideration, however, the rational evaluation of a theory is not limited to the estimation of its present acceptability. This evaluation has also to take into consideration the fruitfulness of the theory (or research tradition), its "rate of progress": and "it is always rational to pursue any research tradition which has a higher rate of progress than its rivals" (Ibid., p. 111). From this point of view, it was not necessary to know what Mill could not know in 1873 to be able to contrast the sterility of natural theology and the prospects of a research program in evolutionary biology as the one that can be derived from *The Origin of Species*. Mill's position on Darwin, for all its subtlety, mirrors also the intrinsic limitations of the framework he adopts in his philosophical examination of science.

Acknowledgements All my gratitude goes to Jean Gayon for his many suggestions. I also thank Anita Conrade for her translation, which I have revised.

References

Darwin, C. (1998) [1868]. *The variation of animals and plants under domestication , Volume 1*. The Johns Hopkins University Press.

Forest, D. (2017). Mill, Darwin et l'argument du Dessein. *Cahiers Philosophiques, 148*, 9–25.

Forest, D. (2018). Darwin est-il résistible? Mill sur le Dessein et la sélection naturelle. In F. Merlin & P. Huneman (Eds.), *Philosophie, histoire, biologie: Mélanges offerts à Jean Gayon* (pp. 183–194). Éditions Matériologiques.

Gayon, J. (1992). *Darwin et l'après-Darwin : Une histoire de l 'hypothèse de sélection naturelle*. Kimé. Translated as Gayon, J. (1998). *Darwinism's struggle for survival. Heredity and the hypothesis of natural selection*. Cambridge, Cambridge University Press.

Gayon, J. (2011). History of evolutionary theories. In G. Auletta, M. Leclerc, & R. A. Martinez (Eds.), *Biological evolution: facts and theories. A critical appraisal 150 years after "The Origin of Species"* (pp. 49–71). Gregorian Biblical Press.

Hull, D. L. (2000). Why did Darwin fail ? The role of John Stuart Mill. In R. Creath & J. Maienschein (Eds.), *Biology and epistemology* (pp. 48–63). Cambridge University Press.

Hume, D. (1995) [1779]. *Dialogues concerning natural religion*. Penguin Classics.

Laudan, L. (1977). *Progress and its problems*. University of California Press.

Mill, J.S. (1973) [1843]. *A system of logic*. In *Collected works of John Stuart Mil*, (Vols. VII and VIII). Routledge.

Mill, J.S. (1979) [1865]. *An examination of the philosophy of sir William Hamilton*. In *Collected works* (Vol. IX).

Mill, J.S. (1969) [1874]. *Three essays on religion*. In *Collected works* (Vol. X).

Ruse, M. (1975). Darwin's debt to philosophy. *Studies in the History and Philosophy of Science, 6*, 159–181.
Reader, L. C. (2002). *John Stuart Mill and the religion of humanity*. University of Missouri Press.
Skorupski, J. (1989). *John Stuart Mill*. Routledge.
Snyder, L. (2006). *Reforming philosophy*. University of Chicago Press.
Whewell, W. (2001) [1837]. *The history of the inductive sciences*. John W. Parker (ed.). London: Thoemmes Press.

Part IV
Jean Gayon: Personal Memories

Chapter 17
Jean Gayon on the History of French Genetics: A Personal View

Richard M. Burian

Abstract This essay highlights some special strengths of Jean Gayon's approach to the history of French genetics by providing a partial account of our collaboration. I begin with our first meeting and the work it initiated. I then illustrate key aspects of Jean's approach to understanding complex histories involving conceptual conflict and conceptual change like those we encountered in our studies of the history of French genetics. In my view, Gayon's great skill in teasing out conceptual connections and analyzing conceptual revisions has sharpened our thinking on topics far beyond those on which any one of us has worked on. Today, his exemplary efforts to explore an enormous range of topics in both breadth and depth provide a model of the role of a scholar, or of a historian-philosopher.

Keywords French genetics · Jean Gayon · Georges Teissier · Philippe L'Héritier · Non-Mendelian inheritance · Deep history

Jean Gayon and Marjorie Grene are the two people who most influenced my intellectual life and career since I left graduate school at the University of Pittsburgh. Beyond that, Jean and Elisabeth Gayon have been my wife's (Anne McNabb's) and my best friends for more than thirty years. They and their children are, in effect, our French family and we their American family. It is a great delight to participate with many of Professor Gayon's colleagues, friends, and former students in this *Festschrift* honoring his work and career. This essay highlights some special strengths of Jean Gayon's approach to the history of French genetics by providing a partial account of our collaboration.

I begin with our first meeting and the work it initiated. It will illustrate key aspects of Jean's approach to understanding complex histories involving conceptual conflicts

---

From Burian (2018).

R. M. Burian (✉)
Virginia Tech University, Blacksburg, USA
e-mail: rmburian@vt.edu

Virginia Tech, Blacksburg, Virginia, USA

P.-O. Méthot (ed.), *Philosophy, History and Biology: Essays in Honour of Jean Gayon*, History, Philosophy and Theory of the Life Sciences 30,
https://doi.org/10.1007/978-3-031-28157-0_17

and conceptual changes like those we encountered in our studies of the history of French genetics. In 1984 Jean had a Fulbright Summer Fellowship to work on the history of evolutionary biology with Ernst Mayr and others.[1] As part of his Fulbright trip, he planned a quick visit to the late Will Provine at Cornell to learn more about interrelations between population genetics and evolutionary biology. That same summer I began a project on the history and epistemology of gene concepts. For a month I worked in Provine's enormous collection of reprints, acquired from leading ecologists, evolutionists, geneticists, developmental biologists etc. That summer, Provine was working—both intellectually and physically—at his farm, about 50 km from Cornell. One afternoon he telephoned to tell me that a young Frenchman had arrived to visit him; he asked me to introduce the visitor to the reprint collection and then bring him to the farm for dinner and conversation. I was particularly pleased to do so, because I had just encountered some articles written in the 1930s by Philippe L'Héritier and Georges Teissier, two mathematically sophisticated zoologists. These articles were difficult: their terminology was unsettled, their project did not fit with what little I knew about the history of French genetics, and it pointed in unexpected and puzzling directions regarding ways in which French scientists had contributed to genetics. I hoped that Provine's guest, namely, Jean Gayon, could help me better understand those articles.

Here's a bit about the puzzles raised by L'Héritier and Teissier's papers. In 1932 or 1933 L'Héritier, with help from Teissier, invented devices to conduct demographic experiments on populations of drosophila. He called them *démomètres*, i.e., devices to measure demography, but they are now known as *population cages*.[2] How did he happen on such a project? After all, after World War I, despite the excellent start made by Lucien Cuénot at the beginning of the twentieth century, there was almost no university-level teaching or research in genetics in France and absolutely no experimental or theoretical work in population genetics. The reprints showed that L'Héritier and Teissier used *démomètres* in 1933–1937 to study demographics and growth of competing strains of flies in determinate circumstances. Multigenerational experiments showed that competing strains reached stable equilibria, usually supporting about 3–4 thousand flies. A startling result was that strains carrying certain seriously deleterious mutants (e.g., *bar* and *ebony*), were not eliminated when competing with wild-type flies; instead, they reached a stable equilibrium at a low frequency. This implied that the fitness of such mutants increased as they became rarer in the population, i.e., that the fitness of the mutants was frequency-dependent, a result that contradicted some of R. A. Fisher's basic commitments.

As Mendelian papers go, these early papers were quite unusual. They drew on Mendelian findings and used strains of flies with specific Mendelian mutations to determine whether certain predictions of standard Mendelian population genetics about the evolving structure of those populations were correct, but they did no breeding experiments and did not directly study the inheritance of the alternative traits. In that respect, their central methodology was extra-Mendelian. When some

[1] On Gayon's research in the United States, see Méthot and Depew, this volume.

[2] On this invention by L'Héritier and Teissier, see the chapter by Veuille, this volume.

predictions proved wrong, they sought physiological or selective explanations at the phenotypic level for their findings without recourse to explicitly Mendelian considerations, but they also used their results to correct mistaken assumptions and excessively simple abstractions embodied in standard Mendelian population genetic models. One of these was R. A. Fisher's assumption that selection coefficients for given mutations are independent of the frequency of the mutants in the population. In the limited sense that L'Héritier's and Teissier's earliest experiments did not require them to make Mendelian crosses, those experiments might be classified as experiments in population *biology* that tested *consequences* of Mendelian population genetic models rather than as Mendelian *experiments* in *population genetics* strictly speaking.

Another startling finding was particularly intriguing. Their experiments required counting mutants and wild type flies for many generations. One technique they used for this purpose was to anaesthetize the flies briefly with CO_2, photograph them spread out on photographic paper, and return them to the population cage, a procedure that allowed careful counts to be conducted at convenience with no harm to the flies. In 1937, they discovered an anomalous phenomenon. The females of a strain that carried the *ebony* mutation produced offspring with remarkable sensitivity to CO_2: they were not killed outright by exposure to the gas, but when its narcotic effects wore off the offspring were either totally paralyzed or their movements were so spastic that they could not walk, fly, or feed—and before long they died. Stranger yet, the trait was inherited in a non-Mendelian fashion. At first the inheritance appeared to be strictly maternal and, thus, probably cytoplasmic. However, they soon showed that some sperm could cause some offspring to become sensitive, albeit (in later terminology) with incomplete penetrance and with several confusing complications. Due to the exigencies of their careers and the war L'Héritier and Teissier worked in different places by 1938 and, although they co-authored several more papers, they never resumed their collaborative experiments. L'Héritier took over the research on CO_2 sensitivity and Teissier the research on the kinematics and equilibria among competing strains.

It proved quite difficult to understand the details of the inheritance of CO_2 sensitivity, which became a major puzzle. For this purpose, L'Héritier had to resort to serious breeding experiments to show that the causative agent was not carried on any of the male or female chromosomes in his reciprocal crosses of resistant and sensitive strains. After much work on the physiology of the phenomenon L'Héritier ultimately identified the causal agent of the sensitivity first as a viroid (a particulate component of the cytoplasm carrying some gene or gene complex) and later as a virus, a result that led him to focus his research on the genetics of viruses. As became clear to us in our studies of the beginnings of molecular genetics in France, L'Héritier provides but one of many examples in France of fascination with maternal and/or cytoplasmic inheritance and of phenomena that were easily interpreted in a neo-Lamarckian framework but might, arguably, and initially with great difficulty, be reinterpreted within a Mendelian framework.

Let's turn from L'Héritier and Teissier to discuss how Jean Gayon shaped our work together and how he approached major puzzles concerning large-scale scientific processes. Jean was nearly as puzzled as I was about how their papers fit into the development of French genetics, so we decided to investigate further and see what role that work played in the French biology's turn toward genetics. He led the way since he knew far more than I about the history of biology, and especially population genetics, in France. We wanted to understand how L'Héritier and Teissier came to this line of inquiry, what they each brought to it, the influence of their work on further developments both in France and internationally, and the interactions between French and 'mainstream' genetics. Jean took our research in two directions: first to a deep history of teaching and research regarding heredity in France and, second, an examination of all the main starting points from which the French gained entry into international genetics. The strategy of examining deep history is found throughout Jean's innovative work in history and philosophy of science. One cannot easily set boundaries on such historical work; to do so requires understanding the sources of scientists' problems and the influences affecting their research objectives and their standards for solving problems. For the French scientists who sought to understand biological heredity in the first half of the twentieth century, especially those who came to focus on genetics, it quickly became clear that embryology, physiology, and microbiology provided extra-Mendelian traditions that quickly became entangled with Mendelian genetics or with the transformation of genetics from classical Mendelism into molecular genetics.

It was soon obvious that institutional and disciplinary histories should be included in the relevant deep history. But it was also evident that epistemic and social histories are also relevant—what are the standards of scientific proof, the merits and demerits ascribed to the use of this technique or instrument rather than that, and the standards for evaluating scientists' work? Again, what are the social and institutional standards of promotion, of funding, of reputation? Etc. We had to take all these issues into account to understand scientists' formulation of problems, deployment of research tools, epistemological commitments, and their standing with their colleagues. And we also had to take account of several 'external' contingencies, such as the loss of a generation in World War I.

Deep history of this sort broadens and sharpens the tools used to analyze the events of concern. It facilitates analysis of the impact of differences in disciplinary background and the ways in which alternative problem formulations lead protagonists in strongly different directions and to revise their evaluations of familiar research results. Let me exemplify some of these points by reference to our early work. We began by asking which disciplines were salient in providing accounts of heredity in France. By the end of the nineteenth century, France's biological culture was strongly influenced by Comtean positivism, neo-Lamarckism, and the work of Claude Bernard and Louis Pasteur. The neo-Lamarckian program was anti-speculative and mechanistic; it sought evolutionary explanations with no role for Darwinian selection. Indeed, nearly all French scientists who accepted evolution in the first half of the twentieth century dismissed natural selection as sheer speculation or as only guiding microevolution and minor organismal traits, but not macroevolution or species-determining characters. This position was reinforced by the strong development of physiology, causal embryology, and microbiology (including cytology), the dominant biological disciplines relevant to the detailed study of heredity in France. Pasteurian microbiology showed that microbes respond to environmental change by altering various hereditary morphological and physiological properties, some of which, such as virulence, were humanly important. Although Pasteurians were not heavily involved in neo-Lamarckian polemics, they were dominantly persuaded that acquired characters could be inherited. Further reinforcement came from the strong tradition of causal embryology, according to which all the major action in differentiation and determination takes place in the cytoplasm, not the nucleus, and its commitment to the holistic view that cells interact as wholes in shaping the main steps of development. These claims support the view that nuclear particles are impotent to guide development. At the same time, biology was struggling to match the prestige of physics and engineering, sciences whose reputations were greatly enhanced by the rigor of mathematics, another of France's disciplinary crown jewels. Mathematics provided the French with a second entry to the international stage: population genetics is the most easily mathematized of the subdisciplines dealing with heredity.

The situation of Mendelian genetics was somewhat different than that of Darwinian selectionism. Immediately after the rediscovery of Mendel, Lucien Cuénot, working with mice, made several major contributions to genetics. He showed that animals, like plants, exhibit clearcut Mendelian inheritance. His pioneering research on coat color patterns in mice demonstrated polygeny (i.e., that multiple genes are required to produce many, perhaps most, standard Mendelian traits), epistasis (i.e., that there are interactions among genetic determinants), the existence of multiple alleles of some genes, and the existence of genes that are lethal when homozygous. He also was one of the first to suggest that genes are responsible for the formation of enzymes. Finally, starting in 1908, he did pioneering experiments in physiological genetics, studying the complex interrelations of physiology and genetics in the formation of cancers in mice. This work was widely recognized both in France and abroad. But with regard to genetics Cuénot's story is rather tragic. During World War I his strains of mice were lost. In France, in spite of strong recognition of his accomplishments in genetics, he was strongly discouraged from continuing to work in that discipline and he refused to encourage his students to take up that work for fear that they would not be able to obtain a faculty position if they did so. At least part of the explanation for this is sociological. It centers on the consequences of the loss of an entire generation in World War I, the institutional conservatism on the universities, whose key mission was to train teachers for Lycées, and the power of the mandarins in key faculty positions, many of whom retained great power into their old age with no younger replacements in sight. They saw Mendelian genetics as unable to answer the key questions of "general biology". They did not a totally reject Mendelian genetics, which was considered valid for "special heredity", i.e., for inheritance of "minor" and "ornamental" traits of organisms; rather, they rejected its relevance to "general heredity", which dealt with fundamental traits, such as organisms' *Baupläne* or ability (vs. inability) to respire. The commitment to physiological issues like this last one is emblematic: Genetic control of respiration in yeast was crucial to Boris Ephrussi's role in securing physiological genetics as a core component of the French contribution to molecular biology and molecular genetics.

The point just made about the importance of physiology plus microbiology and embryology is critical, for it feeds into a summary of the take-home message of this paper. Thanks to Jean Gayon's leadership in doing deep history, we understood that all the major figures on whom we focused who helped bring genetics into French biology before the end of World War II—Lucien Cuénot, Boris Ephrussi, Émile Guyénot, Philippe L'Héritier, André Lwoff, Jacques Monod, Georges Teissier, and Eugène and Elisabeth Wollman—formulated their research problems in partly physiological terms. Several of them were also committed to the techniques connected to experimental embryology and thought of organisms in terms of developmental processes. And several of them were primarily microbiologists with strong allegiances to or contacts in the Institut Pasteur. They all worked in research institutions outside the university system and, before WW II, only Cuénot, L'Héritier and Teissier held professorships in French universities (all in zoology). None of them had significant opportunity to teach genetics except, in Cuénot's case, in the limited confines of a course in Biologie Générale, aimed at the education of teachers in Lycées. In the context of the deep history that we developed, this brief, rather abstract summary indicates how impervious mainstream French biology was in the interwar years to Mendelian thinking, let alone Darwinian selectionism. At the same time, the sheer fact that, with the exception of Cuénot, who abandoned his research on genetics, and Guyénot, who took a professorship outside France in Geneva, all of these figures enjoyed protective environments within which to develop a physiologically, embryologically, or microbiologically-based theory of heredity while working outside of the traditions of Mendelian genetics, an opportunity that was not readily available in Great Britain, the US, or even Germany. Thus, there was opportunity in France to develop the elements of what became regulatory and/or microbial genetics after WW II, when it became possible to interact with Mendelian traditions during the founding era of molecular biology and microbial genetics. French microbial genetics had the means to take this path thanks to its deep tradition of dealing with heredity in microbial physiology and biochemistry. The success of our studies of the French entry into genetics (Burian et al., 1988) depended on the deep history that we were able to develop thanks to Jean Gayon's leadership in constructing that history.

Let me conclude with a brief coda. This presentation falls far short of portraying all that owe to Jean Gayon on both personal and professional fronts. Most relevant in this regard is the enormous range of his professional accomplishments, well represented in this volume. Beyond that, his institutional leadership, the mentoring that he provided for his students and many other young colleagues, and the breadth and depth of his acquaintanceship and his interests, have advanced the level of scholarship and collegiality in history and philosophy of science (and of biology) in France and internationally. His great skill in teasing out conceptual connections and analyzing conceptual revisions has sharpened our thinking on topics far beyond those that any one of us has worked on. His exemplary efforts to explore an enormous range of topics in both breadth and depth provide a model of the role of a scholar, or of a historian-philosopher. His example is exceptionally worthy of our emulation. Many thanks to you, Jean, for what you have done for us all.

References

Burian, R. M. (2018). Jean Gayon on the history of French genetics. In F. Merlin & P. Huneman (Eds.), *Philosophie, histoire, biologie: Mélanges offerts à Jean Gayon* (pp. 111–119). Éditions Matériologiques.

Burian, R. M., Gayon, J., & Zallen, D. (1988). The singular fate of genetics in the history of French biology, 1900–1940. *Journal of the History of Biology, 21*(3), 357–402.

Chapter 18
Historical Epistemology and Analytic Philosophy: The Path of a Singular Synthesis. Testimony of a Former Doctoral Student

Élodie Giroux

Abstract The goal of this chapter is to describe the philosophical style of Jean Gayon through my own experience as a Ph.D. student. I show how Gayon's work represents a unique synthesis of two main traditions in philosophy of science: analytical Anglo-American philosophy of science and biology and French historical epistemology. His openness and curiosity towards the Anglo-American tradition led him to believe that a genuine dialogue between these traditions could benefit both. Jean also encouraged his graduate students to draw on multiple philosophical approaches, with an eye toward the sciences. In this chapter, I propose to characterize his own style as a *historically* grounded *analysis* of concepts. The analytical style he promoted is underpinned by historical reflection in which it finds a concrete and living substratum.

Keywords Risk factor epidemiology · Causation · Health and disease concepts · Precision medicine

18.1 Introduction

In the philosophy of science, the traditions of historical epistemology (mainly continental) and analytic philosophy (mainly Anglo-American) have for a long time been opposed, either ignoring one another or considering themselves to be incompatible. The difficulties and criticisms encountered by the philosophy of science deriving from logical positivism in the 1960s and 70s undermined the desire for unity and therewith also a certain vision of science. This situation was propitious to the development of the philosophy of the special sciences on the Anglo-American side with the emergence of philosophy of biology, in parallel to the development of a criticism of logical positivism from the history of the sciences (notably Kuhn). The French

Translated by Henry Dicks from Giroux (2018).

É. Giroux (✉)
Institut de Recherches Philosophiques de Lyon (EA 4187), Université Jean Moulin Lyon 3, Lyon, France
e-mail: elodie.giroux@univ-lyon3.fr

P.-O. Méthot (ed.), *Philosophy, History and Biology: Essays in Honour of Jean Gayon*, History, Philosophy and Theory of the Life Sciences 30,
https://doi.org/10.1007/978-3-031-28157-0_18

tradition of historical epistemology (Bachelard, Koyré, Canguilhem, Foucault) has already made the choice of regional approaches to the sciences and of an integration of history into philosophizing about the sciences. The evolution of Anglo-American philosophy of the sciences created the space for a dialogue and a rapprochement between these two traditions that had for a long time been kept apart, and even held each other in mutual contempt.

It was in this context of the birth of analytically oriented philosophy of biology that Jean Gayon entered the philosophical scene. One of the principal specificities of his work, and which he transmitted to his doctoral students, was this openness towards the international and to the Anglo-American tradition. This openness led him to believe profoundly in the need for a dialogue. It seems to me that his work represents a form of singular synthesis of these two main traditions in the philosophy of science, a way of taking the best of each of them. It is not a question here of a superposition or an artificial articulation but rather of a specific way of philosophizing about science, born from the encounter between these two traditions. This encounter was first experienced in his own academic studies, which combined a ground in historical epistemology with strong knowledge of current developments in philosophy of biology in the analytic style thanks to his research visits in the United States and his numerous contacts with North American philosophers of biology.[1]

In this brief testimony, I aim to show that it is the encounter and dialogue of these two traditions of philosophy of science that he transmitted and brought to life for his doctoral students, before detailing my own view of the nature of the synthesis that he achieved. It is as having personally experienced the encounter of these two traditions as a doctoral student, guided by Jean in a process of what one might call intellectual initiation, that I have been able most clearly to perceive the originality of his work.

18.2 Jean, Thesis Supervisor

In what follows I will take the liberty of talking about myself, but many doctoral students will no doubt recognize themselves in my story. It is in this form of testimony that it seems to me both possible and pertinent to speak of Jean as a thesis supervisor. In so doing, I wish to do justice to what he transmits and has transmitted, of the traces he left in some of us and in the philosophy and the history of the sciences that we seek to keep alive and pursue.

If it is difficult to know and to identify what one transmits, perhaps it is slightly less difficult to identity what one has received. I received from Jean the invitation to live this encounter between historical and analytic approaches in the philosophy of science through people, institutions, and texts, but also to find for myself, following Jean, though not necessarily in a way that resembles his, the path for an articulation

---

[1] On Gayon's contacts with American philosophers of biology, see Méthot, Depew, and Burian, this volume.

and a reciprocal enrichment. Nothing is determined in advance. But a rich, stimulating, and fertile path is opened up here. Showing respect and confidence in the ability of his doctoral students to discover their own originality and to find their own path, Jean put down several markers, leaving them free to choose the forms and modes of their journey.

One usually meets a thesis supervisor at a key moment in one's intellectual journey, and more generally, in one's own life story. It is a stage and a moment that will often strongly determine the following stages of our intellectual and personal life and the type of researcher that we will become. I remember my first meeting with Jean in the context of discussing my project of doing a *Diplôme d'études approfondies* (DEA), the old equivalent of a second year Master's programme, in history and philosophy of the sciences. He was in a period of transition between what was at the time University Paris 7 and University Paris 1, where he had only just been transferred, succeeding Anne Fagot-Largeault in the post of professor of philosophy of science. I was interested in questions relating to medicine; I had notably discovered Canguilhem's *The Normal and the Pathological* (1991 [1966/1943]) in the context of the oral study programme for the "*agrégation*".[2] I expressed to him my desire to study this domain as a current concern and from the perspective of epistemology, rather than bioethics, telling him that my conviction that the latter requires the former. It quickly became apparent that we would get on well.

I would not want to hide the fact that there was a small attempt on his part to encourage me to work on philosophy of biology. And it is true that I presented myself in a way that indicated my interest in issues related to the concept of the individual in biology and medicine. But my resolve to undertake a reflection on the concepts of health and disease was greater. Jean then related to me his own experience of a silent illness that had led him to undergo some important surgery. The question of the status of chronic and asymptomatic diseases opened an important field of reflection that Canguilhem had little explored. The context was different from that of his 1943 thesis (*Essai sur quelques problèmes concernant le normal et le pathologique*) because of change to morbidity that had since arisen and of the importance of preventive, statistical and probabilistic approach towards these chronic and asymptomatic diseases. Anne Fagot-Largeault, with whom I was also in contact, invited me to start a dialogue with Pierre Corvol, an expert on hypertension and holder of the Chair in Experimental Medicine at the Collège de France. A common preoccupation animated them in relation to the concept of "risk of disease", which derived from epidemiology, and its impact on our conceptions of the normal and the pathological. Jean and I decided that there was scope here for a DEA, and certainly for a thesis, on the subject of the silent disease (or the disease that has not yet taken the form of an illness) and an epistemological study of the emergent risk factor approach to disease. This was confirmed by my DEA study.

[2] The *agrégation* is a nationwide exam that one must pass to become a high school teacher.

18.3 Doing History

It was at this point that the first initiatory stage took place in the journey of Jean's doctoral student: the invitation to do history. He advised me to talk to the historian, Jean-Paul Gaudillière. The message was clear: to confront and familiarize myself with the history of historians and not settle for the history of philosophers. I became acquainted with the team of historians at the CERMES, today CERMES3 (*Centre de recherche médecine sciences, santé, santé mentale, société*) centred on Jean-Paul Gaudillière, Ilana Löwy et Luc Berlivet, who, by a fortunate coincidence—or convergence of interests around emerging subjects and questions in medicine—had just launched a research project on the theme of risk and disease.

Doing history did trouble me slightly. I neither felt myself to be a historian, nor competent to work in the archives. But I climbed on board, confident in Jean's invitation and in the support of the CERMES team. I threw myself into archival research into a prospective study in cardiovascular epidemiology, during which the concept of "risk factor" of disease was in part established: the Framingham Heart Study (1947–), Framingham being a city close to Boston. The archives were spread out between Boston and Washington (the National Institute for Heart Lung and Blood Diseases at Bethesda). I met historians of science and medicine from the Department of the history of science at Harvard University, and then from the National Institutes of Health's Office of History, where I was subsequently welcomed. This contact with archives and with the community of historians confronted me with what was for me an enigma throughout the whole of my doctorate: how can a philosopher do history? Can one simply convert? Are historical epistemology and epistemological history the only ways of resolving this enigma? Are there no other possible paths for articulation?

18.4 The Encounter with North American Philosophy of Medicine

In France, at that time, philosophy of medicine was only just getting off the ground, and Anglo-American philosophical studies of the concepts of health and disease, which had been important since the end of the 1970s, were little, if at all known. It was also difficult to get hold of the Anglophone literature on these concepts. Jean helped me to acquire the imposing collection of texts co-directed by Caplan, Engelhardt, and McCartney (1981) and suggested that I carry out, based on this volume, a study of the state of the art with respect to these concepts. I thus discovered the debate between the "naturalist" and "normative" schools. From there, many other things began; it was another stage in my intellectual initiation. The style of the texts present in this collection was a mix of historical, sociological, and analytic reflection, depending on the author. There was nothing on Canguilhem. I thus became aware of an astonishing reciprocal ignorance on both sides of the Atlantic in philosophy of medicine. Jean played a key role here in encouraging me to go and find out what was happening in this field in North America and to carry out research trips to the United States. The name, Jean Gayon, means something like "open sesame" in many North American universities, and among many researchers, whether historians or philosophers of science. These research visits have had a determining influence on my career, and it is in this context that I gained access to a philosophy of medicine literature little known in France.

Moreover, in the framework of the French Institut d'histoire et de philosophie des sciences et des techniques, Jean had launched a research programme on the concept of function in biological and medical sciences and had made me an associate member. This framework strongly contributed to familiarizing me with the style of conceptual analysis and, more generally, to the analytic literature on the concept of function in philosophy of biology. And it is also in this context that I started working on analytic philosophy of medicine, whose precise point of departure are conceptual analyses in philosophy of biology, to develop a functional and theoretical concept of health. I am thinking of the works of Christopher Boorse (1977) in particular.

18.5 The Max Planck Institute for the History of Science

After these initiatory years working on the one hand on the history of medicine and on the other on analytic philosophy of medicine, Jean suggested to me a research trip at the Max Planck Institute for the History of Science, in the department of Hans-Jörg Rheinberger. There, I discovered other ways of doing the history of the sciences and of articulating history and philosophy than that of historical epistemology. Researchers at the Max Planck Institute adopt an international and multidisciplinary approach to the study of sciences. It was at that point that I was writing up my doctoral thesis and trying to resolve the issue of how to articulate philosophy and history. But the realization of an original synthesis between history and philosophy of medicine, but also of the style of historical epistemology and the analytic style, was not yet fully formed in my thinking of the time. I thus anchored myself in the French tradition to attempt an "epistemological history" of the Framingham epidemiological study, while also sketching out the beginnings of a dialogue between the theses of Canguilhem on the normal and the pathological and the controversies surrounding the concepts of health and disease in the analytic style.

18.6 In Search of a Synthesis

Part of the research I did following my doctorate consisted in proposing articulations between different ways of philosophizing within the philosophy of medicine, and notably establishing bridges between the reflections of Canguilhem and those of analytic philosophers of medicine like Christopher Boorse or Lennart Nordenfelt

(Giroux, 2009). I showed that there were proximities, despite important differences. It was after immersion in the analytic style and having developed a critical analysis of its limits for the philosophical analysis of the concept of disease, that I discovered an intermediary research path. And it was at that point that I discovered that that was the path already taken by Jean.

I may also say that Jean enabled me to equip with tools for situating myself in a contemporary approach of philosophizing *on* and *with* the sciences, in an international context. Above all, he gave me the means to invent new forms of conjoining these two traditions of the philosophy of science. It is the retrospective understanding of this path, for which Jean has been my guide, which enables me better to understand the originality of his way of philosophizing about the sciences, and which I will now attempt to explain.

18.7 A Singular Way of Articulating Two Traditions in the Philosophy of Science: Historical Epistemology and Analytic Philosophy

Jean is not the only figure to carry out what I describe here as a marriage, a conjunction of those two traditions. It isn't simple to combine history and logic. But Jean has his own method, which I propose to characterize as a *historically* anchored *analysis* of concepts. More generally, he has a concern for clarity of reasoning which will not settle simply for logic and which takes the risk to enrich itself with *history*. The latter is even what often makes it possible to signal or neutralize certain stumbling blocks in *conceptual analysis* in the strict sense of the term.

I have always been impressed by the clarity of Jean's presentations, whether at conferences or in his teaching. And his love for etymology is for me one of the key aspects of this original articulation. The concept is important, but its history is essential to understanding its meaning. This is a departure from the analytic tradition. But, unlike in the tradition of historical epistemology, a much greater place is made for analysing concepts, on the one hand, and for clarity of argumentation and logical reasoning, on the other. I would say that Jean, more so than Canguilhem and the authors one habitually associates with this style, is concerned with conceptual clarification and logical argumentation. This is surely due to his conscientiousness and an intellectual honesty, which integrate transatlantic analytical questions and debates. But this clarification and analysis are woven by history, where they find a concrete and living substratum.

References

Boorse, C. (1977). Health as a theoretical concept. *Philosophy of Science, 44*(4), 542–573.
Canguilhem, G. (1991) [1966/1943]. *The normal and the pathological*. Zone Books.

Caplan, A. L., Engelhardt, H. T., & McCartney, H. T. (Eds.). (1981). *Concepts of health and disease: Interdisciplinary perspectives*. Addison-Wesley.

Giroux, É. (2009). *Après Canguilhem. Définir la santé et la maladie*. Presses Universitaires de France.

Giroux, É. (2018). Épistémologie historique et philosophie analytique: la voie d'une synthèse singulière. Témoignage d'une ancienne doctorante. In F. Merlin & P. Huneman (Eds.), *Philosophie, histoire, biologie: mélanges offerts à Jean Gayon* (pp. 267–273). Éditions Matériologiques.

Chapter 19
Doing Philosophy of Evolutionary Biology with Jean Gayon

Philippe Huneman

Abstract Throughout my university career, and since I began my Ph.D., Jean Gayon was there. Unlike many contributors to this volume, to the early or mid-career researchers who do French philosophy of biology today, I did not know Jean as a dissertation supervisor or a professor, but as a dissertation examiner, as expert witness to the beginning of my career and as indisputable scientific authority. For fifteen years I have been doing philosophy of evolutionary biology with Jean Gayon. In this chapter, I do not offer an analysis of "evolutionary biology according to Gayon" but I aim to pay tribute to a philosopher who was at once a "major contemporary", a colleague, a friend, and an inspiration.

Keywords Evolutionary theory · Modern Synthesis · Philosophy of biology · Function · Darwin · History of biology · Historical epistemology

19.1 Introduction

Throughout my university career, in the broad sense, or since I stopped being only a student, since I began my Ph.D., Jean Gayon was there. We got to know each other when he was at Université Paris 7 Diderot, in the laboratory Sphere (called REHSEIS at the time[1]) of epistemology and history of the sciences, where he was already following my work with a benevolent eye, over 20 years ago now. I hope the reader will understand that it is difficult for me to synthesize all this, to say something quite precise about it.

---

[1] Sphere: Sciences, philosophy, history; REHSEIS: Epistemological and historical research on exact sciences and scientific institutions.

Translated by Henry Dicks from Huneman (2018).

P. Huneman (✉)
IHPST, CNRS, UMR 8590, Paris 1 Panthéon-Sorbonne, Paris, France
e-mail: philippe.huneman@gmail.com

P.-O. Méthot (ed.), *Philosophy, History and Biology: Essays in Honour of Jean Gayon*, History, Philosophy and Theory of the Life Sciences 30,
https://doi.org/10.1007/978-3-031-28157-0_19

Unlike many contributors to this volume, to the early or mid-career researchers who do French philosophy of biology today, I did not know Jean as a dissertation supervisor or a professor, but as a dissertation examiner, as expert witness to the beginning of my career and as indisputable scientific authority.

If I consider myself today above all a philosopher of evolutionary biology, it is certain that it was with Jean that I set out on the path towards the biology of evolution. The first philosophical book I read in French on evolution, a long time before knowing him in person, was *Darwin et l'après-Darwin*. I learnt in it all sort of things I knew nothing about: the existence of a "*théorie synthétique de l'évolution*" (the "Modern Synthesis"), the complexities surrounding the apparently simple idea of natural selection, the fundamental theorem that Ronald Fisher developed on the subject, the subtleties of the relationships between chance, neutralism, and selection. I had never read the founding texts, other than the *Origin of Species*, and this long reading was therefore a discovery. Retrospectively, I can see that my own current research themes are already there; and often, when I reopen this book to find this or that analysis, I realize that my own limited intuitions were already there, explored, developed, or contradicted, in scientifically detailed analyses of this or that work on evolutionism prior to 1970…

If I had to be more precise in the exercise of my memory, I think my first encounter with Jean's work was a less well-known text, an off-print I picked up (it was at the time when one still read things printed on paper) from a table corner where he'd left it after a REHSEIS meeting: an article on Nietzsche and Darwin (Gayon, 1999). I had been impressed by the finesse of his reading of the philosopher of the transmutation of values; at the time of a weird leftist Nietzscheanism, nowadays somewhat subdued, Jean recalled the eugenicist project of Nietzsche, his arguments in favour of political ideals that many found unpleasant. I sympathized with this way of demolishing, by erudition and analysis, sacred cows.

But of course, when I think back to this period when I was simultaneously discovering the work of Jean and Jean at work, the figure of "evolutionary biology" stands out, and it belongs under the aegis of Darwin, rather than that of Nietzsche. This stands out even more clearly when I recall how all my work, starting from my arrival at the IHPST in 2002, after my dissertation, focussed on the philosophy of evolutionary biology, and the extent to which I did it *with Jean*. I will therefore try to spell out what this latter expression means. I will conclude by trying to set out how I saw Jean, throughout this history or process, bring together philosophy and history of science and thus illustrate a possible response to the question that every philosopher of science at some point asks about the relation between philosophy and history in her work.

19.2 About Evolutionary Biology

When I am called on to say what exactly it is that I do, "philosophy of evolutionary biology" comes straight to mind. Does that constitute an identity? I know not, though I could discuss the issue, but that is not the question. People sometimes react, and those who do not belong to the field express surprise ("Really, that exists?"). But if this "identity" is possible in our country—in France, in French philosophy—I think that we owe that essentially to Jean Gayon. That is why even if some of my research themes are no longer those of Jean, even if I invest in projects that are not his, I always have the impression of doing philosophy of evolutionary biology *with* Jean Gayon. It is with unpacking this notion that the present text will be concerned. Of course, *philosophy* of evolutionary biology is not the only thing Jean does—the list of his publication subjects is long and astonishing, going well beyond biology and even the philosophy of science—but it seems to me that it provides the core of his work, even when Jean tells the "true" story of genetics or of Darwinism, with all the characteristics that differentially define the history of science (archives, correspondence, original texts, etc.).

In this respect he is a bit like Jean Cocteau, who wrote in every genre, but who, at the end of his books, arranged the list of his other publications in the form of "theatrical *poetry*", "romantic *poetry*", "cinematographic *poetry*", etc., to get across that whatever form his writings took, they always revolved around a common centre, which was poetry.

So, if one accepts my proposal that this is the core of his work, what does it mean to "do philosophy of evolutionary biology with Jean Gayon"?

To begin, to do philosophy of evolutionary biology "with Jean Gayon" means a practice: doing philosophy together, discussing texts, writing with one another, but also in my case developing a framework for this activity concretely: constructing institutional philosophy of biology at the IHPST. Then, philosophizing "with Jean Gayon", in the sense in which one sometimes says "think with Descartes", is to philosophize reading Jean, listening to him, discussing with oneself the positions he defended or which are attributed to him. It is to philosophize, in some sense, in his wake, which implies being able to represent to oneself at the time the nature of this wake. I will now set out these two aspects in order.

19.3 Working with Jean Gayon: The Philosophy of Evolutionary Biology

For fifteen years I have been doing philosophy of evolutionary biology with Jean Gayon. It would be difficult to ignore it in the sense that, wherever I go for my seminars, conferences, etc., as soon as one learns that I come from Paris or even from France, I am asked "do you work with Jean Gayon?", to which I have the pleasure of answering affirmatively. And in fact, I can no longer count the seminars, symposiums, conferences, I have organized or co-organized with Jean—including that great adventure that was the organization of the large congress of the International Society for History, Philosophy and Social Sciences of Biology (ISHPSSB) at Montpellier in 2013—or where we were both speakers, in France, in Europe, in America, or in Australia.

But beyond these touristic memories of talks and conferences, I can clearly discern today the continuity of a common project on which Jean and I had embarked. Indeed, I had the good fortune, arriving at the IHPST, to be immediately thrown into a collective adventure consisting in making this Institute, already famous for its tradition in the history of biology and medicine, a prestigious centre in the philosophy of biology at the international level. This was not the explicit initial intention, but, in the end, that is what progressively occurred. With Jean, several research programmes were defined, determining the theoretical orientations of the lab as well as the noteworthy collaborations, and, as far as I was concerned, certain research questions. The team grew naturally, adding doctoral students—some of whom are contributing to this collection—postdocs, associates; it established links with research institutions in biology, some of which had been collaborating with Jean for quite a while, while others recognized themselves in the type of work that Jean had initiated with us, at the frontier between philosophy, history of biology, and theoretical biology.

These research programmes have always maintained a strong link with evolutionary biology: the "function" collective research project (discussed by Françoise Longy (2018) in the French version of this volume), first of all, and its reading seminar, which was the occasion for impassioned discussions featuring contributions from, amongst others, the sorely missed Marie-Claude Lorne[2] the psychologist Françoise Parot, the historian of biology Jean-Claude Dupont, the doctoral students of Jean's first Parisian wave, like Élodie Giroux. Then there came the projects we had developed around the general question of cooperation, which allowed me to discover an entire continent within contemporary evolutionary biology, and which through showing us the economic tools used by biologists to think the enigma of altruism, gave rise to the ANR-funded research programme on economy and evolution, recently led by Jean.

A word about this officially closed project, but which in reality is only now coming to completion, because its most emblematic achievement represents an aspect of working with Jean: in order for philosophers, biologists, and economists collectively to write a double economics-evolution lexicon of key concepts, considered from the perspective of the two disciplines, we met up over the course of several years with the biologists Silvia de Monte and Jean-Baptiste André, the economist Bernard Walliser, and the philosopher of economics Mikaël Cozic, to discuss notes, each of which was written by one or the other of us, on key concepts (market, competition, chance, strategy, etc.) in one of the two disciplines. The premise was in some sense to adopt a position of complete naivety: in the most general way possible, what do words like "competition", "adaptation", or "cycle" mean to a biologist, or to

[2] For a tribute to Marie-Claude Lorne, see Huneman and Barberousse (2011).

an economist?—While explicitly leaving on the side the empirical and theoretical controversies surrounding these concepts. The wager was to keep each entry down to two or three pages (see André et al., 2022).

Jean's impressive familiarity with the details of evolutionary biology was a source of wonder. What has left the most enduring mark on me, however, was discussing with him things that he had no doubt thought about thousands of times, to which he had often devoted an article or a decisive passage in his book, and which he treated as if he had only recently discovered, as if the conceptual difficulties were new to him.

Even more recently, Jean launched, with his friends, firstly David Depew and Richard Burian, but also me, a vast project of re-reading the synthetic theory of evolution, or Modern Synthesis, which today brings together the contradictory arguments of biologists, often split into two opposed camps, "for" and "against". But as Jean noted, we still don't really know what this Modern Synthesis was, for the testimonies and descriptions very often come from actors, each of whom has their own vision, defending their own point of view—primarily Ernst Mayr.

Of course, by embarking us on this project, Jean was meeting up again with his initial programme—as Pierre-Olivier Méthot explains so well here. His original CNRS application project, which became *Darwin et l'après-Darwin* (2019 [1992]), was indeed to understand the Modern Synthesis, in its philosophical meaning, structure, and genesis. His research, by deepening its roots, by also noting the irreducible character of the Modern Synthesis as a social construction (as François Duchesneau reminds us in this volume), had thus deviated towards a reference work in philosophy on the conceptual history of the "hypothesis of natural selection", as Jean names the exact object of the conceptual thread he brings to light.

For several years now, this has therefore been a collective project which takes up again Jean Gayon's initial intention. It has been published as a special issue of *the Journal of the History of Biology* that I edited.[3]

19.4 A Way of Doing Philosophy of Biology

Next up, "doing philosophy of evolutionary biology with Jean Gayon", is to do it oneself, in dialogue and in concert with that of Jean, some traits of which I will now present and which I find striking. At the outset, I wanted to grasp the intrinsic unity of Jean's approach—to synthesize it all if necessary, in a table, an image, with text boxes and arrows. But in truth this is evidently not possible.

What I can make out is rather a constellation of questions:

- Eugenics—with a great interest for Fisher, to whom *Darwin et l'après-Darwin* devotes landmark analyses, but also to whom Jean regularly returns, right up to the beautiful paper on Fisher's economics, written in the context of the projects

[3] See, *Journal of the History of Biology*, "Special Issue on Revisiting the Modern Synthesis", *52*(4), 2019.

on economics and evolution that I discussed above (see Gayon, 2014). And in fact, eugenics is a constant in the work of the English biologist: his main book, *The Genetical Theory of Natural Selection*, devotes five chapters out of twelve to eugenics, though the work's still lively scientific legacy concerns only the other seven. Moreover, Jean remarked to me a few years ago that there is still no intellectual biography of Fisher, as there is for Sewall Wright, George Price, David Lack, or J.B.S. Haldane.

- Human enhancement, which clearly resonates with the old eugenicist projects from before the war, to which it adds a technological and medical dimension.
- Races, which were in their time a major preoccupation for the architects of the Modern Synthesis, who were often engaged in the enterprise of providing a scientific refutation of racisms. This is evidently a point of interface with social Darwinism.
- The notion of heredity—both from the point of view of the history of genetics, with respect to which Richard Burian and David Depew remind us of certain aspects of Jean's contribution—and in a general manner concerning the philosophical meaning and long-term history of the idea of hereditary transmission. Stéphane Tirard mentions in this context a contribution on Lamarck, but I would also refer to some pages of *Darwin et l'après-Darwin* that have left a lasting impression on me about the way that Mendelism undid the common presupposition that heredity is an immemorial force, at the time often called atavism.
- The history of transformism and its important figures since Lamarck; and in general, it goes without saying, the various Darwinisms.
- French philosophy of science (the so-called 'historical epistemology'), with the often-discussed figures of Canguilhem and especially Bachelard, but probably more profoundly the examination, dear to Jean, of the relations between the history of the sciences and the philosophy of the sciences, addressed in this volume by François Duchesneau. His recent implication in the new international dynamic developing around the phrase Integrated History and Philosophy of Science (IHPS), discussed by Anastasios Brenner in this volume, was a way of institutionalizing a question which has never left him and which is in some ways meta-reflexive (if one accepts the idea that the philosophy of science is itself already reflexive). It is plausible to think that Jean's own work sketches out paths for thinking about these relations, and Jean himself has certainly never ceased trying to formulate them explicitly, both for himself and for others.

These objects of questioning are loosely connected. If they are arranged into a discursive constellation or nebula, it seems to me (as François Duchesneau also notes in his contribution) that *Darwin et l'après-Darwin* would in a sense constitute the kernel—both as initial inquiry which uncovered the possible areas of questioning and as the material of the meta-reflexive investigation mentioned above.

Of course, along these last few years at the IHPST, Jean has made some new suggestions for research, which have provided the occasion for him to write some new articles on various questions: economics, with the project I mentioned on economics and evolution, animal and human enhancement, cultural evolution, the definition of life and synthetic life, or even time in evolutionary biology. On this last theme, Jean wrote a remarkable text with Maël Montévil, containing entirely new material, that I had the pleasure to edit, and which shows that, in opposition to what has been thought since Fisher, there is no reversibility in (determinist) population genetics, and so no parallel with classical mechanics (Gayon & Montévil, 2017). These new paths appear to me as extensions of the constellation mentioned above, organized around a core of questions about evolutionary biology.

What I find striking in this way of doing philosophy is a sort of ethos of return or reprisal: coming back to something to which you have already devoted a considerable amount of in-depth study, that one could believe to be definitively known, and that in any case amounts to knowledge that others believe you to possess, and discovering there a new path, a new layer of problems, a new question. Jean had for example written from *Darwin et l'après-Darwin* (2019 [1992]) onwards about the "neutralist" theory of molecular evolution advanced by Motoo Kimura in the period 1960–70. However, at the occasion of a meeting of the ANR working group "Explabio" in which I asked him for a presentation on the subject, he set out, to our surprise, on a fresh inquiry into the profound conceptual meaning of this theory. Similarly, when he recently came back to Lamarck and the heredity of acquired traits, or even to the notions of genetic and epigenetic heredity (see Gayon, 2016, p. 2) I often had the impression that he was finding a way to take a fresh look at a topic that had already been studied by him at length.

A literary image come to my mind here—the *Turn of the Screw* by Henry James, whose title comes from an image in which by turning and returning to the same place one nevertheless takes a step forward with respect to where one was before, and which defines for this novel the dynamic of the plot but also of the text itself. Jean's theoretical relation to certain conceptual objects seems to me to belong to this process of "turning the screw".

The best example of this for me is the recent work on the synthetic theory of evolution. It is thought that he knows the issue well, that he has own vision of it, set out in the book *Darwin et l'après-Darwin* and numerous other texts, but he actually decides to re-think it all. Indeed, it was Jean who found a way to articulate the project of revisiting the Modern Synthesis, to which we were in 2018 putting the finishing touches, as we ourselves were banging our head against the wall working out how to construct our approach. One fine day Jean thus came to find us with this quotation from Julian Huxley's *Evolution: The Modern Synthesis*: "The time is ripe for a rapid advance in our understanding of evolution. Genetics, developmental physiology, ecology, systematics, paleontology, cytology, mathematical analysis, have all provided new facts or new tools of research; The need today is for a concerted attack and synthesis" (1942, p. 8).

After reading this passage, it became evident that the simple and natural way to proceed and to share the workload was to examine one by one each of the domains that Huxley said the Modern Synthesis must take on, to evaluate to what extent that has in fact occurred, and how the theoretical nature of the synthesis had in turn been modified.

Of course, discovering what the first research project written by Jean was—cited here by Méthot—one clearly perceives the resonance between this question, and his programme of the time, in 1984, which announced: “The inquiry will examine [with respect to the Modern Synthesis] the ‘theoretical’ nature of this theory, and the meaning of ‘synthesis’.”

A second image, to characterize this art of the reprisal comes to me from biology itself: the duality between exploitation (searching for resources in an area in which one knows there are plenty) and exploration (heading out into uncertain terrains which may yield new resources but may also yield nothing at all). Behavioural ecologists study how different species make trade-offs between exploration and exploitation in the search for resources or the choice of partners. Thus, the history of genetics and evolution, as thought by Jean, could be understood in terms of this duality: whereas one field is exploited, new adjacent domains are regularly opened and explored, such as human enhancement, evolutionary economics, eugenics, etc.

In any case, a remarkable strength of this approach resides in its capacity for synthesis. This is often organized around the statement of a misleadingly simple proposition—which imposes itself as the self-evident fact around which everyone gravitates without being able to recognize or formulate it. This is the case for the discussions we had about evolutionary economics, economics in evolution, and their different variants. In the end, Jean succeeded in discerning, as the principle of differentiation between two economic transpositions of evolution, the type of role conferred to heredity in the vision of natural selection at play in each case. This is what in fact allows him to assert a clear conceptual difference between the “evolutionary theory of games” and “evolutionary economics” (Gayon, 2012), and to bring to light an ordering principle, which retrospectively seems obvious, in the jumbled comings and goings between economics and evolutionary biology.

In an analogous way, I can also recall a moment in our collective exchanges about the notion of function and functional explanation. In the middle of endless debates in the discussion of more or less byzantine refinements of a theory interpreting the proposition “the function of X is Y”, Jean suddenly reminded us that in reality biologists use the word “function” in *two* ways: “small functions”, as it were, like the “function of the kidney”, the “function of the gene Pax 6”—and “big functions”, those which define a field of investigation: the “circulatory function”, the “digestive function”. A large part of the discussion of analytic philosophy thus came to appear as blindly focussed on the former. Such a remark thus put in perspective the entire philosophical debate (which took place between the 1980s and now) and returned it to the context of the very general question “What does it mean to ‘do biology’?”.

19.5 On a Certain Style of Relations Between Philosophy and History of the Sciences

Jean Gayon has directly reflected on the notion of *style* in the sciences, following Alistair Crombie or Ian Hacking. He has often thought about the relation between history and philosophy of the sciences, which he developed in his own style of philosophy. What synthesis or what alliance holds between these two domains?

When I re-read *Darwin et l'après-Darwin*, I am left with the impression of a remarkable equilibrium between the two: the conceptual history is present, appreciated by connoisseurs as a major contribution of unrivalled systematic scope; but Fisher's theorem, the nature of heredity, the question of validation, the establishment of causality are also all there as objects of philosophical reflection. At the same time, I know many biologists, in France, in the USA, or in the UK, who work on this book and recognize it as important for them—an infinitely rare accolade held only by a few books in philosophy of biology, since philosophy is generally considered by the scientist in accordance with Feynman's witticism that it is as necessary to the researcher as ornithology is to birds.

How can we understand this? The book begins with Darwin and Wallace, and ends with Kimura (incidentally, Jean once told me about meeting Kimura at the time he was writing the book). Kimura's neutral theory gave rise to an exceptionally fierce controversy around natural selection, and indeed, Jean's entire book led up to the adoption of a position on this controversy; this position is characterized by a significant philosophical distance. In fact, Jean Gayon dispels the appearance of conflict: neutrality and selection do not concern the same things, because the scales of the processes in play establish an important difference. Of course, selection remains omnipresent and explanatory, its "paramount power", to take up a Darwinian expression that Jean made into the title of an article (Gayon, 1997), remaining uncontested. The "hypothesis of selection", the touchstone of Darwinism, remains corroborated, but Kimura is shown to displace the question towards "At what scale is it corroborated exactly?". (The problem here is not the question of the levels of selection, which Thomas Pradeu rightly suggests here has been the focus of too much attention by philosophers of biology, but rather that of spatial and temporal scales.) Thus, Jean may conclude his discussion of the neutral theory with the following paradox: "Far from belonging to an anti-Darwinian philosophy of biology, it is one of the most remarkable modern interpretations of the Darwinian theory of natural selection."

Let us generalize about this style of thinking. When faced with a philosophical question, Jean Gayon certainly likes to come to a decision—I often remember having heard him, at a conference or in private, ask the speaker after a complex and nuanced argumentation: "Ok, but what are you ultimately defending?"—But he only comes to a decision right at the very end, after the advocates of the parties have fulfilled their roles. (And the unfolding of the trial is precisely where one learns so much.)

On the other hand, these philosophical questions with respect to which decisions must ultimately be made are often constructed around concepts that in the last analysis lack clarity, and this is where the necessity of doing the history of science comes in.

Against an analytic philosophy which prides itself on leaving history to one side, Jean ceaselessly defends the necessity of tracking down historically constituted concepts, because only their genealogy can indicate something about their meaning which is not simply the assumed position of the day. The best example for evolutionary biology is probably the word "fitness". Even if one always recognizes in it a certain combination of survival and reproduction, this term, for which numerous interpretations have been advanced (lifetime fitness, trait fitness, organism fitness, inclusive fitness, Malthusian parameter, invasion fitness), has divided biologists since Fisher, whereas philosophers of biology find in it (as a result) one of their favourite objects: is it a propension, a disposition, a purely statistical measure, etc.? But it would be incorrect to think that simply carrying out a well-researched inquiry into contemporary uses of the concept in the models would settle the debate about what fitness is, for these uses themselves contain interpretative choices. It is therefore necessary to take a look into the past and ask oneself where, when, and how the concept of fitness was introduced in evolutionary biology, via an adaptation of the Spencerian expression "survival of the fittest" adopted in the final analysis by Darwin, as well as why this term exactly. Thus, in *Darwin et l'après-Darwin*, Jean devotes a few pages to the word itself, before embarking on an investigation of the concept.

One thus sees with this example how conceptual analysis, at the heart of supposedly pure philosophy of science, cannot do without the semantic depth of the history of science. More precisely, there appears here the strange historicity specific to evolutionary biology, on which Jean Gayon had from the outset wanted to reflect when he was planning to tackle the synthetic character—or otherwise—of the Modern Synthesis, which is the following

Ultimately, everyone recognizes that this synthesis is only a synthesis in an extraordinary sense of the word, because its protagonists were rarely in agreement. One could even describe the Modern Synthesis by enumerating the debates with which it was concerned: Fisher versus Wright on the respective roles of drift and natural selection—the debate that *Darwin et l'après-Darwin* qualified as "constitutive"; Müller versus Dobzhansky on the explanation of polymorphism; selectionism vs saltationism in its diverse forms, from De Vries to Goldschmidt, etc. The distance afforded by history here can give the impression of an eternal re-start, with contemporary biologists repeating over and over again these conflicts over the significance of selection, the random nature (or otherwise) of variation, the gradualism or discontinuity of evolution, etc. Far from being linear, the history of evolutionary biology seems to progress by ceaselessly reactivating a small handful of aptly named "constitutive debates", changing the arguments and the precise meaning of the terms in the light of empirical and theoretical advances (macromutations become certain types of mutations in networks of regulator genes, Kimura's neutrality transforms the meaning of "drift", etc.).

Are we in the presence here of what Gerald Holton calls "themata"? (Holton, 1981) No, these are less finely grained objects, such as: preformism vs epigenesis, vitalism and the machine analogy, etc.; they nourish the whole of biological thought well beyond evolutionary biology. Of course, neither is it a question here of Kuhnian "paradigms", for however it is that one takes this equivocal notion, what is at issue in our case is rather that which remains constant across presumed changes of paradigms.

But precisely because the history of evolutionary biology takes on the appearance of a re-start, the philosopher, if she wishes to knows what is repeating itself here—like the biologist if she doesn't want to engage in endless repetitions of the same—must understand the matrix of this re-start.

How are we to understand this "matrix"? It is a space of historically mobile interrelated conceptual determinations, in which both controversies and advances unfold, where radically new conceptions appear as palimpsests of old ones. One thinks here of something close to Foucault—an author of which Jean was never a great fan. But if one can certainly think of the matrix as a conceptual space, it is something more general than Foucauldian epistemes, whose character as ruptures is always emphasized, whereas these matrices Jean reconstructs establish continuities; and at the same time it is also something more regional, for the Foucauldian episteme aims to go beyond disciplinary frontiers, including grammar, economics, or natural history, as *The Order of Things* so brilliantly sets out.

In any case, to explore the matrices of evolutionary biology, which have been developed at least since Darwin, shifts slightly the methodological opposition separating philosophers of science between historical and analytic approaches to science. The elements that belong to a conceptual matrix cannot be considered in terms of causal relations, as if an etiological explanation was necessary for the appearance of this or that conceptual novelty in this or that author or school, but rather in terms of logical conditioning (in the broad sense).

It is therefore less a question of retracing the influences and conquests than of grasping the movements of the driving antagonisms: thus, Jenkins' criticism of the hypothesis of natural selection (on the grounds that it forgets the dilution of variation occurring in the mixing of traits during reproduction) reappears several decades later in the opposition between biometricians and Mendelism as two possible answers to Jenkins' dilemma—which Jean calls the "foundational crisis" (*crise principielle*) of Darwinism—while the "constitutive debate" between Fisher and Wright is prolonged into the neutralist challenge of Kimura; the discussion around Fisher's theorem informs right up to the present discussions around the optimizing character of natural selection and ultimately also the adaptationism criticized by Gould and Lewontin in a famous article, or the "phenotypic gambit" that the English biologist, Alan Grafen, places at the foundation of the whole of behavioural ecology (Grafen, 1984).

A question of fundamental importance in the analysis of this theoretical matrix, and which differentiates it from the conceptual analysis of philosophers working in the analytic tradition in the strict sense, is: "Who is speaking?", "Who exactly is advancing this thesis, looking at this object, employing this term?" "Who?", not in the sense of the person, but rather the disciplinary background, the tradition, or the scientists' network. Around Darwinism, in fact, there are quite different populations, whose importance for this field of knowledge changes over the course of the decades: breeders (one of the fundamental equations of evolution, which links selection differential and heritability, is called the "breeder's equation"); naturalists;

geneticists; statisticians; cytologists, etc. So, in seminars or informal discussions, Jean is often fast to point out that physicians were the first in each period to speak of heredity, physiologists to speak of function (rather than naturalists, for example), etc. The origin of a concept or a conceptual development must be identified to understand its role in the disciplinary matrix, its possible relations with other versions of the same concept and with other concepts, and ultimately the contingencies of its dynamics.

The distance appears clearly here with analytic programmes in philosophy of science, marked by the very "Vienna Circle" ideal, if not of axiomatization, at least of rational reconstruction[4] One often speaks today in the Anglo-Saxon context of a supposed "practical turn" in philosophy of science; there is even a young *Society for philosophy of science in practice.* What one is after here is a philosophy attentive to the practice of science—without however doing sociology, which studies science in action, according to the Latourian watchword—as much as or even more than the study of texts, dogmas, or models. A nice programme of course, but in a sense, Jean Gayon had already been doing something like that for a long time, and, it should be noted, did so under the auspices of Georges Canguilhem, Jacques Roger, and François Dagognet (to whom one may add Marjorie Grene for a "heretical" philosopher of biology).

19.6 Conclusion

In a sense, if one wants to call "historical epistemology" the inspiration behind his way of doing philosophy of biology, one gets an idea of the extent to which Jean Gayon was able, by turning to evolution and its problems, to operate a transition that no one in France had done before him towards Anglo-Saxon analytic philosophy of biology. This could even be the model for whosoever wants to undertake this transition, as in a way I did two decades later. But above all, this transition is less a passage than a *reinvention*, through which the terms of the domain of inspiration (historical epistemology) see themselves transformed into a way of philosophizing that is also not the one defined by the initiators of the analytic tradition.

This reinvention is the way that Jean Gayon links philosophy of science and history of concepts, which in turn constitutes an inspiration for those following in his footsteps, the pupils, former students, colleagues, and friends.

By means of these few remarks, it should be obvious that I did not want to offer an analysis of "evolutionary biology according to Jean Gayon" or even of the "the relations between philosophy and history of the sciences according to Jean Gayon", but rather to salute a philosopher who was at once a "major contemporary", a colleague, a friend, and an inspiration.

[4] *Darwin et l'après-Darwin* (2019 [1992]) does lay claim to a certain idea of "rational reconstruction", but this would nevertheless integrate the realities of conceptual history. On Gayon and Lakatos, see Duchesneau and by Loison, this volume.

References

André, J. B., Cozic, M., De Monte, S., Gayon, J., Huneman, P., Martens, J., & Walliser, B. (2022). *From evolutionary biology to economics and back. Parallels and crossings between economics and evolution*. Springer.

Gayon, J. (1997). The paramount power of selection: from Darwin to Kauffman. In M. L. Dalla Chiara, M. L., Doets, K., Mundici, D., & Benthem, J. (Eds.). (1997). *Structures and norms in science. Volume Two of the Tenth International congress of logic, methodology and philosophy of science, Florence, August 1995* (pp. 265–282). Kluwer.

Gayon, J. (1999). Nietzsche and Darwin. In M. Ruse & J. Maienschein (Eds.), *Biology and the foundations of ethics* (pp. 154–197). Cambridge University Press.

Gayon, J. (2012). Economic natural selection: What concept of selection? *Biological Theory, 6*(4), 320–325.

Gayon, J. (2014). Population genetics, economic theory, and eugenics in R.A. Fisher. In P. Thompson & D. Walsh (Eds.), *Evolutionary biology: Conceptual, ethical, and religious issues* (pp. 137–150). Cambridge University Press.

Gayon, J. (2016). From Mendel to epigenetics: History of genetics. *Comptes Rendus Biologies, 339*(6–7), 225–230.

Gayon, J. (2019) [1992]. *Darwin et l'après-Darwin. Une histoire de l'hypothèse de sélection naturelle*. Matériologiques.

Gayon, J., & Huneman, P. (2019). The modern synthesis: Theoretical or institutional event? *Journal of the History of Biology, 52*(2), 519–535.

Gayon, J., & Montévil, M. (2017). Repetition and reversibility in evolution: Theoretical population genetics. In P. Huneman & C. Bouton (Eds.), *Time of nature and the nature of time. Philosophical perspectives on time in natural sciences* (pp. 275–313). Springer.

Grafen, A. (1984). Natural selection, kin selection and group selection. In J. R. Krebs & N. B. Davies (Eds.), *Behavioural ecology* (pp. 62–84). Blackwell Scientific Publications.

Holton, G. (1981). *L'imagination scientifique*. Gallimard.

Huneman, P. (2018). Faire de la biologie évolutive avec Jean Gayon. In F. Merlin & P. Huneman (Eds.), *Philosophie, histoire, biologie: Mélanges offerts à Jean Gayon* (pp. 347–362). Éditions Matériologiques.

Huneman, P., & Barberousse, A. (2011). In memoriam, Hommage à Marie-Claude Lorne (1969–2008). In T. Heams, P. Huneman, G. Lecointre, & M. Silberstein (Eds.), *Les mondes darwiniens. L'évolution de l'évolution* (pp. 1559–1567). Éditions Matériologiques.

Huxley, J. (1942). *Evolution: The modern synthesis*. George Allen & Unwin Ltd.

Longy, F. (2018). Le pilote de l'ACI 'Fonction.' In F. Merlin & P. Huneman (Eds.), *Philosophie, histoire, biologie: Mélanges offerts à Jean Gayon* (pp. 275–281). Éditions Matériologiques.

References

Chapter 20
Intellectual Fertilizers

Pascal Engel

Abstract In a fine essay on agriculture in Flaubert's *Bouvard et Pécuchet,* Jean Gayon shows how much knowledge of the biology and agronomy of his time Flaubert is able to display. He also suggests that Flaubert's interest in Pouchet's views on spontaneous generation is connected to romantic theories of life and matter, and to his lifelong interest in the nature of stupidity.

Keywords Flaubert · *Bouvard et Pécuchet* · Agronomy · Biology · Pouchet · Spontaneous generation

Jean Gayon has scarcely written on literature. At first sight, his article on "Agriculture et agronomie dans *Bouvard and Pécuchet* de Gustave Flaubert" (Gayon, 1998) seems to be marginal within his output as an historian of science, but it is very significant in terms of his writing style. Gayon liked to focus on seemingly minor issues, as if he were adding a footnote or an appendix, but he meant to get important points across about science and philosophy. This article, though published in the journal *Littérature*, is typical of this ambitiously modest strategy, as we might call it, in the history of science.

For a Normand like Gayon,[1] Raymond Queneau's hometown ("*mon père était mercier, ma mère était mercière, ils trépignaient de joie*" (Queneau, 1937, p. 31)), close to Honfleur (where Alphonse Allais started his apprenticeship as a pharmacist and invented instant coffee), it was quite natural to write on Flaubert. It is also noteworthy that the biologist Félix-Archimède Pouchet, Pasteur's enemy in the quarrel about spontaneous generation, was a Normand too. Gayon himself possessed a master's degree in plant biology and has always been fond of botany. So it is not

[1] Editor's note: for his college (high school) studies, Gayon attended the Institution Saint-Joseph in Le Havre (Normandy), prior to entering the prestigious Lycée Henry IV where he stayed only a few weeks. Then, Gayon took philosophy classes for one year in Rouen before entering the Sorbonne. He was born in the Paris area.

P. Engel (✉)
École des Hautes Études en Sciences Sociales, Paris, France
e-mail: pascal.engel@ehess.fr

P.-O. Méthot (ed.), *Philosophy, History and Biology: Essays in Honour of Jean Gayon*, History, Philosophy and Theory of the Life Sciences 30,
https://doi.org/10.1007/978-3-031-28157-0_20

surprising that he became interested in Bouvard's and Pécuchet's agricultural adventures. Through the episodes of the settlement of the two "morons" ("*les deux crétins*", as he calls them in his correspondence) near Falaise, Flaubert evokes the gardening and agronomy manuals upon which, as with all the topics raised in the novel, he has accumulated abundant documentation.

In the second chapter of *Bouvard and Pécuchet*, Gayon is mostly interested in three topics. The first is Flaubert's references to agronomical knowledge of his time, and the fact that agronomy is to plants as medicine is to the human body. The book consequently turns into an occasion to review work in natural history, agricultural science, and chemistry of his times, but also the biological sciences of the nineteenth century.

The second topic is the narrative pattern of the novel, which is centered on various enterprises of the two fellows, invariably followed by their abject failures. Bouvard and Pécuchet reproduce faithfully the mistakes registered in the agriculture manuals of the time, such as Gasparin's and Gressent's. This prompts Gayon to reflect, as a disciple of Gaston Bachelard, on the relationship between theory and experience and on scientific knowledge as the correction of errors. The two friends make systematic errors, repeated at each attempt. Gayon quotes Bouvard's method to get rid of maggots: "In order to destroy the maggots, he thought of shutting up the fowls in a hen coop on rollers, which two men had to push along behind the plough—a thing which had only the effect of breaking the claws of the fowls" (Flaubert, 1896, p. 50).

The method used by Bouvard was orthodox, but its implementation was inept.[2] Stupidity, the central theme of the novel, is not a matter of error or of ignorance, as Queneau and Deleuze have noted (Deleuze, 1995; Queneau, 2007). Bouvard and Pécuchet are not ignorant; on the contrary, they know a lot of things and are eager to learn. They do not just make mistakes, but also fail to perform the correct judgments when they move from theory to practice. Flaubert is fully Kantian in his theory of stupidity as lack of judgment.[3]

Jean Gayon's third interest concerns the theories of fertilizers and composts which Bouvard and Pécuchet find in manuals, in particular the one by Girardin, a professor at the school for agronomy in Rouen, who was a specialist in compost. At this point we reach the metaphysics of the vegetal world, to which Gayon, as a student of Canguilhem studying scientific ideologies, could not have been indifferent.

> Stimulated by Pécuchet, he began to rave about pasture. In the pit for composts were heaped up branches of trees, blood, guts, feathers—everything that he could find. He used Belgian cordial, Swiss wash, lye, red herrings, wrack, rags; sent for guano, tried to manufacture it himself; and, pushing his principles to the farthest point, he would not suffer even urine or other refuse to be lost. Into his farmyard were carried carcasses of animals, with which he manured his lands. Their cut-up carrion strewed the fields. Bouvard smiled in the midst of this stench. A pump fixed to a dung-cart spattered the liquid manure over the crops. To those

---

[2] The failure reminds us of the episode in *Madame Bovary*, when Charles Bovary tries to remove a boy's club-foot.

[3] See Kant (1998; I 179—Rc 234). See Deleuze and Guattari's comments, in *Difference and repetition*, characterizing stupidity in terms congenial to *Bouvard and Pécuchet*: "digestive and vegetable like background" (1995, p. 151).

who assumed an air of disgust, he used to say, "But 'tis gold! 'tis gold!" And he was sorry that he had not still more manures. Happy the land where natural grottoes are found full of the excrements of birds! (Flaubert, 1896, p. 48)

Gayon informs us that lye (*Da Olmi wash*) was a fertilizer coming from a mixture of animal compost and various salty substances: saltpeter, cinder, and plaster. Bouvard subscribes to the romantic theories of fertilizers as the regeneration of life. Although Flaubert criticizes Bouvard's delusions about compost, guano, and excrement, he had himself some sympathy for this theory. He had read in Gasparin that a good fertilizer must be a recycling of human excrements in order to renew the life cycle. Other advocates of these theories included Victor Hugo and the positivist philosopher Pierre Leroux.

In a recent article on compost in Flaubert, Florence Vatan (2019; see also Vatan, 2015) has explored these questions. She quotes a passage from Hugo's *Miserables*, which directly echoes Bouvard's exclamation cited above:

Science, after having long groped about, now knows that the most fecund and the most efficacious of fertilizers is human manure. The Chinese, let us confess it to our shame, knew it before us. Not a Chinese peasant – it is Eckberg who says this – goes to town without bringing back with him, at the two extremities of his bamboo pole, two full buckets of what we designate as filth. Thanks to human dung, the earth in China is still as young as in the days of Abraham. Chinese wheat yields a hundred-fold of the seed. There is no guano comparable in fertility with the detritus of a capital. A great city is the most mighty of dung-makers. Certain success would attend the experiment of employing the city to manure the plain. If our gold is manure, our manure, on the other hand, is gold. (1887, V, p. 87)

Pierre Leroux argues against what Malthus has called "the law of nature". When one knows the latter's influence on Darwin, one sees what Gayon was after, when he comments on this romantic dream. He does not quote such texts, but certainly knew Flaubert's fad for Félix Archimède Pouchet and his theory of spontaneous generation. Flaubert was himself a friend of his son Georges Pouchet, who was also a naturalist in Rouen. "Fermentation and putrefaction", said Félix Pouchet in his *Hétérogénie*, "are almost indispensable to spontaneous generation" (1859, p. 358). And Flaubert, who wrote to Louise Collet in 1846, said the following:

Life comes to superimpose itself on Death: it makes plants grow in petrified skulls; and on the stone where one of us has carved his dream the Life Principle reasserts itself with every new blossoming of the yellow wallflowers. I find it a sweet thought that one day I'll help the tulips grow. Who knows? The tree at whose feet I'll be laid may bear splendid fruit. I'll perhaps make superb manure, a superior kind of guano. (Flaubert, 1980, pp. 70–71)

I wonder whether Flaubert's commentators have noted the proximity of the names of Pouchet and Pécuchet. Flaubert was so imbued with his views that he called his adversaries, like, Pasteur, "imposters" and "morons" (see Pecheron, 2015, quoted in Vatan, 2019). In fact, things are a little more complex, since Flaubert also knew the naturalist Georges Pennetier, who was favorable to Darwin's ideas, and a reader of Ernest Haeckel.[4] Gayon brings his study to a close by placing the two fellows' alleged stupidity into perspective:

[4] In a letter of 1878 to Mrs. Roger de Genettes, Flaubert praises Spencer, and in a number of other texts and letters he declares himself in favor of Darwin. See Dors-Crouslé (2008).

> All this suggests that every nineteenth century bourgeois, with the same project, the same means, and acting in the same scientific and technical context would have reached the same point, even if he did not commit the innumerable picturesque blunders which shine through the second chapter of Flaubert's novel. (1998, p. 73)

The moral which can be drawn from Gayon's subtle and elliptical remarks is that the stupidity of a time, even within the domain of empirical science, is widely shared, even by those who intend to react against the stupidity of masses, on account of the artistic power of individual creation. Flaubert indeed was close to adopting the same views as those of the two morons:

> Not forgetting even the toilet, especially not the toilets! The chemistry that goes on there is marvelous, such a prolific process of decomposition. Who knows what excremental broth nourishes the scent of the roe and the taste of the melon? Has anyone ever reckoned up the number of despicable actions which must have been abstained from in order to constitute one great soul? All the nauseating vapours that have to be inhaled, all the sorrows and the agonies to be endured in order to write one decent page? That is what we do, our kind, sewage-sorting and garden-digging. From human putrefaction we distill enchantments. We grow cluster of flowers all over the accumulation of misery. The fact is transmuted in the Form, and it raises up to the sky, like a pure incense of the Spirit, towards the Eternal, the immutable, the absolute, the ideal. (Flaubert, 1980, p. 234)

Today Bouvard's and Flaubert's ideal of regeneration through excrements and animal putrefaction seems to be a fact of everyday life. In the newspaper *Le Monde* of November 27, 2018, we learn that: "In Villetaneuse, the pedagogical farm is the setting, in partnership with a lab of the Center of National Scientific Research, of various experiments in order to produce energy from animal excrements" (Gourdon, 2018).

Let us further Gayon's and Flaubert's reflection by moving from shit in the literal sense to shit in the intellectual sense. Today, the term *bullshit*—namely, the shit of bulls—serves to designate various products of our information society, as well as the discourse of politicians (Frankfurt, 2005; see also Engel, 2016, 2021). The term has also been applied to the pseudo-scientific productions of philosophers (Sokal & Bricmont, 1999). One could dream, in Flaubert's manner, that this massive amount of stupidity could, through spontaneous regeneration, in the end produce some amount of intelligence and cosmically be fudged into it, leading to a new Advancement of Learning and a Renewal of Spiritual Life. Who knows whether bullshit will not one day produce deep truths? Isn't it the fate of science, as the sum of collective errors, to produce correct theories?

Acknowledgements This essay was read in December 2018, in an homage to Jean Gayon at the École normale, Paris. I thank Elisabeth Gayon and Stéphane Deligeorges for their invitation, and Florence Vatan for having let me see her remarkable article and provided me with the English translations of Flaubert's works

References

Deleuze, G. (1995). *Difference and repetition* (Patton, Trans.). Columbia University Press.
Dors-Crouslé, S. (2003). 'Le darwinisme de Flaubert': L'idée de 'race' dans la littérature et les sciences humaines (XVIIIe-XIXe siècles). In S. Moussa (Ed.), *Histoire des sciences humaines* (pp. 283–297). L'Harmattan.
Engel, P. (2016, November 17). Trump of la pathologie du pragmatism. *Le Monde.*
Engel, P. (2021). Post-truth is an assertion crisis. *Revue Internationale De Philosophie, 297*(3), 27–41.
Flaubert, G. (1896). *Bouvard and Pécuchet* (Hannigan, D. F., Trans.). S. Nichols.
Flaubert, G. (1980). *The letters of Gustave Flaubert* (Steegmuller, F., Transl.). Harvard University Press.
Frankfurt, H. G. (2005). *On bullshit.* Princeton University Press.
Gayon, J. (1998). Agriculture et agronomie dans *Bouvard and Pécuchet* de Gustave Flaubert. *Littérature, 109*, 59–73.
Gourdon, J. (2018, November 27). L'agriculture urbaine s'enracine dans les universités. *Le Monde.*
Hugo, V. (1887). *Les misérables* (Hapgood, I. F., Trans.). Crowell & Co.
Kant, I. (1998). *Critique of pure reason* (Guyer, P., & Wood, A. W., Eds., Trans.). Cambridge University Press.
Percheron, B. (2015). Flaubert, les naturalistes rouennais et les théories biologiques de 1865 à 1889. *Flaubert* [En ligne], *13.*
Pouchet, F. A. (1859). *Hétérogénie ou traité de la génération spontanée, basé sur de nouvelles expériences.* Baillière.
Queneau, R. (1969) [1937]. Chêne et chien. *Chêne et chien suivi de Petite cosmogonie portative et de Le chant du Styrène.* Gallimard.
Queneau, R. (2007). *Letters, numbers, forms: Essays, 1928–1970* (Stump, J., Trans.). University of Illinois Press.
Sokal, A., & Bricmont, J. (1999). *Intellectual impostures. Postmodern philosophers' abuse of science.* Profile Books.
Vatan, F. (2015). Le vivant, l'informe et le dégoût: Baudelaire, Flaubert et l'art de la (dé)composition. *Flaubert* [En ligne], *13.*
Vatan, F. (2019), 'Cultiver son jardin': rêves et délires de l'engrais dans Bouvard et Pécuchet. *Revue Flaubert, 18.*

Chapter 21
In Memoriam: Jean Gayon

Hans-Jörg Rheinberger

Abstract In this personal note, I wish to recall my first encounters with Jean Gayon in the context of a conference of the International Society for the History, Philosophy, and Social Studies of Biology, an International Colloquium in honor of Marjorie Grene and our collaborations at the Max Planck Institute for the History of Science in Berlin in the context of the Cultural History of Heredity project.

Keywords Bachelard · Heredity · Model organism · Historical epistemology

What I can contribute to this volume dedicated to the memory of Jean Gayon is a personal, if not peripheral note. I met Jean for the first time, about 25 years ago, during one of the European meetings of the International Society for the History, Philosophy, and Social Studies of Biology in Leuven in the summer of 1995. It might also have been a few months earlier in the spring of that year in Dijon, on the occasion of an International Colloquium in honor of Marjorie Grene, organized by Jean and by Richard Burian. Back then, Jean was engaged, in close collaboration with Dick Burian and Doris Zallen, in writing a couple of decisive papers on the development of genetics in the first half of the twentieth century and beyond in France (Burian & Gayon, 1990, 1999[1]; Burian et al., 1988; Méthot, 2018). I was deeply impressed by these papers both because of their precision in argumentation and their richness in historical detail. This has also been, by the way, a hallmark of Jean's interventions in discussions. They were always discursively disciplined and polite but challenging and at the same time undergirded by historical detail. I do not know of many people who were able to integrate historical and epistemological thinking in such a reflexive manner.

When I assumed a directorship at the newly founded Max Planck Institute for the History of Science in Berlin in 1996, we had to form an Advisory Board for the

[1] On the encounter between Jean Gayon, Richard Burian and Marjorie Grene, see Méthot and Burian, this volume.

H.-J. Rheinberger (✉)
Max Planck Institute for the History of Science, Berlin, Germany
e-mail: rheinb@mpiwg-berlin.mpg.de

P.-O. Méthot (ed.), *Philosophy, History and Biology: Essays in Honour of Jean Gayon*, History, Philosophy and Theory of the Life Sciences 30,
https://doi.org/10.1007/978-3-031-28157-0_21

Institute whose function was not only to critically accompany our work, but also to do justice to its name and give advice. I could think of no one better suited to serve in this function concerning the work of my department than Jean. He became a member of our first board in 1998 and served in this function for the coming six years until 2004.

In 2001, my colleague Staffan Müller-Wille and I set up a long-term research project on the Cultural History of Heredity from the early modern period to present-day post-genomics. From its very inception, Jean Gayon was one of those scholars among the international community of historians and philosophers of biology who not only closely accompanied the project, but also significantly contributed to it, in participating in our regular annual conferences as well as in writing (see Gayon, 2016). Over the whole period of this project that lasted for a decade, we could always count on Jean's input and suggestions.

Another collaboration in which Jean participated was one between the MPI for the History of Science and the Archive of the Institut Pasteur. The driving force behind the three meetings in the context of this cooperation was the late Charles Galperin (1929–2019), whose constant efforts to generate and entertain a dialogue between historians of science and scientists must also not be forgotten. I vividly remember a meeting of our group, including Jean Gayon, at the Musée Claude Bernard in Saint-Julien-en-Beaujolais in the fall of 2003, out of which a collected volume on *Model Organisms in Medical Research* resulted (Gachelin, 2006). Today it is barely remembered anymore that over a quarter of a century ago, Dick Burian was one of the first to argue for paying close attention to the role of model organisms in the practice of twentieth century biology and the medical sciences (Burian, 1993; Gayon, 2006).

As well around that time, I began again to pay closer attention to the epistemological work of Gaston Bachelard. The notion of phenomenotechnique was at the center of my interest then. It goes without saying that in Jean, I found someone who was intimately familiar with Bachelard's historical epistemology, Jean's own formation having been shaped by Georges Canguilhem, the successor to Bachelard at the Institut d'histoire et de philosophie des sciences et des techniques and at the Sorbonne, and by François Dagognet (Méthot, 2018, pp. 17–22), who actually had happened to be a young medical doctor to Bachelard during his summer vacations in Dijon in the early years after World War II. When I told Jean about my intention to follow more closely Bachelard's notion of phenomenotechnique, he gave me a lengthy manuscript he had written on Gaston Bachelard's *Le rationalisme appliqué*. It was the most succinct work on Bachelard's historical epistemology that I had come across, and yet Jean had never published it in a readily visible fashion (Gayon, 1994). I think this episode gives us some of the essence of Jean Gayon's character in a nutshell. He was always willing to share his insights, even if unpublished, with others, and he successfully withstood the pressure for what would have been, according to his judgement, a premature publication.

And Jean was a generous host. When my doctoral student Sandra Pravica was working on her dissertation on *Gaston Bachelard's Tentative Philosophy of Science* (in German: *Bachelards tentative Wissenschaftsphilosophie*, Pravica, 2015), Jean

Gayon was immediately willing to take her up for a stay in Paris at the IHPST at rue Du Four from which she immensely profited.

Although our contacts became less frequent over the past several years, our encounters remained always memorable. On the fiftieth anniversary of the Nobel Prize for François Jacob, André Lwoff, and Jacques Monod, at the Institut Pasteur in September 2015, Jean's long-lasting friendship with Dick Burian bore fruit once more, when they presented their joint paper on Eugène and Elisabeth Wollman's work regarding the question of lysogeny (Gayon & Burian, 2017). And when I saw Jean for the last time, about three years ago, at his home in rue Vaugirard, he advised me on how to find a French publisher for my new little book on the encounter between the copper engraver Albert Flocon and the philosopher of science Gaston Bachelard (Rheinberger, 2017). Now, much too prematurely, time has come to say goodbye forever to a distinguished friend and companion.

References

Burian, R. M. (1993). How the choice of experimental organism matters: Epistemological reflections on an aspect of biological practice. *Journal of the History of Biology, 26*, 351–367.

Burian, R. M., & Gayon, J. (1990). Genetics after World War II: The laboratories at Gif. *Cahiers d'histoire du CNRS, 7*, 25–48.

Burian, R. M., Gayon, J., & Zallen, D. (1988). The singular fate of genetics in the history of French biology, 1900–1940. *Journal of the History of Biology, 21*, 357–402.

Burian, R. M., & Gayon, J. (1999). The French school of genetics: From physiological and population genetics to regulatory molecular genetics. *Annual Review of Genetics, 33*, 321–322.

Gachelin, G. (Ed.). (2006). *Les organismes modèles dans la recherche médicale*. Presses Universitaires de France.

Gayon, J. (1994). *Gaston Bachelard: Le rationalisme appliqué*. Centre National d'Enseignement à Distance.

Gayon, J. (2006). Les organismes modèles en biologie et en médecine. In G. Gachelin (Ed.), *Les organismes modèles dans la recherche médicale* (pp. 9–43). Presses Universitaires de France.

Gayon, J. (2016). Natural selection, regression, and heredity in Darwinian and post-Darwinian evolutionary theory. In S.W. Müller-Wille & C. Brandt (Eds.), *Heredity Explored. Between Public Domain and Experimental Science, 1850–1930* (pp. 167–188). MIT Press.

Gayon, J., & Burian, R.M. (2017). Eugène et Elisabeth Wollman: la question de la lysogénie. In L. Loison & M. Morange (Eds.), *L'invention de la régulation génétique. Les Nobel 1965 (Jacob, Lwoff, Monod) et le modèle de l'opéron dans l'histoire de la biologie* (pp. 31–65). Éditions Rue d'Ulm.

Méthot, P.-O. (2018). De l'épistémologie historique à la philosophie de la biologie: le double héritage philosophique de Jean Gayon. In F. Merlin & P. Huneman (eds.), *Philosophie, histoire, biologie. Mélanges offerts à Jean Gayon* (pp. 15–50). Editions Matériologiques.

Pravica, S. (2015). *Gaston Bachelards tentative Wissenschaftsphilosophie*. Passagen.

Rheinberger, H.-J. (2017). *Le graveur et le philosophe. Albert Flocon rencontre Gaston Bachelard* (Lochmann, A., Trans.). Hermann.

Index

P.-O Méthot (ed.), *Philosophy, History and Biology: Essays in Honour of Jean Gayon*, History, Philosophy and Theory of the Life Sciences 30,
https://doi.org/10.1007/978-3-031-28157-0

K

L

M

Q

R

S

MIX
Papier aus verantwortungsvollen Quellen
Paper from responsible sources
FSC® C105338

If you have any concerns about our products,
you can contact us on
ProductSafety@springernature.com

In case Publisher is established outside the EU,
the EU authorized representative is:
Springer Nature Customer Service Center GmbH
Europaplatz 3, 69115 Heidelberg, Germany

Printed by Libri Plureos GmbH
in Hamburg, Germany